A TREATISE

ON THE

LINE COMPLEX

A TREATISE

ON THE

LINE COMPLEX

BY

C. M. JESSOP, M.A.

FORMERLY FELLOW OF CLARE COLLEGE, CAMBRIDGE,
ASSISTANT PROFESSOR OF MATHEMATICS IN THE DURHAM
COLLEGE OF SCIENCE, NEWCASTLE-ON-TYNE.

CAMBRIDGE
AT THE UNIVERSITY PRESS
1903

CAMBRIDGE
UNIVERSITY PRESS

University Printing House, Cambridge CB2 8BS, United Kingdom

Cambridge University Press is part of the University of Cambridge.

It furthers the University's mission by disseminating knowledge in the pursuit of education, learning and research at the highest international levels of excellence.

www.cambridge.org
Information on this title: www.cambridge.org/9781107457997

First published 1903
First paperback edition 2014

A catalogue record for this publication is available from the British Library

ISBN 978-1-107-45799-7 Paperback

A TREATISE

ON THE

LINE COMPLEX

BY

C. M. JESSOP, M.A.

FORMERLY FELLOW OF CLARE COLLEGE, CAMBRIDGE,
ASSISTANT PROFESSOR OF MATHEMATICS IN THE DURHAM
COLLEGE OF SCIENCE, NEWCASTLE-ON-TYNE.

CAMBRIDGE
AT THE UNIVERSITY PRESS
1903

CAMBRIDGE
UNIVERSITY PRESS

University Printing House, Cambridge CB2 8BS, United Kingdom

Cambridge University Press is part of the University of Cambridge.

It furthers the University's mission by disseminating knowledge in the pursuit of education, learning and research at the highest international levels of excellence.

www.cambridge.org
Information on this title: www.cambridge.org/9781107457997

First published 1903
First paperback edition 2014

A catalogue record for this publication is available from the British Library

ISBN 978-1-107-45799-7 Paperback

PREFACE.

THE important character of the extensive investigations into the theory of line-geometry renders it desirable that a treatise should exist for the purpose of presenting these investigations in a form easily accessible to the English student of mathematics. With this end in view, the present work on the Line Complex has been written.

The subject owes its origin to Plücker, who suggested the idea of taking the straight line as the element of space*. The straight line thus holds to the present subject the relationship in which the point and the plane stand to the older geometry. Types of coordinates of the line were introduced by Cayley and Grassmann; Plücker adopted a coordinate system which is a special form of them.

In his work the *Neue Geometrie des Raumes*, Plücker introduced the conception of a *complex of lines*, *i.e.* the ∞^3 lines which satisfy one given condition, so that one equation exists between the four coordinates of each line of a complex. He investigated in detail the linear and the quadratic complex; his work contains most of the chief properties of such complexes; in particular he shows that if any screw motion about a certain axis be given to the lines forming a linear complex, these lines still remain within the complex. He discovered the *polar properties*, viz. that the lines of a linear complex in any plane pass through a point, the *pole* of the plane; that the lines of the complex through any point lie in a plane, the *polar plane* of the point; and that if a point moves along any given line, its polar plane turns round another line called the *polar line* of the first line, the relationship between the two lines being reciprocal.

The greater part of the *Neue Geometrie* is concerned with the quadratic complex, of which it contains many of the leading properties; in particular, Plücker shows that while the lines of

* *System der Geometrie des Raumes*, Düsseldorf, (1846).

such a complex through any point form in general a quadric cone, there is a certain surface, the *Singular Surface* of the complex, for whose points these cones break up into two planes. Likewise the lines of the complex in any plane, which in general touch a conic, in the case of any tangent plane of the singular surface form two pencils. This surface is the one known as Kummer's surface; it is of the 4th degree and class and possesses 16 nodes and 16 singular tangent planes.

The next investigator in this field was Battaglini, who pursued still further the ideas of Plücker. He adopted as the general quadratic complex one which was afterwards shown to be a special case, viz. the complex formed by the lines for each of which the points of intersection with two given quadrics form a harmonic range; but many of his results apply also to the general complex.

The success of Plücker's researches was limited by the unsuitable (Cartesian) analysis he employed. The second important step in the development of the subject was due to Prof. Felix Klein, who, in his celebrated memoir in volume II. of the *Mathematische Annalen*, introduced the coordinate-system determined by six linear complexes in mutual involution*. By its adoption a simple and elegant analytical mode of treatment of line-geometry is rendered possible.

Klein further revealed the existence of a singly infinite series of quadratic complexes which have the same singular surface as any given quadratic complex. In his Dissertation (Bonn, 1868) he pointed to the method of Weierstrass for the canonization of two quadratic forms, as the appropriate instrument for classifying the quadratic complex; and this classification was carried out by Weiler. Another service rendered by Klein was his discovery of the analogue existing between the *lines* of three-dimensional space and the *points* of four-dimensional space, together with the equations embodying this relationship. His enunciation of the fact that line-geometry is point-geometry on a quadric contained

* On any line common to two linear complexes a (1, 1) correspondence of points is determined by the planes through the line, viz. by taking the poles of each plane for the two complexes. If a certain condition, connecting the constants of the equations of the two complexes, is satisfied, these pairs of points form an *involution*.

in point-space of five dimensions, offers a new point of view of the subject.

Other important contributions to the theory are introduced from time to time in the text: of these the most fundamental are contained in the investigations of Lie, in which he showed the connexion of line-geometry with sphere-geometry. He established a relationship between the lines and spheres of three-dimensional space of such a nature, that to two intersecting lines there correspond two spheres in contact; and he applied the ideas of both varieties of geometry to the investigation of various types of differential equations.

In the present work the analytical method of treatment with Klein coordinates has been generally adopted; but as it frequently happens that synthetic methods are appropriate, recourse to such has been occasionally made. Since the study of synthetic geometry has been less widely followed in this country than on the Continent, I have not thought it superfluous to insert, by way of Introduction, a short sketch of the simpler portions of that subject which have bearing on the context of the work.

The main object of investigation is, as has been stated, the properties of the line complex, and, in connexion with it, the characteristics of the system of ∞^2 lines common to any two complexes. To any set of ∞^2 lines the name *congruence* is attached; the study of such systems was extensively pursued at a period considerably before Plücker's discoveries took place. The chief property of a congruence is that each of its lines is bitangent to a surface, (including as a special case *two surfuces, a surface and a curve,* etc.). Through any point there pass a definite number m of the lines of a given congruence, and in any plane there lie a definite number n of its lines. *If the congruence is the complete intersection of two complexes,* $m = n$.

Though not necessarily included in the scope of this treatise, nevertheless, on account of its close connexion with the theory of the complex, a discussion has been given in Chapters XIV.—XVI. of the congruence (m, n), and in particular, of the congruences $(2, n)$, so elegantly treated by Kummer.

As regards the various authorities on this subject, the student is referred to the work of Prof. Gino Loria *Il passato ed il presente delle principali teorie geometriche*, which contains detailed references to the chief memoirs. A useful summary with references is given in Prof. E. Pascal's *Repertorio di mathematiche superiori*. The comprehensive treatise of Prof. R. Sturm, *Die Gebilde ersten und zweiten Grades der Liniengeometrie*, is a storehouse of information; his method is, however, *exclusively* synthetic. An introduction to most of the leading ideas is given by Prof. G. Koenigs in his work *La géométrie réglée et ses applications*.

An interesting general account of Line Geometry given by Mr J. H. Grace in the Supplement to the *Encyclopaedia Britannica*, will be found very serviceable by the student of this subject.

I have thought it not desirable to include in this treatise a description of the important investigations of Prof. E. Study, on account of their distinctness in aim and method from those of the other writers who have built up this subject. I rather refer the reader to Prof. Study's work *Geometrie der Dynamen*.

It gives me much pleasure to express my gratitude to several friends for assistance generously given me; and especially to Mr J. H. Grace, M.A., Fellow of Peterhouse, Cambridge, who read the manuscript, and who, by his criticisms and suggestions, has greatly increased the value of the work. My colleague Mr G. W. Caunt, M.A., late Scholar of St Catharine's College, has read all the proofs; such accuracy as the book possesses is largely due to his carefulness. I am also under obligations to Mr P. W. Wood, B.A., Scholar of Emmanuel College, who has read the proofs and verified many of the examples.

Professor T. J. I'A. Bromwich, Fellow of St John's College, has kindly put at my disposal a collection of examples, most of which were made by him; they have been incorporated in the Miscellaneous Results and Exercises, and add greatly to the book's usefulness.

Finally, I feel it a pleasant duty to express my appreciation of the admirable manner in which the staff of the University Press have carried out the onerous task involved in the printing.

C. M. JESSOP.

September 1903.

CONTENTS.

INTRODUCTION.

CHAPTER I.

SYSTEMS OF COORDINATES.

CHAPTER II.

THE LINEAR COMPLEX.

CHAPTER III.

SYNTHESIS OF THE LINEAR COMPLEX.

CHAPTER IV.

SYSTEMS OF LINEAR COMPLEXES.

CHAPTER V.

RULED CUBIC AND QUARTIC SURFACES.

CHAPTER VI.

THE QUADRATIC COMPLEX.

CHAPTER VII.

SPECIAL VARIETIES OF THE QUADRATIC COMPLEX.

CHAPTER VIII.

THE COSINGULAR COMPLEXES.

CHAPTER IX.

POLAR LINES, POINTS, AND PLANES.

CHAPTER X.

REPRESENTATION OF A COMPLEX BY POINTS OF SPACE.

CHAPTER XI.

THE GENERAL EQUATION OF THE SECOND DEGREE.

CHAPTER XII.

CONNEXION OF LINE-GEOMETRY WITH SPHERE-GEOMETRY.

CHAPTER XIII.

CONNEXION OF LINE-GEOMETRY WITH HYPERGEOMETRY.

CHAPTER XIV.

CONGRUENCES OF LINES.

CHAPTER XV.

CONGRUENCES OF THE SECOND ORDER WITHOUT SINGULAR CURVES.

CHAPTER XVI.

THE CONGRUENCE OF THE SECOND ORDER AND SECOND CLASS.

CHAPTER XVII.

THE GENERAL COMPLEX.

CHAPTER XVIII.

DIFFERENTIAL EQUATIONS CONNECTED WITH THE LINE COMPLEX.

CORRIGENDA.

Art. 39 *for* 'which meet such a pair' *read* 'which meet such pairs.'

Page 184 *for* $e_2 p_{34} p_{42}$ *read* $e_2 p_{34} p_{14}$.

INTRODUCTION.

BEFORE entering upon the subject proper of the present work, a short preliminary discussion of those parts of synthetic geometry which have the closest connexion with line geometry, has been inserted here for convenience of reference.

i. **Double Ratio.** For four points $ABCD$ lying on a straight line the number $\frac{AB}{BC} \div \frac{AD}{DC}$ is called the Double Sectional Ratio or Double Ratio of the points, the *sense* of the segments AB &c. being taken into consideration; the terms Anharmonic Ratio and Cross Ratio are also used to designate this quantity, which is usually denoted by $(ABCD)$.

The orders in which the points may be taken are 24 in number, but there are only six different Double Ratios of the four points, for we find that any two orders which differ by a double interchange of two points have their double ratios equal, *e.g.*

$$(BADC) = \frac{BA}{AD} \div \frac{BC}{CD} = \frac{AB}{BC} \div \frac{AD}{DC} = (ABCD);$$

so that $\quad (ABCD) = (BADC) = (CDAB) = (DCBA).$

Secondly, if two non-consecutive members of a double ratio be interchanged the double ratio is *inverted, e.g.*

$$(ADCB) = \frac{AD}{DC} \div \frac{AB}{BC} = \frac{1}{(ABCD)}.$$

Thirdly, the sum of the double ratios for two orders which differ in their second and third members is *unity, e.g.*

$$(ABCD) = \frac{AB \cdot DC}{BC \cdot AD}, \quad (ACBD) = \frac{AC \cdot DB}{CB \cdot AD} = \frac{CA \cdot DB}{BC \cdot AD},$$

and since

$$BC + CA + AB = 0, \quad AD = BD + AB, \quad AD = CD - CA,$$

therefore

$$BC.AD + CA(BD + AB) + AB(CD - CA) = 0,$$

or $$BC.AD + CA.BD + AB.CD = 0,$$

hence $$(ABCD) + (ACBD) = 1.$$

So that denoting $(ABCD)$ by λ, we have

$$(ADCB) = \frac{1}{\lambda}, \quad (ACBD) = 1 - \lambda, \quad (ADBC) = \frac{1}{1-\lambda},$$

$$(ABDC) = 1 - \frac{1}{1-\lambda} = \frac{\lambda}{\lambda - 1}, \quad (ACDB) = \frac{\lambda - 1}{\lambda}.$$

All the other double ratios have one of these six values. If the value of the double ratio is -1 the points are said to be Harmonic; in this case since

$$\frac{AB}{BC} + \frac{AD}{DC} = 0,$$

hence $$AB(AC - AD) + AD(AC - AB) = 0,$$

or $$\frac{2}{AC} = \frac{1}{AB} + \frac{1}{AD};$$

whence it is easily found that if O is the mid-point of AC,

$$OC^2 = OB.OD.$$

[It should be noticed that in this case the points $ABCD$ are arranged consecutively.]

ii. **Correspondence.** If between the points of a straight line a connexion is established such that to each point of the line corresponds one and only one point of the line, there is said to exist a "one-one" correspondence, or correlation, between its points. If x is the distance of any point P of the line from a fixed point O of the line, this correlation is defined by an equation of the form

$$Axx' + Bx + Cx' + D = 0,$$

where x' is the distance from O of the point P' which corresponds to P. It follows from this equation that

$$x = -\frac{Cx' + D}{Ax' + B}$$

and hence if P, Q, R, S are four points and P', Q', R', S' their corresponding points,

$$(PQRS) = (P'Q'R'S');$$

for $$PQ = x_2 - x_1 = (x_2' - x_1')(AD - BC)/(Ax_1' + B)(Ax_2' + B),$$

$$SR = x_3 - x_4 = (x_3' - x_4')(AD - BC)/(Ax_3' + B)(Ax_4' + B), \text{ \&c.},$$

hence

$$(PQRS) = \frac{(x_1 - x_2)(x_4 - x_3)}{(x_2 - x_3)(x_1 - x_4)} = \frac{(x_1' - x_2')(x_4' - x_3')}{(x_2' - x_3')(x_1' - x_4')} = (P'Q'R'S').$$

It follows that if three pairs of points of the line be associated the correspondence is determined, for if P and P', Q and Q', R and R' be made to correspond, then the point S' which corresponds to any fourth point S is determined by the equality

$$(PQRS) = (P'Q'R'S').$$

iii. **United Points.** The coincidence of a point and its corresponding point will occur twice, for putting $x' = x$ we have

$$Ax^2 + (B + C)x + D = 0,$$

thus in every (1, 1) correspondence there are two "united" points (real or imaginary).

If the point midway between the united points (E, E') be the point O from which the distances are measured, we must have $B + C = 0$, and the equation defining the correspondence is of the form $Axx' + B(x - x') + D = 0$, while the distance α of either united point from O is given by the equation $A\alpha^2 + D = 0$; combining these two equations and writing κ for $\frac{B}{A}$, the equation of correspondence becomes

$$xx' + \kappa(x - x') - \alpha^2 = 0,$$

which may be written in the form

$$(x + \alpha)(x' - \alpha) = (\alpha + \kappa)(x' - x),$$

hence $$\frac{2\alpha}{\alpha + \kappa} = \frac{(x - x')(-2\alpha)}{(x' - \alpha)(x + \alpha)} = \frac{PP' \,.\, E'E}{P'E \,.\, PE'} = (PP'EE'),$$

thus the double ratio of a point, its corresponding point, and the united points is *constant.* The correspondence is therefore determined if its united points and one pair of corresponding points are given.

iv. **Involution.** If $B = C$ the relation between x and x' is symmetrical, and hence if P' corresponds to any point P then will P correspond to P', and the points of the line form "closed systems" of two points. The correspondence is in this case called an Involution. The equation which connects corresponding points being now

$$Axx' + B(x + x') + D = 0,$$

it may be written

$$A\left(x + \frac{B}{A}\right)\left(x' + \frac{B}{A}\right) = \frac{B^2 - AD}{A},$$

or if y and y' are the respective distances of P and P' from the point whose distance from O is $-\frac{B}{A}$,

$$yy' = \frac{B^2 - AD}{A^2}.$$

If $B^2 > AD$ there are two *real* points each of which coincides with its corresponding point, viz. those given by the equation

$$y = \pm \frac{\sqrt{B^2 - AD}}{A},$$

so that if these points be E and E' and M their middle point (the origin for the y's) and P, P' any pair of corresponding points

$$MP \,.\, MP' = ME^2.$$

This shows that the two "double" points E and E' of the involution form with any pair of corresponding points *a harmonic range*.

v. **Harmonic Involutions.** If in the two involutions on the same line determined respectively by

$$xx' + A(x + x') + B = 0,$$
$$yy' + C(y + y') + D = 0,$$

the double points of one form a pair in the other, *i.e.* are harmonic conjugates to the double points of the other, it is clear that

$$B + D - 2AC = 0 \text{..........................} (a).$$

The Involutions are then said to be "harmonic" to each other. In this case, if to two points P and P' which are conjugate in the first Involution the conjugate points in the second Involution are Q and Q' respectively, Q and Q' are themselves conjugate in the first Involution; for by hypothesis

$$OQ = -\frac{C \,.\, OP + D}{OP + C}, \quad OQ' = -\frac{C \,.\, OP' + D}{OP' + C},$$

hence $\quad (OP + C)(OP' + C)(OQ \,.\, OQ' + A \,.\, \overline{OQ + OQ'} + B)$

$= (C \,.\, OP + D)(C \,.\, OP' + D) - A(\overline{C \,.\, OP + D} \,.\, \overline{OP' + C}$

$\qquad + \overline{C \,.\, OP' + D} \,.\, \overline{OP + C}) + B(OP + C)(OP' + C)$

$= (C^2 - D)(OP \,.\, OP' + A \,.\, \overline{OP + OP'} + B)$; from (a),

$= 0$, which proves the result stated.

vi. **Correspondences on different lines.** We shall now consider correspondences between the points of two different lines and the ruled surfaces (or plane curves) obtained as loci or envelopes of lines joining corresponding points. In what follows use will be made of the obvious fact that when a correspondence is established between the points of a line and these points are joined to any external point by a plane pencil of lines, a similar correspondence is thereby established between the lines of the pencil; similarly a correspondence established on a line gives rise to a correspondence between the planes of *any* pencil of planes (*i.e.* planes having a common line of intersection).

When a (1, 1) *correspondence exists between the points of two lines* in the same plane, the joins of pairs of corresponding points envelope a curve of the second class;* for joining any point P of the plane to the points of the two lines a (1, 1) correspondence is established between the lines of the pencil centre P; since there are two *united lines* in this correspondence, through P will pass two and only two lines which connect a pair of corresponding points. To this envelope the two given lines are themselves tangents. A special case arises when the point of intersection O of the two given lines *corresponds to itself, i.e.* regarded as a point of the first line has itself as corresponding point in the second line; in this case, of the two lines through P which join corresponding points *one* coincides with PO and therefore passes through the fixed point O; the envelope of lines joining corresponding points breaks up into two points, O and one other point C, the two rows of points are said to be in *perspective*, and the point C through which pass all lines joining corresponding points is called the "centre of perspective."

The corresponding theorem afforded by the Principle of Duality is, *if between the lines of two plane pencils a* (1, 1) *correspondence exists, the locus of intersection of corresponding lines is a curve of the second order passing through the centres of the pencils;* for on any line the two pencils determine a (1, 1) correspondence of points, the two double points of which are the points of intersection of the line with the required locus. A special case arises when the line joining the centres of the pencils corresponds to itself; in this case, of the two double points on any line, *one* lies on the line joining the centres of the pencils, *i.e.* the locus of intersection of corresponding lines of the two pencils breaks up into the line joining the centres of the pencils and one other line c, the two pencils are said to be in *perspective*, and the line c which contains

* The x of Art. ii here refers to a point P of one line, and x' to its corresponding point P' on the other line.

the intersections of corresponding lines is called the *axis of perspective.*

vii. *A* (1, 1) *correspondence between the lines of two pencils* (*or the points of two lines*) *in one plane is established when to three elements of one are assigned as correspondents three elements of the other;* for if three lines of one pencil SA, SB, SC meet any given line p in A, B, and C, and three corresponding lines $S'A'$, $S'B'$, $S'C'$ of the other pencil meet the same line in A', B', C', then (Art. ii) the three pairs of corresponding points AA', BB', CC' determine a correlation on p, hence if any other line of the first pencil meets p in P the corresponding line $S'P'$ of the other pencil is determined.

viii. *The joins of corresponding points on two non-intersecting lines form one set of generators of a quadric, that is, a Regulus**; for the points of the two lines u and v establish on the pencil of planes whose axis is any line l a (1, 1) correspondence, viz., if P and P' are corresponding points on u and v, to the plane (P, l) corresponds the plane (P', l), each of the two double planes of this correspondence will meet the lines u, v in a pair of corresponding points, hence *two and only two* lines joining corresponding points on u and v will meet l, and any line l will meet the locus of lines which join corresponding points on u and v in *two* points. The two given lines belong to the other system of generators of the quadric determined by the Regulus. It is to be noticed that the lines of a Regulus determine on *any* two lines of the other system two rows of points having a (1, 1) correspondence; and four given generators determine on a variable generator of the opposite system four points having a constant Double Ratio.

The Principle of Duality gives the theorem, *the locus of intersection of corresponding planes of two pencils of planes connected by a* (1, 1) *correspondence is a Regulus;* for the two pencils of planes determine on any line l a (1, 1) correspondence of points, hence corresponding planes will only meet on l at the double points of this correspondence.

The following properties of a quadric should be observed:

First, if A, B, C, D are any four points on a generator and a, b, c, d the tangent planes thereat, $(ABCD) = (abcd)$.

* In the sequel the word Regulus is restricted to mean 'one set of generators of a quadric surface,' the other set of generators is called, in reference to it, the 'complementary' regulus. The word 'demi-quadrique' is used by Koenigs in this sense.

This follows from taking any other generator of the same system which meets a, b, c, d in A', B', C', D' respectively, then AA', BB', CC', DD' are generators and therefore from what precedes $(ABCD)=(A'B'C'D')$, while $(A'B'C'D')=(abcd)$, hence $(abcd)=(ABCD)$;

Second, if $ABCD$ are any given points on the quadric and x any generator, the double ratio of the four planes xA, xB, xC, xD is *constant*; for if the generators through A, B, C, D of the opposite system to x meet x in $PQRS$ respectively, xA, xB, xC, xD are the tangent planes at P, Q, R, S respectively, therefore $(PQRS)=(xA, xB, xC, xD)$, but $(PQRS)$ being the double ratio of points of section by a generator of the four given generators through A, B, C, D is *constant*.

ix. **Correspondence between the points of a conic and the lines of a plane pencil.** *If a* (1, 1) *correspondence exists between the points of a conic and the lines of a plane pencil centre* S, *there are three points on the conic, of which one at least is real, through which pass their corresponding lines;* for take any point S' on the conic f^2 and join it to the points of f^2, then between the lines of the pencils centres S and S', a (1, 1) correspondence is established and the intersection of corresponding lines being a conic ϕ^2 which meets f^2 in four points, of which S' is one, it is seen that there are three points on f^2 through which pass the respective corresponding lines of the pencil centre S.

If between a pencil of planes and the lines of a regulus a (1, 1) correspondence exists, and if the section of the pencil and the regulus by any plane be taken, the last result shows that in this plane there are three points in which a plane of the pencil meets its corresponding line of the regulus, hence the locus of the intersection of corresponding generators and planes is a curve of which three points lie in any plane or a *twisted cubic*.

x. **Involution on a conic.** If a (1, 1) correspondence exists between the points of a conic, to a point P will correspond a point Q and to Q a point R in general different from P, thus through Q pass two lines which join corresponding points, and this being the case for each point of the conic, the envelope of lines joining corresponding points is a curve of the *second class.* If however the correspondence is involutory, *i.e.* if to Q corresponds P, the envelope becomes of the first class or the lines joining corresponding points are concurrent; if U is their common point, U is called the *centre* of the involution.

Condition for Involution. A (1, 1) correspondence on the same conic is an Involution when to a point A *there corresponds doubly* a point A_1; for let B and B_1 be two other corresponding points, so that to AA_1B correspond A_1AB_1, let U be the intersection of AA_1 and BB_1 and u its polar line with respect to the conic.

The pencils $B_1(A_1AB...)$, $B(AA_1B_1...)$ are in (1, 1) correspondence and since they have the line BB_1 as self-corresponding line the pencils are in perspective and the locus of intersection of corresponding lines of the pencils is a straight line which must be u.

So that any line BC meets its corresponding line B_1C_1 in u, hence C, U, and C_1 must be collinear and therefore since to C corresponds C_1, to C_1 will correspond C, or the correspondence is an *Involution*.

xi. **Corresponding Sheaves.** The assemblage of ∞^2 lines which pass through one point are said to form a *sheaf*; the same name is given to all the planes through a point. If the sheaf of planes through any point S is connected by a (1, 1) correspondence with the sheaf of planes through a point S', so that to the intersection of two planes through S corresponds the intersection of the corresponding planes through S', the sheaves are said to be *collinear*.

xii. **Systems of Lines.** *The system of lines formed by the intersection of pairs of corresponding planes of two collinear sheaves is of the* first order, *i.e. through any point P there passes* one *line of the system;* for join S to P, then through S' there is one corresponding line $S'P'$ and the two pencils of planes for SP and $S'P'$, being connected by a (1, 1) correspondence, have as locus of intersection of corresponding pairs a *regulus* (Art. viii), of which *one line* passes through P; SP and $S'P'$ intersect all the lines of this regulus.

But if the line $S'P$ corresponds to SP, i.e. if the two corresponding lines intersect in P, the above regulus becomes a quadric cone of vertex P and *all* the generators of the cone belong to the system of lines; P is then called a "singular" point of the system of lines.

[It should be noticed that SP and $S'P$ are both generators of this cone, for to the plane SPS' of the pencil whose axis is SP corresponds a plane through $S'P$, hence $S'P$ is a line of intersection of two corresponding planes; similarly for SP.]

If the plane SPS' corresponds to itself the cone becomes a plane; for take any plane α through SS', the two pencils of planes through SP and $S'P$ meet α in corresponding lines SQ, $S'Q$ and the locus of Q is a *line* p, since for the pencils (S, α), (S', α) the line SS' corresponds to itself, (because to the plane SPS' the plane $S'PS$ corresponds), hence the lines of the system through P, being the lines PQ, lie in the plane (P, p).

Every regulus or cone of the system passes through each singular point, since to a plane through such a point P corresponds a plane which must also pass through P.

It has been shown that through each point which is not a singular point one line of the system passes; it remains to determine the *class* of the system of lines, *i.e.* the number in any plane.

I. *When the correspondence is such that SS' is a self-corresponding line the class is unity;* for if P is a singular point, to the plane PSS' corresponds the plane $PS'S$, *i.e.* this plane corresponds to itself and since SS' corresponds to itself the locus of intersection of corresponding lines in it is a line p of which every point is a singular point, therefore corresponding planes through S and S' meet p in the same point. Singular points not on p are seen for a similar reason to lie on another line p' which is intersected by all lines of the system. Thus the system of lines is formed by all the lines which meet the two lines p and p'; in any given plane there is one line of the system, viz. the join of the points in which the plane meets p and p'. There are two self-corresponding planes in the sheaves, viz. (SS', p), (SS', p').

II. *When the correspondence is such that the sheaves have only one self-corresponding plane, the class of the system is two;* for in this plane there is a set of singular points lying on a conic c^2 through S and S', and at a point P of c^2 the cone of P is a plane (see above); the planes of two such points P and Q meet in a line v every point of which is singular, for through any point R of v there are *two* lines, *i.e.* RP, RQ, belonging to the system and therefore an infinite number, also since v meets the given plane it intersects c^2. The lines of the system are hence the lines joining the points of v to those of c^2, the lines in any plane α are the joins of the point (v, α) to the points (c^2, α), *i.e.* two in number.

III. *When the collinear sheaves have no self-corresponding element the class of the system is three;* for if p is the intersection of two corresponding planes, to the pencil of lines (S, p) corresponds the pencil of lines (S', p) and these two pencils determine on p a

(1, 1) correspondence; if P and Q are its united points they are singular points of the system of lines and on P there are *no other* singular points; also if P is any singular point the plane PSS' contains the three singular points P, S, S' and no other, since if in it another singular point existed the plane PSS' would be self-corresponding as containing two pairs of corresponding lines; again since the cones of any two singular points P and P' each contain all the singular points, the locus of singular points consists of the partial intersection of two quadric cones having the generator PP' in common, *i.e.* a *twisted cubic**. The lines of the system consist of the chords of this cubic and those in a plane are the joins of the intersections of the plane and the cubic, three in number.

The following property of the twisted cubic is of importance:

If x is any chord of the curve and A, B, C, D given points on the curve the four planes xA, xB, xC, xD have a constant double ratio; for if any point P of the curve be joined to the other points of the curve we have a "cone of the system" and if x be any generator of this cone (xA, xB, xC, xD) is constant, since this merely expresses the property that the double ratio of the lines joining a variable point on a conic to four fixed points on the conic is constant, hence for *all* the chords of the cubic through P the theorem is true and if Q be *any other* point of the cubic since the cones for P and Q have one generator common, the theorem is also true for Q.

The locus of intersection of three corresponding planes of three pencils of planes which have mutually a (1, 1) *correspondence, is a twisted cubic of which the axes of the pencils are chords;* for the lines of intersection of corresponding planes of two of the pencils form a regulus and to each pair of corresponding planes there corresponds a plane of the third pencil, hence to each line of the regulus one plane of the third pencil corresponds, and it was seen (Art. ix) that the locus of intersection of corresponding members of a regulus and pencil of planes is a twisted cubic.

xiii. **Collinear Plane Systems.** By aid of two collinear sheaves a (1, 1) correspondence can be established between the points of any two planes, *i.e.* if any line of the sheaf centre S meets one plane in P and the corresponding line of the sheaf centre S' meets the other plane in P', then P and P' are a pair of corresponding points in the two planes. Further if the two planes are superposed on each other a (1, 1) correspondence of points of this plane arises, so that to each point of the plane correspond two points P', P'' according as P is supposed to belong to *one* of these (indefinitely near) planes or to the *other*; the points of the plane

* See Salmon's *Geometry of Three Dimensions*, third edition, p. 304.

are then said to form a *collinear* system; conversely if for each pair of corresponding points P, P' of a collinear plane system the point P is joined to S and the point P' to S', when S and S' are any two points of space, two collinear sheaves are obtained.

Thus a collinear plane system has three self-corresponding points, viz. the points where the twisted cubic of the sheaves (S) and (S') meet the plane.

The collineation of a plane system is determined if to any four points P, Q, R, S are assigned as corresponding points four other points P', Q', R', S'; which may be shown as follows:—to the intersection O of the lines PQ and RS will correspond the intersection O' of the corresponding lines $P'Q'$ and $R'S'$, and if X be any point on PQ, the corresponding point X' on $P'Q'$ is determined from the equation $(OPQX) = (O'P'Q'X')$, similarly for corresponding points on PR, &c.; also if x be any line through P the corresponding line x' through P' is determined from the equation

$$(x, PQ, PR, PS) = (x', P'Q', P'R', P'S');$$

similarly for corresponding lines through Q, R, S; hence to any line which meets PQ, RS in X, Y will correspond the line which meets $P'Q'$, $R'S'$ in the corresponding points X', Y'; and to any point whose joins to P and Q are x and y will correspond the point which is the intersection of the corresponding lines through P' and Q'.

It follows that the collineation of two sheaves is determined when to four lines of one are assigned as respective correlatives four lines of the other.

Hence through six points no four of which are in one plane one and only one twisted cubic can be described, for the points being S, S', A, B, C, D if the lines SA, $S'A$; SB, $S'B$; SC, $S'C$; SD, $S'D$ be made to respectively correspond, the collineation of the sheaves (S) and (S') is determined and hence a cubic which passes through the six points; also if there were another cubic through them we should again obtain by the same means a collineation of the sheaves (S) and (S') which must be the *same* collineation, and hence the cubics must be the *same*.

xiv. **Collineation of systems of space.** A collineation between two three-dimensional spaces Σ and Σ' is a relation such that to any point of Σ corresponds one point of Σ' and to a plane π of Σ through a point P of Σ corresponds one plane π' of Σ' through the corresponding point P' of Σ' and π' contains all points of Σ' corresponding to points of Σ in π; hence to a pencil of planes in Σ will correspond a pencil of planes in Σ' and to a line of Σ corresponds a line of Σ'. The collineation is determined when to five points of Σ (no four of which are coplanar) are assigned as correlatives five points of Σ' (no four of which are coplanar); for if A, B, C, D, E are the given points of Σ and A', B', C', D', E' their corresponding points in Σ' the collineation of the sheaves (A) and (A') is established, since to four lines of one correspond four lines of the other; similarly the collineation of the other pairs of sheaves (B), (B'), &c. is determined; thus to three planes of the sheaves (A), (B), (C) through any point P of Σ will correspond three definite planes of the sheaves (A'), (B'), (C') respectively and the intersection of these latter three gives the point P' corresponding to P.

If the spaces Σ and Σ' are *the same*, we obtain a collineation of the points of one space Σ; in this case it can easily be shown that there are four (real or imaginary) self-corresponding points; for denote by ρ_{AB} the quadric determined by the corresponding pencils of planes whose axes are AB and $A'B'$ and by p_{ABC} the intersection of the planes through ABC, $A'B'C'$ respectively, and by $k_A{}^3$ the twisted cubic corresponding to the two sheaves (A) and (A'), then ρ_{AB} contains both $k_A{}^3$ and $k_B{}^3$ (Art. xii), also the quadrics ρ_{AB}, ρ_{AC} intersect in p_{ABC} and $k_A{}^3$ while the quadrics ρ_{BA}, ρ_{BD} intersect in p_{ABD} and $k_B{}^3$; moreover ρ_{AB}, ρ_{AC}, ρ_{BD} meet in eight points (real or imaginary) and of these, four are the intersections of p_{ABC} and $k_B{}^3$, p_{ABD} and $k_A{}^3$, so that the remaining four are the intersections of $k_A{}^3$ and $k_B{}^3$, hence the twisted cubics $k_A{}^3$ and $k_B{}^3$ meet in four points each of which corresponds to itself, for if K be such a point, to the point of intersection of the lines KA and KB will correspond the point of intersection of the corresponding lines KA' and KB', *i.e.* K itself.

If in a space collineation there are five self-corresponding points *every* point will correspond to itself, for it has been seen that when five pairs of corresponding points are given the collineation is determined uniquely; hence, if five self-corresponding points are given, the *identical* collineation being a *possible* is also the *only* one.

xv. **General Involution.** If a correspondence be established on a curve such that to each of its points P correspond n points of the curve $P_1P_2 \ldots P_n$ and if *one* of the points corresponding to

P_1 be P, one point corresponding to P_2 be P and so on, the correspondence is said to be *Involutory* and is denoted by $[n]$; if in addition the points $P_1 \ldots P_n$ correspond to each other so that the points $PP_1 \ldots P_n$ form a closed system we are said to have an Involution of the $n+1^{\text{th}}$ degree. If on a plane curve there is an Involutory correspondence $[n]$, the lines joining corresponding points envelope a curve, called the *Direction Curve,* which is of class n since from each point of the curve n tangents can be drawn to the curve. If there are two Involutory correspondences $[n]$ and $[n']$ on the same curve, since the two direction curves have nn' common tangents it is clear that the two correspondences must have nn' pairs of corresponding points in common.

If there are two sets of points on a curve defined respectively by coordinates x and y, such that to each point x there correspond m points y, and to each point y there correspond n points x, we have an (m, n) point correspondence on the curve. Such a correspondence is expressed by an algebraic equation of the form $\phi(x, y)=0$, where ϕ is a rational polynomial of degree m in y and n in x. Putting $x=y$ we obtain in general an equation of degree $m+n$ in x, which gives $m+n$ united points or 'coincidences' of the correspondence. This result is known as the Correspondence Principle of Chasles.

If the correspondence is an Involution $m=n$, and if the coefficient of $x^r y^s$ is a_{rs} we have $a_{rs}=a_{sr}$. Should the correspondence be such that one point x coincides with *two* of its corresponding points y, taking this point as the zero point of both x and y we notice that the coefficients a_{00}, a_{01}, a_{10} all disappear, hence the equation which gives the coincidences must have two zero roots, or, if in an Involution $[n]$ a point coincides with two of its corresponding points, this is to be counted as a double coincidence.

xvi. **Involution on a twisted cubic.** Take the quadric cone formed by the chords joining any point of a twisted cubic to its other points, then an Involution [2] on a plane section of this cone is projected into an Involution [2] on the cubic. Another Involution [2] on the cubic is made by the pencil of planes through any line l; this is a *cubic* Involution, and is projected into a cubic Involution on the given conic; moreover since the two Involutions on the conic have in common four pairs of corresponding points so also will the two Involutions on the cubic; it follows that four lines joining corresponding pairs of points in the first Involution on the cubic meet l. Hence the lines joining corresponding points in any Involution [2] on a twisted cubic form a ruled surface of the fourth degree; through each point of the cubic pass two generators of this surface for which therefore the cubic is a *double curve.*

If it once occurs in this Involution that the points Q and R which correspond to P also correspond to each other the Involution will be cubic; for the plane PQR since it contains three lines of a ruled quartic must cut this surface in another line l which does not meet the cubic curve; any plane through l will meet the cubic in three points P', Q', R' and the line joining any two of these latter points, *e.g.* P' and Q', meets the surface in five points, viz. twice in both P' and Q' and once in the point $(P'Q', l)$, and hence must lie altogether in the surface; the generators of the surface therefore consist of all the chords of the cubic which meet l.

xvii. **[2, 2] Correspondences.**

A [2, 2] correspondence of points upon two lines is given by an equation of the form

$$x^2u_1+xu_2+u_3=0 \quad\text{.............................(i)},$$

where u_1, u_2, u_3 are quadratic expressions in y, the points P of one line being determined by x, the points Q of the other by y. There are four points x for each of which the two corresponding values of y coincide; such a point is called a Branch Point, so that there are four branch points X_1, X_2, X_3, X_4 on one line, and four Y_1, Y_2, Y_3, Y_4 on the other line; it will now be shown that

$$(X_1X_2X_3X_4)=(Y_1Y_2Y_3Y_4).$$

For writing $x'=(x-X_1)/(x-X_2)$, a (1, 1) correspondence is established between points P and P' of one line, hence, (Art. ii),

$$(P_1P_2P_3P_4)=(P_1'P_2'P_3'P_4');$$

similarly if $y'=(y-Y_1)/(y-Y_2)$, substituting for x and y in (i), we obtain another [2, 2] correspondence between points P' and Q' of the two lines. It is obvious that when P is a branch point of the *first*, the corresponding point P' is a branch point of the *second* [2, 2] correspondence; it follows that when x' is zero or infinity we obtain branch points for P', similarly for Q', therefore the [2, 2] correspondence between P' and Q' must be determined by an equation of the form

$$x'^2(y'+a)^2+2x'(by'^2+2cy'+d)+b^2(y'+e)^2=0,$$

in which $d^2=a^2b^2e^2$.

The equations to determine the two remaining branch points of the second correspondence on either line are quadratic: if their roots are x_3', x_4'; y_3', y_4'; it is seen by inspection of these quadratics that

$$\text{if } d=abe,\ \frac{x_3'}{x_4'}=\frac{y_3'}{y_4'};\ \text{or, } (x_3', 0, x_4', \infty)=(y_3', 0, y_4', \infty);$$

$$\text{if } d=-abe,\ \frac{x_3'}{x_4'}+\frac{y_3'}{y_4'}=1;\ \text{or, } (x_3', 0, x_4', \infty)=(y_3', y_4', 0, \infty);$$

so that for a definite assignment of the suffixes

$$(X_1X_2X_3X_4)=(Y_1Y_2Y_3Y_4).$$

CHAPTER I.

SYSTEMS OF COORDINATES.

1. In the analytical treatment by Des Cartes of the Geometry of Space, the point is the space-element; the researches of Poncelet and other geometers, and in particular the Principle of Duality, lead naturally to the plane as a space-element; finally to Plücker* is due the conception of a geometry in which the line serves as the element of space. Just as the point and plane are defined by coordinates and the investigation of loci of points and envelopes of planes is conducted by algebraical methods in the older geometry, so in that with which this work is concerned, line-coordinates are employed, and loci of lines are by their means discussed. The object of the present treatise is then, mainly, the investigation of the properties of the assemblage of lines which satisfy one or more given conditions, *i.e.* of lines whose coordinates satisfy one or more equations of given form.

If only one condition is imposed the lines which fulfil it are said to form a Complex, and since a line has four coordinates, it is clear that the lines of a complex are triply infinite, or ∞^3, in number. If a double condition or two conditions are imposed we have a Congruence†, this is seen to consist of ∞^2 lines. If three conditions are specified we have ∞^1 lines forming a ruled surface. A fact which has most important bearings upon our subject is that since a straight line may be regarded either as *the locus of its points* or as *the envelope of its planes*, it is found that the propositions of line geometry stand in the same relationship to *points* as to *planes*, or the subject is *dual*.

The present chapter is mainly given to the description of various kinds of coordinates of the straight line.

* *Neue Geometrie des Raumes.*

† The term 'system of lines' is also sometimes employed.

2. The line of intersection of the planes

$$x = rz + \rho,$$
$$y = sz + \sigma,$$

is known, if the values of r, s, ρ, σ are given; these quantities may be taken as four of the coordinates of the line. Another coordinate is then also taken for the following reason; on linearly transforming the Cartesian coordinates, we get the same line represented by the equations

$$x' = r'z' + \rho',$$
$$y' = s'z' + \sigma',$$

where r', s', ρ' and σ' are fractions of the same denominator, and whose numerators and denominators are linear expressions in r, s, σ and ρ, and also $r\sigma - s\rho$. Thus an equation of degree n in r, s, σ, ρ would in general be transformed into one of degree $2n$ in those quantities. To obviate this Plücker introduced a fifth coordinate η, where $\eta = r\sigma - s\rho$. Thus an equation of degree n in r, s, σ, ρ, η is transformed into an equation of the same degree in $r', s', \sigma', \rho', \eta'$.

3. Homogeneous Coordinates. The introduction of homogeneous line-coordinates greatly assists the study of this branch of geometry. Such systems will now be discussed.

Two points of a line being $(\alpha_1\alpha_2\alpha_3\alpha_4)$, $(\beta_1\beta_2\beta_3\beta_4)$ or as will be written in future, α and β, and two planes through the line being similarly u and v, we have the following four equations

$$\left.\begin{array}{l} u_1\alpha_1 + u_2\alpha_2 + u_3\alpha_3 + u_4\alpha_4 = 0, \\ u_1\beta_1 + u_2\beta_2 + u_3\beta_3 + u_4\beta_4 = 0, \\ v_1\alpha_1 + v_2\alpha_2 + v_3\alpha_3 + v_4\alpha_4 = 0, \\ v_1\beta_1 + v_2\beta_2 + v_3\beta_3 + v_4\beta_4 = 0. \end{array}\right\} \text{..............(i).}$$

Eliminating successively u_1, u_2, u_3, and u_4 from the first two equations we obtain

$$(\alpha_1\beta_2 - \alpha_2\beta_1)\, u_2 + (\alpha_1\beta_3 - \alpha_3\beta_1)\, u_3 + (\alpha_1\beta_4 - \alpha_4\beta_1)\, u_4 = 0;$$

or if

$$\alpha_i\beta_k - \alpha_k\beta_i = p_{ik},$$

we have

$$\left.\begin{array}{l} \cdot \quad p_{12}u_2 + p_{13}u_3 + p_{14}u_4 = 0, \\ p_{21}u_1 + \quad \cdot \quad + p_{23}u_3 + p_{24}u_4 = 0, \\ p_{31}u_1 + p_{32}u_2 + \quad \cdot \quad + p_{34}u_4 = 0, \\ p_{41}u_1 + p_{42}u_2 + p_{43}u_3 + \quad \cdot \quad = 0. \end{array}\right\} \text{..............(ii).}$$

The ratios of the quantities p_{ik} thus defined are taken as the coordinates of the line joining α and β. Observe that if any two other points on the line (α, β) be taken in place of α and β, the ratios of the p_{ik} are not altered.

It should be noticed that for an edge of the tetrahedron of reference all the p_{ik} are zero except the one having the same suffixes, *e.g.* for the edge A_1A_2,

$$p_{13}=p_{14}=p_{34}=p_{42}=p_{23}=0.$$

If the equations (ii) *are satisfied the line "p_{ik}" will lie in the plane u.* The quantities p_{ik} are connected by an equation, for on eliminating the u_i from equations (ii) we have

$$\begin{vmatrix} 0 & p_{12} & p_{13} & p_{14} \\ p_{21} & 0 & p_{23} & p_{24} \\ p_{31} & p_{32} & 0 & p_{34} \\ p_{41} & p_{42} & p_{43} & 0 \end{vmatrix} = 0;$$

whence observing that $p_{ik} = -p_{ki}$ we obtain

$$(p_{12}p_{34} + p_{13}p_{42} + p_{14}p_{23})^2 = 0.$$

The same result is obtained by developing the zero determinant

$$\begin{vmatrix} \alpha_1 & \alpha_2 & \alpha_3 & \alpha_4 \\ \beta_1 & \beta_2 & \beta_3 & \beta_4 \\ \alpha_1 & \alpha_2 & \alpha_3 & \alpha_4 \\ \beta_1 & \beta_2 & \beta_3 & \beta_4 \end{vmatrix},$$

thus, $$2(p_{12}p_{34} + p_{13}p_{42} + p_{14}p_{23}) = 0 \quad\text{..................(iii).}$$

This then is an identical relation connecting the six coordinates of a line. It is usual to denote the left side of the last equation by $2P$. By aid of this relation any one of the equations (ii) may be deduced from two of the others.

Again, if we eliminate the α_i in the same manner from the first and third of the equations (i), we have on writing

$$\pi_{ik} = u_iv_k - u_kv_i,$$

the equations

$$\left.\begin{array}{l} \cdot \quad \pi_{12}\alpha_2 + \pi_{13}\alpha_3 + \pi_{14}\alpha_4 = 0 \\ \pi_{21}\alpha_1 + \quad \cdot \quad + \pi_{23}\alpha_3 + \pi_{24}\alpha_4 = 0 \\ \pi_{31}\alpha_1 + \pi_{32}\alpha_2 + \quad \cdot \quad + \pi_{34}\alpha_4 = 0 \\ \pi_{41}\alpha_1 + \pi_{42}\alpha_2 + \pi_{43}\alpha_3 + \quad \cdot \quad = 0 \end{array}\right\} \quad\text{..............(iv).}$$

4. A fact connected with the duality of the subject is that the quantities p and π are proportional, so that we may take either the p_{ik} or the π_{ik} as coordinates. For since u and v both pass through p we have

$$p_{12}u_2 + p_{13}u_3 + p_{14}u_4 = 0,$$
$$p_{12}v_2 + p_{13}v_3 + p_{14}v_4 = 0.$$

Now eliminating p_{12} it follows that

$$\frac{p_{13}}{p_{14}} = \frac{\pi_{42}}{\pi_{23}},$$

and proceeding similarly we obtain,

$$p_{12} : p_{23} : p_{31} : p_{14} : p_{24} : p_{34} = \pi_{34} : \pi_{14} : \pi_{24} : \pi_{23} : \pi_{31} : \pi_{12} \ldots\ldots \text{(v)}.$$

The equations (iv) are those of the four planes through the line and the different vertices of the tetrahedron of reference; so that a geometrical interpretation is given to the π_{ik}, for π_{12}, π_{13}, π_{14} are proportional to the coordinates of the plane through the line and the vertex A_1, and so on. In like manner from (ii) we notice that p_{12}, p_{13}, p_{14} are proportional to the coordinates of the point of intersection of the line and the coordinate plane α_1, and so on.

Conversely six quantities p_{ik} connected by an equation $P = 0$, and such that $p_{ik} = -p_{ki}$, are the coordinates of a line. For, in this case, since any one of the equations (ii) is deducible from two of the others, the four points they respectively represent lie in one line, and taking two particular values of the u, *i.e.* two particular planes through this line, we derive the equations (v) showing that the coordinates of this line are the p_{ik}.

5. Intersection of two lines. Two lines p and p' will intersect if

$$p_{12}p'_{34} + p'_{12}p_{34} + p_{13}p'_{42} + p'_{13}p_{42} + p_{14}p'_{23} + p'_{14}p_{23} = 0.$$

For α and β being any two points of p, and α', β' two points of p', these four points are coplanar, hence

$$\begin{vmatrix} \alpha_1 & \alpha_2 & \alpha_3 & \alpha_4 \\ \beta_1 & \beta_2 & \beta_3 & \beta_4 \\ \alpha'_1 & \alpha'_2 & \alpha'_3 & \alpha'_4 \\ \beta'_1 & \beta'_2 & \beta'_3 & \beta'_4 \end{vmatrix} = 0,$$

which gives the condition just stated. Observe that this may be written

$$\Sigma p'_{ik} \frac{\partial P}{\partial p_{ik}} = 0.$$

6. Coordinates of Plücker and Lie. Passing from homogeneous to Cartesian coordinates, or writing x', y', z', 1 respectively for $\alpha_1, \alpha_2, \alpha_3, \alpha_4$, and x'', y'', z'', 1 for $\beta_1, \beta_2, \beta_3, \beta_4$, we obtain

$$p_{12} = x'y'' - x''y', \quad p_{23} = y'z'' - y''z', \quad p_{31} = z'x'' - z''x',$$
$$p_{14} = x' - x'', \quad p_{24} = y' - y'', \quad p_{34} = z' - z''.$$

These are the homogeneous coordinates adopted by Plücker*. If now $x'' = x' + dx'$, *i.e.* if the points are consecutive, omitting accents, Lie's coordinates are obtained, viz.

$$p_{12} = xdy - ydx, \quad p_{23} = ydz - zdy, \quad p_{31} = zdx - xdz.$$
$$p_{14} = -dx \quad , \quad p_{24} = -dy \quad , \quad p_{34} = -dz.$$

If the tetrahedron of reference be formed by three mutually perpendicular planes and the plane at infinity, it is clear that the coordinates p_{ik} are proportional to the components along the axes, and the moments about the axes, of a force whose line of action is the given line.

7. Transformation of Coordinates. If a new tetrahedron of reference be chosen, the coordinates of the line p_{ik} will become p'_{ik}, where

$$p'_{ik} = \alpha'_i\beta'_k - \alpha'_k\beta'_i.$$

If the equations of transformation are

$$\mu . x'_i = a_{i1}x_1 + a_{i2}x_2 + a_{i3}x_3 + a_{i4}x_4,$$

p'_{ik} is obviously a linear function of the p_{ik}, so that

$$p'_{ik} = A_{ik,12}\,p_{12} + A_{ik,13}\,p_{13} + \dots\dots\dots$$

The six coefficients of p_{12} on the right are proportional to the coordinates of the edge A_1A_2 of the old tetrahedron of reference with regard to the new one, as is seen by putting

$$p_{13} = p_{14} = p_{23} = p_{34} = p_{42} = 0$$

on the right-hand side of the equations, and so for the other coefficients.

8. Generalized Coordinates. In place of the p_{ik} we may use any six given linear functions of them as coordinates. Denoting for convenience the p's by $p_1p_2 \dots p_6$ and the new coordinates by $q_1q_2 \dots q_6$, the identity $P = 0$ will be replaced by a new quadratic function of the q's equated to zero. Let the latter be $\omega(q)$† $= 0$; then $P(p) \equiv \omega(q)$, hence

$$\frac{\partial P}{\partial p_i} = \sum_k \frac{\partial \omega}{\partial q_k}\frac{\partial q_k}{\partial p_i};$$

* See *Neue Geometrie des Raumes*, Bd I. S. 2.

† The notation ω, Ω appears to be due to M. Koenigs, see *La Géométrie réglée*, p. 9.

now let p and p' be two lines denoted by q and q' in the second system of coordinates, then

$$\sum_i p'_i \frac{\partial P}{\partial p_i} = \sum_i \left(p'_i \sum_k \frac{\partial \omega}{\partial q_k}\frac{\partial q_k}{\partial p_i}\right)$$
$$= \sum_k \left(\frac{\partial \omega}{\partial q_k} \sum_i p'_i \frac{\partial q_k}{\partial p_i}\right)$$
$$= \sum_k \frac{\partial \omega}{\partial q_k} q'_k.$$

Thus the condition for the intersection of two lines is

$$\sum_1^6 q'_k \frac{\partial \omega}{\partial q_k} = 0.$$

9. Coordinates of Klein. The simplest case included in the last transformation is the coordinate-system of Klein*. This is obtained by writing

$$\left.\begin{aligned} x_1 &= p_{12} + p_{34}, & x_3 &= p_{13} + p_{42}, & x_5 &= p_{14} + p_{23},\\ ix_2 &= p_{12} - p_{34}, & ix_4 &= p_{13} - p_{42}, & ix_6 &= p_{14} - p_{23}, \end{aligned}\right\} \ldots\text{(vi)},$$

where $i = \sqrt{-1}$. These quantities x are adopted as the coordinates of the line.

The equation $P = 0$ becomes

$$\sum_1^6 x_i^2 = 0.$$

The condition of intersection of two lines x, y assumes the simple form

$$\sum_1^6 x_i y_i = 0.$$

The last equation will usually be denoted in future by $(xy) = 0$,

and $\quad \sum_1^6 x_i^2 = 0$ by $(x^2) = 0$.

Conversely six quantities x which satisfy $\Sigma x_i^2 \equiv (x^2) = 0$, may be taken as the coordinates of a line. For if the x_i are given, by equations (vi) we can find six quantities p_{ik} which by virtue of the equation $(x^2) = 0$ satisfy $P = 0$ and are therefore the coordinates of a line.

Intersection of consecutive lines. The condition of intersection of two lines x and y in Klein coordinates is in general $(xy) = 0$; when applied however to *consecutive* lines this condition is satisfied identically for terms of the first order, and the condition of inter-

* See *Math. Ann.* Bd. II., *Zur Theorie der Liniencomplexe des ersten und zweiten Grades.*

section of x and $x+dx$ is $(dx^2)=0$. For let x be determined by its two points α and β, then $x+dx$ is determined by the points $\alpha+d\alpha$, $\beta+d\beta$, thus if x_i is the (bilinear) function of α and β $F_i(\alpha, \beta)$ we see that

$$dx_i = F_i(\alpha, d\beta) + F_i(d\alpha, \beta) + F_i(d\alpha, d\beta)$$

where $\Sigma F_i^2(\alpha, d\beta) = 0$, $\Sigma F_i^2(d\alpha, \beta) = 0$, $\Sigma F_i^2(d\alpha, d\beta) = 0$;

also
$$\Sigma F_i(\alpha, \beta) F_i(\alpha, d\beta) = 0,$$

since the lines have the common point α, and

$$\Sigma F_i(\alpha, \beta) F_i(d\alpha, \beta) = 0,$$

since the lines have the common point β, so that $(xdx)=0$, for terms of the first order.

If x and $x+dx$ have a common point, let it be α, then $d\alpha = 0$, and $dx_i = F_i(\alpha, d\beta)$, hence $(dx^2)=0$.

10. Plane Pencil of Lines. Referring to the coordinates p_{ik} it is observed that they are linear in the coordinates of each of the two points α and β, thus the new coordinates x are so also. It follows that if x and y are the coordinates of two lines which join the point α to the points β and γ respectively, then denoting by z the line joining α to the point $\beta+\lambda\gamma$, we have

$$z_i = x_i + \lambda y_i.$$

By giving all values to λ we obtain the lines of the plane pencil determined by the two intersecting lines x and y. Conversely if x and y are two intersecting lines we have

$$(x^2)=0, \quad (xy)=0, \quad (y^2)=0,$$

hence $x+\lambda y$ is a line, and thus is one of the pencil determined by x and y.

11. Double Ratio of four lines of a pencil. The four lines $x+\lambda_1 y$, $x+\lambda_2 y$, $x+\lambda_3 y$, $x+\lambda_4 y$ will pass respectively through the points $\alpha+\lambda_1\beta$, $\alpha+\lambda_2\beta$, $\alpha+\lambda_3\beta$, $\alpha+\lambda_4\beta$, where α and β are points on x and y respectively; hence the double ratio of the four lines is

$$\frac{(\lambda_1-\lambda_2)(\lambda_4-\lambda_3)}{(\lambda_2-\lambda_3)(\lambda_1-\lambda_4)}.$$

12. Von Staudt's Theorem. If α, β are the points in which a line meets the coordinate planes respectively opposite to the vertices A_1, A_3 of the tetrahedron of reference, then $\alpha+\lambda\beta$, $\alpha+\mu\beta$ are the points in which the line meets the coordinate planes opposite to A_2, A_4 if $\alpha_2+\lambda\beta_2=0$, $\alpha_4+\mu\beta_4=0$.

The Double Ratio of these four points on the line is $\frac{\lambda}{\mu} = \frac{\alpha_2\beta_4}{\alpha_4\beta_2}$; but it is easily seen that for the given line

$$p_{12} = -\alpha_2\beta_1,\quad p_{34} = +\alpha_3\beta_4,\quad p_{14} = -\alpha_4\beta_1,\quad p_{23} = -\alpha_3\beta_2,$$

hence the D.R. of the four points in which it meets the coordinate planes is $-\frac{p_{12} \cdot p_{34}}{p_{14} \cdot p_{23}}$.

By a precisely similar process it is obvious that the D.R. of the four planes through the line and the respective vertices of the tetrahedron is equal to $-\frac{\pi_{12} \cdot \pi_{34}}{\pi_{14} \cdot \pi_{23}} = -\frac{p_{34} \cdot p_{12}}{p_{23} \cdot p_{14}}$, or, *the Double Ratio of the points in which a line meets* any *tetrahedron is equal to the* D.R. *of the planes through the line and the vertices of the tetrahedron.*

13. Sheaf and plane system of lines. If β, γ, δ are respective points on the three lines x, y, z which meet in the point α, then by the reasoning just employed the coordinates of the line joining α to the point $\lambda\beta + \mu\gamma + \nu\delta$ are

$$\lambda x_i + \mu y_i + \nu z_i,$$

i.e. any line through the intersection of three concurrent lines x, y, z is $\lambda x + \mu y + \nu z$. All such lines are said to form a "*sheaf*"*. If the lines x, y, and z, on the other hand, lie in the same plane t, then if u, v, w are planes through x, y, z, respectively, the line of intersection of t and any plane $\lambda u + \mu v + \nu w$ through the point (u, v, w), has for its coordinates again $\lambda x_i + \mu y_i + \nu z_i$. Thus if x, y, z are concurrent, $\lambda x + \mu y + \nu z$ includes all the lines of the sheaf through their point of intersection; if x, y, z are coplanar $\lambda x + \mu y + \nu z$ includes all the lines which lie in their common plane, or the "plane system." Taking the case where x, y and z are concurrent, we see that if P is any point on $\lambda x + \mu y + \nu z$, the coordinates of this line being proportional to linear functions of the coordinates of P and α; λ, μ and ν are each linear functions of the coordinates of P. Now if $\lambda = 0$ it was seen that the line, and hence the point P, will lie in the plane of y and z, or,

$\lambda = 0$ is the equation of the plane (y, z);
$\mu = 0$ " " " " (z, x);
$\nu = 0$ " " " " (x, y).

* See Introduction, Art. xi.

14. Closed system of 16 points and 16 planes*.

From Art. 4 we see that the line x may be defined either by the equations

$$x_1 = p_{12} + p_{34},$$
$$ix_2 = p_{12} - p_{34}, \text{ \&c.},$$

or by the equations

$$\sigma . x_1 = \pi_{12} + \pi_{34},$$
$$-\sigma . ix_2 = \pi_{12} - \pi_{34}, \text{ \&c.; (where } \sigma = \pi_{12}/p_{34}).$$

A comparison of these forms shows us that we may regard the first sets of equations as the condition *either* that the line $(x_1, x_2, x_3, x_4, x_5, x_6)$ should pass through the *point* $(\alpha_1, \alpha_2, \alpha_3, \alpha_4)$ *or* that the line $(x_1, -x_2, x_3, -x_4, x_5, -x_6)$ should lie in the *plane* whose coordinates are $(\alpha_1, \alpha_2, \alpha_3, \alpha_4)$; thus one condition involves the other.

If the squares of the coordinates x of a line have given values, we obtain a set of 32 lines having important connexions; namely the different lines obtained by taking x_1 positively and $x_2 \ldots x_6$ with either sign. It will now be shown that these 32 lines together with 16 points and 16 planes form a closed system.

For if we substitute the x coordinates of the line for the π_{ik} in (iv), (Art. 3), we obtain the conditions that the line x should contain the point α, and from them may be derived the following, (which can easily be directly verified),

$$\left.\begin{aligned} x_1(\alpha_1\alpha_2 - \alpha_3\alpha_4) + ix_2(-\alpha_1\alpha_2 - \alpha_3\alpha_4) + x_3(\alpha_1\alpha_3 + \alpha_2\alpha_4) \\ + ix_4(\alpha_2\alpha_4 - \alpha_1\alpha_3) = 0, \\ x_1(\alpha_1\alpha_2 + \alpha_3\alpha_4) + ix_2(\alpha_3\alpha_4 - \alpha_1\alpha_2) + x_5(\alpha_1\alpha_4 - \alpha_2\alpha_3) \\ + ix_6(-\alpha_1\alpha_4 - \alpha_2\alpha_3) = 0, \end{aligned}\right\} \ldots \text{(vii)}$$

together with two other equations derivable from these.

If in these equations the signs of any two of the α's are changed the coefficients of the x_i are thereby either unaltered or altered only in sign. The same is true if any two pairs of the α's are interchanged (*e.g.* α_1 and α_2, α_3 and α_4); and thus by a combination of these two methods of change the coefficients are altered at most in *sign*, so that by suitably changing the signs of the x's we return to equations (vii). The arrangement of signs for the six quantities x is easily found to be different in the different cases, we thus have 16 points, viz.

$$\begin{array}{llll} \alpha_1, \ \alpha_2, \ \alpha_3, \ \alpha_4; & \alpha_2, \ \alpha_1, \ \alpha_4, \ \alpha_3; & \alpha_4, \ \alpha_3, \ \alpha_2, \ \alpha_1; & \alpha_3, \ \alpha_4, \ \alpha_1, \ \alpha_2 \\ \alpha_1, \ \alpha_2, -\alpha_3, -\alpha_4; & \alpha_2, \ \alpha_1, -\alpha_4, -\alpha_3; & \alpha_4, \ \alpha_3, -\alpha_2, -\alpha_1; & \alpha_3, \ \alpha_4, -\alpha_1, -\alpha_2 \\ \alpha_1, -\alpha_2, -\alpha_3, \ \alpha_4; & \alpha_2, -\alpha_1, -\alpha_4, \ \alpha_3; & \alpha_4, -\alpha_3, -\alpha_2, \ \alpha_1; & \alpha_3, -\alpha_4, -\alpha_1, \ \alpha_2 \\ \alpha_1, -\alpha_2, \ \alpha_3, -\alpha_4; & \alpha_2, -\alpha_1, \ \alpha_4, -\alpha_3; & \alpha_4, -\alpha_3, \ \alpha_2, -\alpha_1; & \alpha_3, -\alpha_4, \ \alpha_1, -\alpha_2 \end{array}$$

through each of which *one* of the 32 lines x passes.

* See Klein, *Math. Ann.* Bd. II.

And, by what was shown above, in each of the 16 *planes* having these coordinates will lie *one* of the 16 remaining lines; and having given one point or one plane the other points and planes are determined, and *the same system of* 16 *points and planes is arrived at, with whichever of the above points or planes we start.* Any point or plane of space determines such a system; so that by aid of the table just given, all the points of space are divided into such sets of 16; similarly for the planes of space.

An important fact connected with the above system is the following:—since the point $(\alpha_1, \alpha_2, \alpha_3, \alpha_4)$ clearly lies in the six planes

$$(\alpha_2, -\alpha_1, \alpha_4, -\alpha_3),\ (\alpha_2, -\alpha_1, -\alpha_4, \alpha_3),\ (\alpha_3, -\alpha_4, -\alpha_1, \alpha_2),$$
$$(\alpha_3, \alpha_4, -\alpha_1, -\alpha_2),\ (\alpha_4, \alpha_3, -\alpha_2, -\alpha_1),\ (\alpha_4, -\alpha_3, \alpha_2, -\alpha_1),$$

and the plane $(\alpha_1, \alpha_2, \alpha_3, \alpha_4)$ passes through the six points having these coordinates, therefore, from the nature of the system, if any other point of the system *e.g.* $(\alpha_2, \alpha_1, \alpha_4, \alpha_3)$ be taken, we should obtain a similar result, whence it follows that—*through each point of the system there pass six planes of the system, and in each plane of the system there lie six points of the system.*

CHAPTER II.

THE LINEAR COMPLEX.

15. A COMPLEX of lines has been defined, (Art. 1), as the assemblage of lines which satisfy one condition. Thus if $q_1 \dots q_6$ are the general coordinates of the last chapter, the lines whose coordinates satisfy the homogeneous equation of degree n $f(q_1 \dots q_6) = 0$, form such a complex. We have seen that if α and β are any two points of a line the quantities $\rho q_1 \dots \rho q_6$ are each homogeneous linear functions of the coordinates of α and of the coordinates of β: thus $f = 0$ is homogeneous of the nth degree in the coordinates of both α and β. Taking α to be any given point, this equation therefore gives the cone formed by the *lines of the complex through the point* α; and *this cone is seen to be of degree* n.

Similarly the q's of a line being proportional to homogeneous linear functions of the coordinates of any two planes u and v through the line, by taking u in $f = 0$ as any given plane, we have *the curve enveloped by the complex lines which lie in the plane* u; and *this curve is seen to be of the nth class.*

16. The Linear Complex. If n is unity the complex is of the first degree, and we see that in a complex of the first degree all the complex lines through any point lie *in a plane*, the "*polar plane*" of the point, all the complex lines in a plane *pass through a point*, the "*pole*" of the plane. If we employ the "p" coordinates the equation of the complex will be of the form

$$a_{12}p_{12} + a_{13}p_{13} + a_{14}p_{14} + a_{34}p_{34} + a_{42}p_{42} + a_{23}p_{23} = 0.$$

Similarly if the coordinates of Klein are used, the equation $\Sigma a_i x_i = 0$, or $(ax) = 0$, represents a linear complex.

The equation $p_{ik} = 0$ represents the complex of lines which intersect the edge of the tetrahedron of reference opposite to

A_iA_k; for if $\alpha_i\beta_k - \alpha_k\beta_i = 0$, where α, β are two points on a line of the complex $p_{ik} = 0$, let the point α be that in which the line meets the face opposite A_i of the tetrahedron of reference, then $\alpha_i = 0$, it follows that either $\alpha_k = 0$ or $\beta_i = 0$ and in either case the line intersects the edge opposite A_iA_k of the tetrahedron of reference.

The equation $x_1 = 0$, *i.e.* $p_{12} + p_{34} = 0$, represents a complex of the utmost importance in this subject, it will be termed a *fundamental* linear complex; there are six fundamental linear complexes, viz.

$$x_1 = 0,\ x_2 = 0,\ x_3 = 0,\ x_4 = 0,\ x_5 = 0,\ x_6 = 0.$$

Now taking $\xi_1, \xi_2, \xi_3, \xi_4$ to be the coordinates of any point on a line of $\Sigma a_{ik}p_{ik} = 0$ through the point α; or writing in this equation $p_{ik} = \alpha_i\xi_k - \alpha_k\xi_i$, and arranging the terms, we observe that the polar plane of α is, (if we write for convenience $a_{ki} = -a_{ik}$),

$$\xi_1(a_{21}\alpha_2 + a_{31}\alpha_3 + a_{41}\alpha_4) + \xi_2(a_{12}\alpha_1 + a_{32}\alpha_3 + a_{42}\alpha_4) + \xi_3(a_{13}\alpha_1 + a_{23}\alpha_2 + a_{43}\alpha_4)$$
$$+ \xi_4(a_{14}\alpha_1 + a_{24}\alpha_2 + a_{34}\alpha_3) = 0 \quad \text{..............(i)},$$

so that if u is the polar plane of α, we have the equations

$$\left.\begin{aligned} \sigma . u_1 &= \quad\cdot\quad + a_{21}\alpha_2 + a_{31}\alpha_3 + a_{41}\alpha_4, \\ \sigma . u_2 &= a_{12}\alpha_1 + \quad\cdot\quad + a_{32}\alpha_3 + a_{42}\alpha_4, \\ \sigma . u_3 &= a_{13}\alpha_1 + a_{23}\alpha_2 + \quad\cdot\quad + a_{43}\alpha_4, \\ \sigma . u_4 &= a_{14}\alpha_1 + a_{24}\alpha_2 + a_{34}\alpha_3 + \quad\cdot \\ 0 &= u_1\alpha_1 + u_2\alpha_2 + u_3\alpha_3 + u_4\alpha_4 \end{aligned}\right\} \text{................(ii)}.$$

If two complex lines intersect, their point of intersection is the pole of their plane. It is clear from the foregoing that to each point of space a unique polar plane is attached; this may be also seen directly, for if the polar planes of the points m and n coincide, it is necessary that

$$(m_2 - \rho n_2)\, a_{21} + (m_3 - \rho n_3)\, a_{31} + (m_4 - \rho n_4)\, a_{41} = 0 \text{ ...... (iii)},$$

together with three other similar equations, and m, n being supposed to be *different* points, it follows that

$$\begin{vmatrix} 0 & a_{21} & a_{31} & a_{41} \\ a_{12} & 0 & a_{32} & a_{42} \\ a_{13} & a_{23} & 0 & a_{43} \\ a_{14} & a_{24} & a_{34} & 0 \end{vmatrix} = 0;$$

the determinant is skew-symmetrical because $a_{ik} = -a_{ki}$, and has the value $(a_{12}a_{34} + a_{13}a_{42} + a_{14}a_{23})^2$. The quantity in brackets is in general different from zero, (see Art. 18), and hence the equations

(iii) cannot coexist, or, the *polar planes of the points of space are all different.*

17. Polar Lines. The equations (ii) connecting a point a and its polar plane u, show that if

$$a_i = m_i + \lambda n_i,$$

i.e. if we suppose a to describe the line joining the fixed points m and n, then $\sigma \,.\, u_i = M_i + \lambda N_i$

where $M_i = \sum_k a_{ki} m_k,\ N_i = \sum_k a_{ki} n_k,\ (a_{ii} = 0),$

i.e. u turns round a *fixed line* (viz. the line of intersection of the planes M and N); conversely, if $\sigma u_i = M_i + \lambda N_i$, it follows from (ii) that $a_i = m_i + \lambda n_i$.

Two lines thus connected are said to be *polar* to each other.

Polar lines do not intersect unless they coincide, for m and n being any two points and p the intersection of their polar planes, the polar plane of m is the plane through m and p, the polar plane of n is the plane through n and p, so that if p met the line mn these two planes would coincide, which we have seen to be impossible.

If the line mn belongs to the complex it lies in the polar planes of both m and n, and hence coincides with its polar line p.

The proposition just established shows that the polar planes of the points of any given line p' pass through the same line p, from which it follows that *any line meeting both p and p' belongs to the complex*; hence *the polar planes of the points of p will all pass through p'.*

The polar lines of the lines through any point P lie in one plane, the polar plane of P. *Any complex line which meets p must also meet p'*, for let a complex line x meet p in P, then since *all* the complex lines through P belong to the plane determined by P and p', x must lie in this plane, and therefore meet p'.

18. The Invariant of a linear complex*. Taking the equation of the complex in the form

$$\sum_1^6 a_i q_i = 0,$$

* See Klein, *Math. Ann.* v., *Differentialgleichungen in der Liniengeometrie.*

and the identical relation (Art. 8) as

$$\omega(q) = 0,$$

the complex has an Invariant.

For writing $\quad \omega(q) = \alpha_{11}q_1^2 + \ldots + 2\alpha_{rs}q_rq_s + \ldots,$

since the a_i are contragredient to the q_i we know that

$$\begin{vmatrix} \alpha_{11} \ldots\ldots \alpha_{16} & a_1 \\ \alpha_{21} \ldots\ldots \alpha_{26} & a_2 \\ \ldots\ldots\ldots\ldots & \\ \alpha_{61} \ldots\ldots \alpha_{66} & a_6 \\ a_1 \ldots\ldots a_6 & 0 \end{vmatrix} \div \begin{vmatrix} \alpha_{11} \ldots\ldots \alpha_{16} \\ \ldots\ldots\ldots\ldots \\ \alpha_{61} \ldots\ldots \alpha_{66} \end{vmatrix}$$

is invariable for linear transformations effected on the q_i. The numerator of this fraction is an Invariant of the complex. It will be denoted by $\Omega(a)$.

19. The Special Complex. If the invariant $\Omega(a)$ is zero we then have the system of coexistent equations

$$\frac{\partial \omega(b)}{\partial b_i} = 2\rho a_i, \quad (i = 1, 2, \ldots 6),$$

$$(ab) = 0,$$

and hence also $\quad \omega(b) = \rho(ab) = 0:$

or the complex may be written

$$\Sigma x_i \frac{\partial \omega}{\partial b_i} = 0,$$

where b is a line.

Thus each line of the complex cuts the line b, (Art. 8), which is called the *directrix* of the complex. The complex is here said to be *special*. The coordinates of b are proportional to

$$\frac{\partial \Omega}{\partial a_i};$$

for since $\quad -\Omega(a) = A_{11}a_1^2 + \ldots + 2A_{rs}a_ra_s + \ldots$

where A_{rs} is the coefficient of α_{rs} in the discriminant Δ of ω,

$$-\frac{1}{2}\frac{\partial \Omega}{\partial a_i} = A_{i1}a_1 + \ldots$$

and from the equations $\quad \dfrac{\partial \omega}{\partial b_i} = 2\rho a_i$

we see that $\quad \Delta . b_i = \rho(A_{i1}a_1 + \ldots)$

or $\quad b_i = \dfrac{-\rho}{2\Delta} \cdot \dfrac{\partial \Omega}{\partial a_i}.$

When the Plücker coordinates are used, since

$$\omega(p) = 2(p_{12}p_{34} + p_{13}p_{42} + p_{14}p_{23}),$$

we see that $\qquad \Omega(a) = 2(a_{12}a_{34} + a_{13}a_{42} + a_{14}a_{23}),$

or, $\Omega(a)$ and $\omega(a)$ have the *same form.*

When Klein coordinates are used

$$\omega(x) = \sum_1^6 x_i^2, \qquad -\Omega(a) = \sum_1^6 a_i^2.$$

Generally, in any system of coordinates in which each coordinate appears in $\omega(q)$ only once, the first minors of Δ are $\dfrac{\Delta}{a_{ik}}$ and thus

$$-\frac{\Omega(a)}{\Delta} = \Sigma \frac{a_i a_k}{a_{ik}}.$$

20. Coordinates of polar lines. If, when Klein coordinates are used, z and z' are polar lines with respect to a linear complex $\Sigma a_i x_i = 0$, it will now be proved that

$$\rho . z_i' = z_i + \lambda a_i;$$

for the coordinates of any line of $(ax) = 0$ which meets z, satisfy the equations $(ax) = 0$, $(zx) = 0$ and hence the equation

$$\Sigma(z_i + \lambda a_i) x_i = 0,$$

for all values of λ.

The last complex is special if $\Sigma(z_i + \lambda a_i)^2 = 0$, which gives two values for λ, viz. zero and $-2\dfrac{(az)}{(a^2)}$; hence, with this value of λ, $z + \lambda a$ is a line which meets every line x of the given complex which meets z, *i.e.* $z + \lambda a$ is the line z' (Art. 17).

An important case arises when $(ax) = 0$ is a *fundamental* complex, *e.g.* $x_1 = 0$, in which case $a_1 = 1$, $a_2 = a_3 = a_4 = a_5 = a_6 = 0$. The polar of any line z has the coordinates $z_1 + \lambda$, z_2, z_3, z_4, z_5, z_6 which requires that $\lambda = -2z_1$, *i.e.* if z and z' are polar for a fundamental complex $x_i = 0$, then $z_k = z_k'$, except for $k = i$ when $z_k + z_k' = 0$. The 32 lines of Art. 14 may be obtained by starting from one of them and taking the successive polars for the 6 fundamental complexes.

21. Relations between the functions ω and Ω. The following identities, which may easily be verified, are useful. Denoting $\dfrac{1}{2}\dfrac{\partial\omega}{\partial z_i}$ by Z_i where $z_1 \ldots z_6$ are any quantities,

$$\Sigma \frac{\partial\Omega}{\partial a_i} \cdot \frac{\partial\omega}{\partial z_i} = -4\Delta . (az),$$

$$\frac{1}{2}\frac{\partial\Omega(Z)}{\partial Z_i} = -\Delta . z_i,$$

$$\Omega(Z) = \tfrac{1}{2}\Sigma Z_i \frac{\partial \Omega}{\partial Z_i} = -\Delta \,.\, \omega(z),$$

$$\omega\left(\frac{1}{2}\frac{\partial \Omega}{\partial Z}\right) = \Delta^2 \,.\, \omega(z).$$

From these equations it follows that if a complex line x meets the line z it will also meet another line z'.

For if $$(ax)=0, \quad \left(\frac{\partial \omega}{\partial z} x\right)=0,$$

x belongs to each complex of the system

$$\left(x\left(\lambda a + \frac{\partial \omega}{\partial z}\right)\right) = 0;$$

of these complexes two are special, viz. those corresponding to the two values of λ given by $$\Omega(\lambda a + 2Z) = 0$$

or $$\lambda^2 \Omega(a) + 2\lambda\Sigma\left(Z_i \frac{\partial \Omega}{\partial a_i}\right) + 4\Omega(Z) = 0,$$

and since z is a line, $\omega(z)=0$, *i.e.*, $\Omega(Z)=0$.

Thus the values of λ are 0 and

$$-2\frac{\Sigma\left(Z_i \dfrac{\partial \Omega}{\partial a_i}\right)}{\Omega(a)} = \frac{4\Delta(az)}{\Omega(a)}.$$

The directrix of a special complex is (Art. 19)

$$\frac{\partial \Omega(\lambda a + 2Z)}{\partial(\lambda a_i + 2Z_i)}, \quad \text{or} \quad \lambda \frac{\partial \Omega(a)}{\partial a_i} + 2\frac{\partial \Omega(Z)}{\partial Z_i},$$

i.e. taking $$\lambda = \frac{4\Delta(az)}{\Omega(a)}$$

we have $$\frac{4\Delta(az)}{\Omega(a)}\frac{\partial \Omega(a)}{\partial a_i} - 4\Delta z_i \equiv \rho z_i'.$$

Multiplying by a_i and adding we obtain

$$4\Delta(az) = \rho(az').$$

Thus finally

$$\frac{z_i}{(az)} + \frac{z_i'}{(az')} = \frac{1}{\Omega(a)}\frac{\partial \Omega(a)}{\partial a_i}$$

is the equation connecting the coordinates of polar lines in general coordinates.

22. Diameters. If a line lie in the plane at infinity its polar line is called a *diameter* and passes through the pole of the plane at infinity for the complex. Taking a series of parallel planes, their poles lie on a diameter, viz. that corresponding to their common line at infinity, and it follows that *all diameters are parallel.* There is one diameter which is perpendicular to the planes through whose poles it passes, viz. that which joins the poles of the planes which are perpendicular to the diameters. This diameter is called the AXIS of the complex, and is perpendicular to the complex lines which it meets.

23. Reduction of the complex to its simplest form. If we take as two opposite edges of the tetrahedron of reference two polar lines, *e.g.* A_1A_2 and A_3A_4, the other edges will belong to the complex; thus for instance the edge A_1A_3 whose coordinates are given by $p_{12}=p_{14}=p_{23}=p_{34}=p_{42}=0$ belongs to it and hence $a_{13}=0$. Similarly $a_{14}=a_{23}=a_{42}=0$ and the equation of the complex reduces to

$$a_{12}p_{12}+a_{34}p_{34}=0.$$

To refer the complex to Cartesian coordinates take the plane at infinity as one face of the tetrahedron of reference, and then the p coordinates will assume the form given in Art. 6.

Let the axis of the complex be chosen for axis of z and as the edge A_4A_3, the edge A_1A_2, the polar of A_3A_4 will then be at infinity; the edges A_4A_1, A_4A_2, being complex lines, are each perpendicular to A_3A_4, we take them as being also at right angles to each other; the complex is now referred to rectangular axes, of origin A_4; then, (Art. 6), $p_{12}=xy'-x'y$, $p_{34}=z-z'$, and since when A_1A_2 and A_3A_4 are polar, the complex is $a_{12}p_{12}+a_{34}p_{34}=0$, its equation is now seen to be

$$a_{12}(xy'-yx')+a_{34}(z-z')=0$$

where xyz, $x'y'z'$ are the coordinates of any two points on a complex line. If in this equation z and z' be increased by any quantity h the equation is not altered, *i.e.* a complex line may be translated in any way parallel to the axis without ceasing to belong to the complex; also $xy'-yx'$ is unaltered by a rotation of the line about the axis, hence if all the lines of a complex are subjected to a screw motion about the axis, the complex itself is not altered.

It is easily seen that if r is the shortest distance between a complex line and the axis, and θ the inclination of this complex line to the axis,

$$-\frac{a_{34}}{a_{12}}=\frac{xy'-yx'}{z-z'}=r\tan\theta,$$

thus for all the lines of a linear complex the quantity $r\tan\theta$ is the same. The quantity $-a_{34}/a_{12}$ is called the Chief Parameter of the complex. Therefore the lines of the complex are the tangents to a series of helices, each helix being on a cylinder of radius r and axis that of the complex, the angle of the helix being

$$\tan^{-1}\left(-\frac{a_{34}}{a_{12}}\cdot\frac{1}{r}\right)^{*}.$$

Using the coordinates of Lie, Art. 6, it is seen that the equation of a linear complex is, for these axes,

$$x\,dy - y\,dx = k\,dz.$$

24. Two complexes have one pair of polar lines in common. Among the complexes of the system

$$\Sigma\,(a_i + \lambda b_i)\,x_i = 0,$$

two are "special," viz. those in which λ is one of the roots of

$$\Omega\,(a_i + \lambda b_i) = 0, \qquad \text{(Art. 18)}$$

or

$$\Omega\,(a) + \lambda \Sigma b_i \frac{\partial \Omega}{\partial a_i} + \lambda^2 \Omega\,(b) = 0.$$

The roots of this equation are, in general, different; let them be denoted by λ_1 and λ_2.

Thus the lines of $\Sigma\,(a_i + \lambda_1 b_i)\,x_i = 0$ meet the line

$$\frac{\partial \Omega}{\partial a_i} + \lambda_1 \frac{\partial \Omega}{\partial b_i}, \text{ or } z; \qquad \text{(Art. 19.)}$$

and the lines of $\Sigma\,(a_i + \lambda_2 b_i)\,x_i = 0$ meet the line $\frac{\partial \Omega}{\partial a_i} + \lambda_2 \frac{\partial \Omega}{\partial b_i}$, or z'.

Now lines which belong to each of these two complexes must also belong to the two $(ax) = 0$, $(bx) = 0$, and conversely. Thus z and z' are polar lines both in $(ax) = 0$, and in $(bx) = 0$; *and all the lines common to* $(ax) = 0$ *and* $(bx) = 0$ *meet* z *and* z'.

25. Complexes in Involution. If $(ax) = 0$ and $(bx) = 0$ undergo the same transformation of coordinates, the coefficients a_i and b_i are cogredient and $\Omega\,(a + \lambda b)$ is an invariant of

$$\Sigma\,(a_i + \lambda b_i)\,x_i = 0.$$

It follows that $\Sigma a_i \frac{\partial \Omega}{\partial b_i}$ is an Invariant of the two complexes.

Let the complexes be referred to a tetrahedron in which their

* From this property the linear complex is called Strahlengewinde by R. Sturm, see *Liniengeometrie*, I. p. 94. The two cases where the chief parameter is positive or negative are distinguished by Plücker as Right-wound or Left-wound, *i.e. right-handed* or *left-handed*.

two common polar lines are opposite edges; their equations then are

$$a_{12}p_{12} + a_{34}p_{34} = 0,$$
$$b_{12}p_{12} + b_{34}p_{34} = 0,$$

and their mutual invariant $\Sigma a_i \frac{\partial \Omega}{\partial b_i}$ becomes $a_{12}b_{34} + a_{34}b_{12}$: we shall examine the consequences of the vanishing of this quantity. Let α be any point on a line common to the two complexes, its polar planes in them have for coordinates, (Art. 16),

$$-a_{12}\alpha_2,\ a_{12}\alpha_1,\ -a_{34}\alpha_4,\ a_{34}\alpha_3;$$
$$-b_{12}\alpha_2,\ b_{12}\alpha_1,\ -b_{34}\alpha_4,\ b_{34}\alpha_3.$$

Also since $a_{12}b_{34} + a_{34}b_{12} = 0$, if $\kappa = \frac{a_{34}}{a_{12}} = -\frac{b_{34}}{b_{12}}$, it follows that

the polar plane of α in $(ax) = 0$ is $(-\alpha_2, \alpha_1, -\kappa\alpha_4, \kappa\alpha_3)$,

" " " " " $(bx) = 0$ is $(-\alpha_2, \alpha_1, \kappa\alpha_4, -\kappa\alpha_3)$;

and if β is any other point on the line,

the polar plane of β in $(ax) = 0$ is $(-\beta_2, \beta_1, -\kappa\beta_4, \kappa\beta_3)$,

" " " " " $(bx) = 0$ is $(-\beta_2, \beta_1, \kappa\beta_4, -\kappa\beta_3)$.

From this it is clear that if the polar plane of β in $(ax) = 0$ is the polar plane of α in $(bx) = 0$, then the polar plane of β in $(bx) = 0$ is the polar plane of α in $(ax) = 0$.

Hence *if p be a line common to two complexes* $(ax) = 0$ *and* $(bx) = 0$ *for which* $\Sigma a_i \frac{\partial \Omega}{\partial b_i} = 0$, *the correlation of planes through p, obtained by taking the polar planes of the points of p for each complex, is an Involution.*

The complexes are said to be themselves in Involution. A similar method of proof shows that the correlation of points on a line p common to the complexes, obtained by taking the poles in the two complexes of the planes through p, is an Involution.

We saw, (Art. 24), that each line common to the two complexes meets the lines z and z' which are polar for each complex. Let N and N' be the points where a common line x meets z and z' respectively; it is important to observe that N and N' are the *double points* of the involution determined on x; for the polar plane of N in *each* complex is the plane (N, z'), thus N corresponds to itself in the involution, similarly for N'.

If one of the complexes, say $(ax) = 0$, is special, its directrix belongs to $(bx) = 0$, for the condition of involution gives

$$\left(b \frac{\partial \Omega}{\partial a}\right) = 0;\ \text{this is sometimes written } \Omega(a \mid b) = 0.$$

If both complexes are special, the directrix of each belongs to the other, *i.e.* the two directrices intersect.

The following properties of complexes in involution should be noticed.

If $(ax)=0$, $(bx)=0$, are any two linear complexes in involution and x is any line of the first complex, the polar of x with regard to the second complex is x', where $\rho \,.\, x_i' = x_i + \lambda b_i$, (Art. 20), Klein coordinates being used; and since $(ax)=0$, $(ab)=0$, it follows that $(ax')=0$, or x' *belongs to the first complex.*

If two lines z and z' are polar with regard to a given complex $(ax)=0$, $\rho \,.\, z_i' = z_i + \lambda a_i$; and if $(bx)=0$ is *any* linear complex which contains z and z', since $(bz)=0$, $(bz')=0$ it follows that $(ab)=0$, *i.e. any* linear complex through a pair of polar lines for a given complex is in involution with the given complex.

Let the equations of two complexes referred to their respective axes, (Art. 22), be $p_{12} - \kappa_1 p_{34} = 0$, $p_{12} - \kappa_2 p_{34} = 0$; then if d is the shortest distance of the axes and $\frac{\pi}{2} - \phi$ the angle between them, the equation of the second complex referred to the same system of coordinates as the first, is obtained by writing in its equation respectively for x, y, and z, the values $x - d$, $y \sin\phi - z\cos\phi$, $y\cos\phi + z\sin\phi$. The equation of the second complex then takes the form

$$p_{12}\sin\phi - p_{34}(\kappa_2 \sin\phi + d\cos\phi) - p_{13}\cos\phi + p_{42}(\kappa_2\cos\phi - d\sin\phi) = 0.$$

This is seen to be in involution with the first complex if

$$(\kappa_1 + \kappa_2)\sin\phi + d\cos\phi = 0\text{*}:$$

which is the condition of involution of two complexes in terms of their Chief Parameters.

The common pair of polar lines of the given complexes are determined as in Art. 24, and it is clear that for each of them $p_{23}=0$, $p_{14}=0$. For if A and B are the given complexes, and α, β these polar lines,

$$\rho \,.\, A \equiv \Sigma\alpha_{ik}p_{ik} + \lambda_1\Sigma\beta_{ik}p_{ik}, \quad \sigma \,.\, B \equiv \Sigma\alpha_{ik}p_{ik} + \lambda_2\Sigma\beta_{ik}p_{ik},$$

and since the variables p_{23}, p_{14} are absent from A and B,

$$\alpha_{14} + \lambda_1\beta_{14} = \alpha_{14} + \lambda_2\beta_{14} = \alpha_{23} + \lambda_1\beta_{23} = \alpha_{23} + \lambda_2\beta_{23} = 0;$$

i.e.

$$\alpha_{14} = \beta_{14} = \alpha_{23} = \beta_{23} = 0.$$

* The quantity on the left side of this equation is important in the Theory of Screws, one half of it is there designated the Virtual Coefficient of two screws, see *Theory of Screws*, Sir R. S. Ball, p. 17.

The equation $p_{23}=0$ asserts that these lines meet the line which has been taken as the x coordinate axis, *i.e.* the common perpendicular d of the axes of the two complexes; the equation $p_{14}=0$ asserts that these common polar lines are each parallel to the coordinate plane yz, *i.e.* they are each perpendicular to d.

26. Three complexes in Involution. If three complexes $(ax)=0$, $(bx)=0$, $(cx)=0$ are mutually in involution, the points of space are divided by them into closed systems of *four*. For through any point O let there be drawn its polar planes in the three complexes; these intersect in three lines through O each of which belongs to two of the complexes, say

$$OO_1 \text{ belongs to } (bx)=0 \text{ and } (cx)=0,$$
$$OO_2 \quad " \quad " \quad (cx)=0 \text{ and } (ax)=0,$$
$$OO_3 \quad " \quad " \quad (ax)=0 \text{ and } (bx)=0;$$

and let O_1, O_2 and O_3 be the points corresponding to O in the involutions determined on these lines (Fig. 1).

Then OO_1O_2 is the polar plane of O in $(cx)=0$,
hence, (Art. 25), OO_1O_2 " " " O_1 in $(bx)=0$,
similarly OO_2O_3 " " " O_3 in $(bx)=0$;

hence both O_2O_1 and O_2O_3 belong to $(bx)=0$, and therefore O_2 is the pole of the plane $O_1O_2O_3$ in $(bx)=0$; similarly, O_1 is the pole of $O_1O_2O_3$ in $(ax)=0$, and O_3 in $(cx)=0$; hence the three planes through O_1 are the polars of O_1 in the three complexes, and so for O_2 and O_3. Thus each vertex of the tetrahedron is the pole of the faces through it in the three complexes. The following table shows the poles of each face of the tetrahedron in the complexes A, B and C.

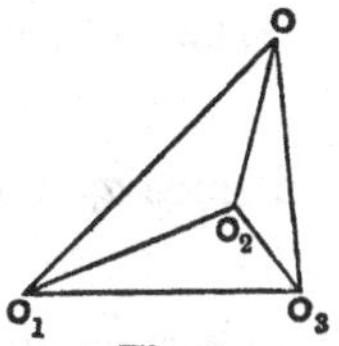

Fig. 1.

	O	O_1	O_2	O_3
A	OO_2O_3	$O_1O_2O_3$	OO_1O_2	OO_1O_3
B	OO_1O_3	OO_1O_2	$O_1O_2O_3$	OO_2O_3
C	OO_1O_2	OO_1O_3	OO_2O_3	$O_1O_2O_3$

27. Six complexes mutually in Involution*. It has been observed that each of the Plücker coordinates of a line equated to zero gives a *special* complex; for the lines which satisfy $p_{ik}=0$ are all the lines which meet the edge opposite to A_iA_k: and in the

* See Klein, *Math. Ann.* II.

coordinates of Klein, $x_k = 0$ represents a fundamental complex, but it is not special, for here $\Omega(a) = -\Sigma a_i^2 = -1$. Thus in this system the line is referred to six *coordinate complexes. These complexes are all in involution with each other*, for here the equation $\Sigma a_i \frac{\partial \Omega}{\partial b_i} = 0$ takes the form $\Sigma a_i b_i = 0$, which is clearly satisfied for any pair of the coördinate complexes.

The six poles of any plane for the fundamental complexes lie on a conic. For, in any given plane, denote by a_{ik} the line joining the pole of $x_i = 0$ to the pole of $x_k = 0$, then if

a_{13} and a_{15} have respectively coordinates a_i and b_i,

a_{23} and a_{25} " " " a_i' and b_i',

the coordinates of a_{14} are $a_i + \lambda b_i$, where λ is determined by expressing that a_{14} belongs to $x_4 = 0$, hence $\lambda = -\frac{a_4}{b_4}$; similarly a_{16} is $a_i + \mu b_i$, where $\mu = -\frac{a_6}{b_6}$.

The double ratio of the pencil formed by $a_{13}, a_{15}, a_{14}, a_{16}$, *i.e.* by $a, b, a + \lambda b, a + \mu b$, is $\frac{\lambda}{\mu}$, (Art. 11), or $\frac{a_4 b_6}{a_6 b_4}$; similarly the double ratio of $a_{23}, a_{25}, a_{24}, a_{26}$, is $\frac{a_4' b_6'}{a_6' b_4'}$; and these double ratios are equal, for since a_{ik} belongs to $x_i = 0$ and $x_k = 0$, we see that

$$a_1 = b_1 = a_3 = b_5 = a_2' = b_2' = a_3' = b_5' = 0,$$

and since a_{15} and a_{23} intersect

$$b_4 a_4' + b_6 a_6' = 0,$$

and since a_{13} and a_{25} intersect

$$a_4 b_4' + a_6 b_6' = 0;$$

therefore

$$\frac{a_4 b_6}{a_6 b_4} = \frac{a_4' b_6'}{a_6' b_4'};$$

hence, since $(a_{13} a_{14} a_{15} a_{16}) = (a_{23} a_{24} a_{25} a_{26})$, the six poles lie on a conic.

28. Transformation of coordinates. Two instances of transformation of coordinates have been met with, viz., the change from one tetrahedron of reference to another, and the change from the coordinates of Plücker to those of Klein. We now consider such transformations in more detail. It has been shown that if for six complexes $x_1 = 0 \ldots x_6 = 0$ we have $\Sigma x_i^2 \equiv 0$, the complexes are in involution. Now let

$$x_1 = a_{12} p_{12} + a_{34} p_{34} + a_{13} p_{13} + a_{42} p_{42} + a_{14} p_{14} + a_{23} p_{23},$$
$$x_2 = b_{12} p_{12} + \ldots\ldots\ldots\ldots$$
$$\ldots\ldots\ldots\ldots\ldots\ldots$$
$$x_6 = f_{12} p_{12} + \ldots\ldots\ldots\ldots$$

then if the complexes $x_1=0 \ldots x_6=0$ are mutually in involution, while

$$\Omega(a)=\ldots=\Omega(f)=k; \text{ it will follow that } \Sigma x_i^2\equiv 0;$$

for, from the six equations

$$a_{12}a_{34}+a_{34}a_{12}+\ldots=k,$$
$$a_{12}b_{34}+a_{34}b_{12}+\ldots=0,$$
$$\ldots\ldots\ldots\ldots\ldots\ldots$$

we obtain $a_{34}\Delta=A_{12}\,.\,k$, where Δ is the determinant of the equations of transformation and A_{12} is the minor of a_{12} in Δ; similarly $a_{12}\Delta=A_{34}\,.\,k$, etc.

Hence solving the equations for $p_{12}\ldots p_{23}$ in terms of $x_1\ldots x_6$, we obtain

$$k\,.\,p_{12}=a_{34}x_1+b_{34}x_2+\ldots+f_{34}x_6,$$
$$k\,.\,p_{34}=a_{12}x_1+b_{12}x_2+\ldots+f_{12}x_6,$$
$$\ldots\ldots\ldots\ldots\ldots\ldots\ldots\ldots$$

thus, since $p_{12}p_{34}+p_{13}p_{42}+p_{14}p_{23}\equiv 0$, we have

$$\Sigma x^2\equiv 0.$$

If the equations of transformation are

$$x_i'=\Sigma a_{ik}x_k,$$

where $\Sigma x'^2\equiv\Sigma x^2=0$, the transformation is "*orthogonal*," the equations connecting the coefficients are

$$\sum_i a_{ik}^2=1,$$
$$\sum_i a_{ik}a_{ik'}=0;$$

from these are derivable

$$\sum_i a_{ki}^2=1,$$
$$\sum_i a_{ki}a_{k'i}=0;$$

from the last equations we learn that the complexes x' are mutually in involution*.

If the equations of transformation are

$$p_{12}'=a_{12}p_{12}+a_{34}p_{34}+a_{13}p_{13}+a_{42}p_{42}+a_{14}p_{14}+a_{23}p_{23},$$
$$p_{34}'=b_{12}p_{12}+\ldots\ldots\ldots\ldots\ldots\ldots\ldots\ldots$$
$$\ldots\ldots\ldots\ldots\ldots\ldots\ldots\ldots\ldots\ldots$$

the coefficients in the same vertical line are the coordinates of the edges of the old tetrahedron of reference with regard to the new tetrahedron of reference. And since the lines for which p_{12}' is zero form a *special* complex of which the edge $A_3'A_4'$ of the new tetrahedron of reference is directrix, it follows that the coefficients of the first row are the coordinates of the edge $A_3'A_4'$ with regard to the old tetrahedron of reference; and so for the other rows. We find that in this case, if Δ is the determinant formed by the coefficients, $b_{12}\Delta=A_{34}, \ldots$ etc.

29. The fifteen principal tetrahedra. We have seen that if $x_1\ldots x_6$ are linear functions of the Plücker coordinates of a line such that $\Sigma x^2\equiv 0$, the complexes $x_1=0, \ldots x_6=0$, are in involution in pairs. Also if we write

$$x_1=p_{12}+p_{34}, \ldots$$
$$ix_2=p_{12}-p_{34}, \ldots$$

we obtain the equations of Art. 9.

* We easily derive that $x_i=\Sigma a_{ki}x_k'$.

These last equations express that the line x has coordinates $p_{12}...$ with regard to a certain tetrahedron $A_1A_2A_3A_4$ whose position in reference to the given complexes $x_1=0...$ will now be determined. The coordinates of the edges A_1A_2, A_3A_4 of this tetrahedron in reference to themselves are got by putting respectively all the p's zero except p_{12} and p_{34}; thus the x coordinates of these edges are $(1, -i, 0, 0, 0, 0)$, $(1, +i, 0, 0, 0, 0)$. But these are the common pair of polar lines for the complexes $x_1=0$ and $x_2=0$, as will now be shown. For let $x_1=0 \ldots x_6=0$ be six complexes mutually in involution: they may be arranged in pairs in 15 ways, take *e.g.*, the pairs x_1, x_2; x_3, x_4; x_5, x_6; then all lines which belong to both $x_1=0$ and $x_2=0$ belong also to the special complexes

$$x_1+ix_2=0, \qquad x_1-ix_2=0,$$

i.e. the directrices of these two complexes are polar lines in both x_1 and x_2, and their coordinates are

$(1, i, 0, 0, 0, 0)$, $(1, -i, 0, 0, 0, 0)$; *i.e.* they are the edges A_3A_4 and A_1A_2.

In the same manner the common polar lines of x_3 and x_4 are

$(0, 0, 1, i, 0, 0)$, $(0, 0, 1, -i, 0, 0)$, *i.e.* the edges A_2A_4, A_1A_3;

while those of x_5 and x_6 are

$(0, 0, 0, 0, 1, i)$, $(0, 0, 0, 0, 1, -i)$, *i.e.* the edges A_2A_3, A_1A_4.

There are 15 tetrahedra which bear this relationship to the coordinate complexes: and starting with one of these tetrahedra we thereby determine six complexes in mutual involution, and thus the 14 other tetrahedra.

For the six fundamental complexes

$p_{12}+p_{34}=0$, $p_{12}-p_{34}=0$, $p_{13}+p_{42}=0$, $p_{13}-p_{42}=0$, $p_{14}+p_{23}=0$, $p_{14}-p_{23}=0$,

the polar planes of a point $(\alpha_1, \alpha_2, \alpha_3, \alpha_4)$ are easily seen to be, (Art. 16),

$$(\alpha_2, -\alpha_1, \alpha_4, -\alpha_3), \quad (\alpha_2, -\alpha_1, -\alpha_4, \alpha_3), \quad (\alpha_3, -\alpha_4, -\alpha_1, \alpha_2),$$
$$(\alpha_3, \alpha_4, -\alpha_1, -\alpha_2), \quad (\alpha_4, \alpha_3, -\alpha_2, -\alpha_1), \quad (\alpha_4, -\alpha_3, \alpha_2, -\alpha_1),$$

i.e. six of the set of the 16 planes of the closed system determined by the point $(\alpha_1, \alpha_2, \alpha_3, \alpha_4)$, (Art. 14), all of which pass through $(\alpha_1, \alpha_2, \alpha_3, \alpha_4)$, and from the properties of this system, it follows therefore that the six polar planes of any point of the system with reference to the six complexes are six planes of the system.

Thus the 16 *points and planes of Art.* 14 *are such that each point is the pole for the six complexes* $p_{12}+p_{34}=0$, *&c. of the six planes of the system through the point, and each plane is the polar plane for these complexes of the six points of the system which lie in the plane.*

CHAPTER III.

SYNTHESIS OF THE LINEAR COMPLEX.

30. THE ratios of the six coefficients of a linear complex are determined if the coordinates of five complex lines are given, or, *five lines determine a linear complex*, since the equation of the complex contains five independent constants, there are ∞^5 linear complexes. The only exception occurs when the coordinates of the five given lines x, y, z, s, t, are connected by the six equations

$$\lambda x_i + \mu y_i + \nu z_i + \rho s_i + \sigma t_i = 0 \text{ ...(i)}, \qquad (i = 1, 2 \dots 6),$$

in which case the complex to which they all belong is not determinate: it is easy to see that here the five lines are intersected by each of two lines, for using the coordinates of Klein, (as will usually be done in future), if α be any line

$$\lambda (\alpha x) + \mu (\alpha y) + \nu (\alpha z) + \rho (\alpha s) + \sigma (\alpha t) = 0,$$

and if α be one of the two lines which meet x, y, z and s, the coefficients of λ, μ, ν and ρ are each zero, and therefore

$$(\alpha t) = 0,$$

so that α also meets t; similarly for the other intersector of x, y, z and s.

If $$\lambda x_i + \mu y_i + \nu z_i + \rho s_i = 0 \dots \qquad (i = 1, \dots 6),$$

x, y, z and s form part of the same system of generating lines of a quadric, *i.e.*, they belong to the same *regulus*, for any line which meets three of them will also meet the fourth. It follows that if three lines x, y, z belong to a given linear complex $(ax) = 0$, then all the lines of the regulus of which x, y, z form part belong to this complex, for since $(ax) = 0$, $(ay) = 0$, $(az) = 0$, it follows from the last equation that $(as) = 0$.

If $$\lambda x_i + \mu y_i + \nu z_i = 0 \dots \qquad (i = 1, 2 \dots 6),$$

the lines x, y, and z form part of the same pencil, so that z, of necessity, forms part of the linear complex to which both x and y

belong. The case when one or more intersections of pairs of the given lines occur, introduces no indeterminateness, but when three of them pass through the same point or lie in the same plane we have, (Art. 13),

$$x_i = \lambda\alpha_i + \mu\beta_i + \nu\gamma_i,$$
$$y_i = \lambda'\alpha_i + \mu'\beta_i + \nu'\gamma_i,$$
$$z_i = \lambda''\alpha_i + \mu''\beta_i + \nu''\gamma_i,$$

where α, β, and γ are three concurrent or coplanar lines.

Then since x, y, z belong to the complex $(ax) = 0$, we have

$$(a\alpha) = (a\beta) = (a\gamma) = 0,$$

so that *all* lines through the point $(\alpha\beta\gamma)$, (or in the plane $(\alpha\beta\gamma)$), belong to the complex which is then special, having for its directrix the line through $(\alpha\beta\gamma)$ which meets s and t.

Observe that the determinant $\begin{vmatrix} \lambda & \mu & \nu \\ \lambda' & \mu' & \nu' \\ \lambda'' & \mu'' & \nu'' \end{vmatrix} \neq 0,$

since it is proportional to the volume of a tetrahedron, for λ, μ, ν are proportional to the perpendiculars from a point of x on the planes $(\beta\gamma)$, (γa) and $(a\beta)$ respectively, (Art. 13), similarly for $\lambda'\mu'\nu'$ and $\lambda''\mu''\nu''$.

If p and p' are the intersectors of the five lines in this exceptional case, the complex is

$$\Sigma (p_i + \lambda p_i') x_i = 0, \text{ where } \lambda \text{ may have any value.}$$

Should the five given lines form a twisted pentagon it is clear they cannot all be intersected by the same line.

31. *The complex is determined by one of its lines and two polar lines p, p'.*

For if l is the given complex line, let a plane through p meet p' in A and l in E, while another plane through p meets p' in C and l in D, then if B is a point on p, the twisted pentagon $ABCDE$ is formed of complex lines and the complex is thereby determined.

Any two pairs of polar lines p, p' and q, q' form part of the same regulus.

For the lines which meet p, p' and q form a regulus ρ and also belong to the complex, hence they all meet q' (Art. 17); thus p, p', q, q' lie on the complementary regulus ρ'.

The lines p, p', q, and q' determine the complex.

For any plane π which cuts p, p' in X and X', and q, q' in Y and Y', has as its pole the intersection of XX' and YY'.

Every line of ρ' has its polar also on ρ' since this polar must meet the lines of ρ, and the lines through the pole of any plane π meet ρ' in pairs of polar lines. Hence the lines of ρ' form an involution, corresponding pairs being polar in the complex, so that *the two double lines of the involution belong to the complex.*

32. Chasles* suggested as a method of generating a complex, to arrange the lines of a regulus in an involution, and to take all the lines which meet pairs of conjugate lines.

Conversely, the complex being given by five of its lines x, y, z, s and t, and the regulus ρ' being taken which is complementary to that to which x, y and z belong, then two lines, α and β, of ρ' meet x, y, z and s; also t will meet two other lines γ and δ of ρ', so that by means of α, β; γ, δ the involution of the lines of ρ' is determined.

33. If three non-intersecting lines are given, as we have seen, the regulus ρ to which they belong forms part of the complex, which may also be seen from the fact that the complementary regulus ρ' is made up of pairs of polar lines, hence each line of ρ belongs to the complex.

Any plane pencil of lines contains one line of a given complex, viz., the line of intersection of the plane of the pencil, and the polar plane of the centre of the pencil.

If a line describes a plane pencil its polar will describe a plane pencil; we notice first that if a point A and a plane π are *united*, the polar plane α of A, and the pole P of π, are also *united*; for the (one) complex line through A in π must clearly lie in α and pass through P, hence α and P are united; so that if a line describe the pencil (A, π), since its polar line must lie in α, (Art. 17), and pass through P, this polar line describes the plane pencil (P, α).

If a line describe a regulus ρ, its polar will describe a regulus ρ_1; for if a line p meet three lines a, b, c of ρ', its polar will meet their three polar lines a', b', c', and hence describes a regulus ρ_1. The six lines a, b, c, a', b', c' are all met by two lines; for a, a', b, b' are met by the lines of a regulus σ, consisting of complex lines; two of these lines meet both c and c'; each of these two (complex) lines belongs to both ρ and ρ_1. Hence *any regulus ρ contains two complex lines.*

This is easily seen analytically; let the coordinates of p be y_i, then $(ay)=0$, $(by)=0$, $(cy)=0$, and there are two lines y which satisfy these equations together with $(Ay)=0$, where $(Ax)=0$ is any given complex.

* See *Liouville*, Série 1, T. IV.

34. Correlations of Space. Collineation and Reciprocity*. The assemblage of lines forming a linear complex has been arrived at by determining the lines which satisfy a linear equation in line coordinates. It may also be reached quite independently from another starting-point, of which correlative systems of space form the base. Such systems will now be discussed.

It has been seen, (Introduction, xiv), that two spaces Σ and Σ' are said to be *collinear*, when to every point P of Σ there corresponds one point P' of Σ', and to every plane π of Σ through P there corresponds one plane π' of Σ' through P'; (then to the join of two points of Σ corresponds the join of the corresponding points of Σ'). To bring about this correlation four equations of the form

$$\mu x_i' = a_{i1}x_1 + a_{i2}x_2 + a_{i3}x_3 + a_{i4}x_4 \ldots \text{(ii)}, \quad (i = 1, 2, 3, 4)$$

are both sufficient and necessary, where x and x' are coordinates of corresponding points in Σ and Σ' respectively, referred in general to different tetrahedra.

As usual, x_i is equal to the ratio of the perpendicular from the point on the face a_i of the tetrahedron of reference to the perpendicular from a fixed point E on the same face.

If the respective correlatives in Σ', of any five given points in Σ, be agreed upon arbitrarily, then the correlative in Σ' of any other point in Σ is determined; for if one pair of given corresponding points be taken as E and E', then inserting in equations (ii) the coordinates of the four other given pairs of points, we obtain 16 equations, from which the ratios of the coefficients a_{ik} are known.

If we take corresponding points in Σ and Σ', as the vertices A_1, A_1', &c. of the two tetrahedra of reference, the equations (ii) reduce to the simple form

$$\mu x_i' = a_{ii} x_i,$$

this may be seen by expressing that $(x_1', 0, 0, 0)$, $(x_1, 0, 0, 0)$, &c. are corresponding points. Observe that in any collineation there are *four* points which coincide with their corresponding points†, for putting $x_i' = x_i$, we have

$$\begin{vmatrix} a_{11} - \mu & a_{12} & a_{13} & a_{14} \\ a_{21} & a_{22} - \mu & a_{23} & a_{24} \\ a_{31} & a_{32} & a_{33} - \mu & a_{34} \\ a_{41} & a_{42} & a_{43} & a_{44} - \mu \end{vmatrix} = 0.$$

* See Reye, *Geom. der Lage*, Bd. II., also Salmon-Fiedler, *Geom. d. Raumes*, Bd. I., of whose work the account of collineation and reciprocity here given is a summary.

† Introd., Art. xiv.

35. Two spaces Σ and Σ' are said to be *reciprocal* when to each point P of Σ there corresponds one plane π' of Σ' and to each plane π through P of Σ there corresponds one point P' on π' of Σ'. Then to the join of two points of Σ corresponds the intersection of two planes of Σ'. To secure this correlation four equations of the form

$$\mu \,.\, u_i' = a_{i1}x_1 + a_{i2}x_2 + a_{i3}x_3 + a_{i4}x_4 \;\ldots\ldots\ldots\ldots \text{(iii)}$$

are both sufficient and necessary.

Also taking any point x' on u' the equation $\Sigma u_i' x_i' = 0$ may be written, by aid of (iii),

$$x_1 \sum_k a_{k1}x_k' + x_2 \sum_k a_{k2}x_k' + x_3 \sum_k a_{k3}x_k' + x_4 \sum_k a_{k4}x_k' = 0,$$

whence, since $\Sigma u_i x_i = 0$, the plane u in Σ corresponding to x' is given by

$$\nu \,.\, u_i = a_{1i}x_1' + a_{2i}x_2' + a_{3i}x_3' + a_{4i}x_4' \ldots\ldots\ldots\ldots \text{(iv)}.$$

The spaces Σ and Σ' need not be distinct, and if not, to any point of space common to them two correlative planes are assigned, because the point may be regarded as belonging either to Σ or to Σ'. These planes (given by (iii) and (iv)) are in general different. We may now take the tetrahedra of reference to which Σ and Σ' have hitherto been respectively referred as *identical*, and may regard the equations (iii) and (iv) as giving the two correlative planes for the *same* point of space, (so that $x_i = x_i'$), then as just stated, these planes u and u' do not in general coincide.

There are four points in the *general* reciprocity, for each of which the corresponding pair of planes u and u' coincide; to obtain them we express that $u_i' = \lambda u_i$, and hence find their coordinates x_i, which satisfy the equations

$$a_{11}x_1 + a_{12}x_2 + a_{13}x_3 + a_{14}x_4 = \lambda\,(a_{11}x_1 + a_{21}x_2 + a_{31}x_3 + a_{41}x_4),$$
$$a_{21}x_1 + a_{22}x_2 + a_{23}x_3 + a_{24}x_4 = \lambda\,(a_{12}x_1 + a_{22}x_2 + a_{32}x_3 + a_{42}x_4),$$
$$a_{31}x_1 + a_{32}x_2 + a_{33}x_3 + a_{34}x_4 = \lambda\,(a_{13}x_1 + a_{23}x_2 + a_{33}x_3 + a_{43}x_4),$$
$$a_{41}x_1 + a_{42}x_2 + a_{43}x_3 + a_{44}x_4 = \lambda\,(a_{14}x_1 + a_{24}x_2 + a_{34}x_3 + a_{44}x_4).$$

The elimination of the x_i leads to a quartic equation for λ, whose roots are in general different and unequal to unity; each root λ_i then gives one point x_i for which the pair of corresponding planes coincide. Each of these four points x_i lies in its corresponding plane, for multiplying the preceding equations by x_1, x_2, x_3, x_4 and adding we obtain $f(x) = \lambda f(x)$, where $f(x) = \frac{1}{2}\Sigma\Sigma\,(a_{ik} + a_{ki})\,x_i x_k$, hence since λ is not unity, $f(x) = 0$, which proves the result; *the quadric* $f(x)=0$ *is the locus of points which lie in their corresponding planes.*

If these special four points be taken as the vertices of the tetrahedron of reference, the preceding four equations must be satisfied by the point $(1, 0, 0, 0)$ for $\lambda=\lambda_1$, by $(0, 1, 0, 0)$ for $\lambda=\lambda_2$, etc.; where we notice that $\lambda_i \neq \lambda_k$.

Hence we obtain the system of equations

$$a_{k1}=\lambda_1 a_{1k},\quad a_{k2}=\lambda_2 a_{2k},\quad a_{k3}=\lambda_3 a_{3k},\quad a_{k4}=\lambda_4 a_{4k},$$

where $k=1, 2, 3, 4$; also $a_{11}=a_{22}=a_{33}=a_{44}=0$, since each of the vertices lies in its corresponding plane.

Moreover if none of the quantities a_{ik} are zero we derive the equations $\lambda_i\lambda_k=1$, where $k=1, 2, 3, 4$; which implies equality between the quantities λ which has been seen not to exist, hence certain of the a_{ik} must vanish; also only *four* of them, such as $a_{ik}, a_{ki}, a_{jl}, a_{lj}$ may not be zero, for if one more were not zero, two pairs of the λ_i would be equal.

Taking $a_{14}, a_{41}, a_{23}, a_{32}$ as being the a_{ik} which are not zero, we obtain the equations of the reciprocity in the form

$$\mu\,.\,u_1'=a_{14}x_4,\quad \mu\,.\,u_2'=a_{23}x_3,\quad \mu\,.\,u_3'=a_{32}x_2,\quad \mu\,.\,u_4'=a_{41}x_1;$$

also

$$\nu\,.\,u_1=a_{41}x_4',\quad \nu\,.\,u_2=a_{32}x_3',\quad \nu\,.\,u_3=a_{23}x_2',\quad \nu\,.\,u_4=a_{14}x_1'.$$

It appears from these equations that the plane corresponding to the vertex A_1 is $x_4=0$, or one of the coordinate planes, and so for the other vertices, hence, the four given points and their corresponding planes form the *same* tetrahedron.

36. Involutory Reciprocity. If for each point of space the two correlative planes coincide, the reciprocity is said to be *involutory*. This is the case if *either* $a_{ik}=a_{ki}$, *or* if $a_{ik}=-a_{ki}$, (which requires that $a_{ii}=0$).

In the latter case we obtain the equations of Art. 16, viz.,

$$\left.\begin{aligned} \nu\,.\,u_1' &= a_{12}x_2+a_{13}x_3+a_{14}x_4,\\ \nu\,.\,u_2' &= a_{21}x_1+a_{23}x_3+a_{24}x_4,\\ \nu\,.\,u_3' &= a_{31}x_1+a_{32}x_2+a_{34}x_4,\\ \nu\,.\,u_4' &= a_{41}x_1+a_{42}x_2+a_{43}x_3,\end{aligned}\right\}\ldots\ldots\ldots\ldots\ldots\ldots\text{(v)},$$

where $a_{ik}=-a_{ki}$. This involves that $\Sigma u_i'x_i=0$, or the point lies in its correlative plane. The distinction between the two spaces Σ and Σ' has now disappeared and we have a $(1, 1)$ correspondence between the points and planes of space in which each point lies in its corresponding plane.

37. Null System*. Such a correspondence is called a *null-system*. The corresponding points and planes are called "null-points" and "null-planes." The join of two null-points corresponds to the intersection of their two null-planes and these lines are called conjugate.

When the join of two points coincides with the intersection of their null-planes we have the coincidence of two conjugate lines. If a line meets two conjugate lines it is such a self-corresponding line, for if it meets them in A and B then the null-plane of A passes through both A and B, likewise the null-plane of B. We see from comparing equations (v) with those of Art. 16, that the null system, and that of poles and polar planes, are identical. Hence the lines of a linear complex are self-corresponding lines of a null system, *i.e. we obtain a linear complex by establishing an involutory reciprocity of two spaces in which corresponding points and planes are united, and then taking the lines which correspond to themselves.* The properties of a linear complex as given in the last and present chapters may be deduced from this involutory reciprocity, a brief sketch is added.

38. A null-system, or linear complex, is seen to be determined by

(i) *any three points and their null-planes,* provided the plane of the three points and the point of intersection of the three planes are *united*; for if A, B, C are the given points and γ any plane through A and B, β any plane through A and C, the null-point of β is known, being the intersection of the complex lines in β through A and C; similarly the null-point of γ is known, hence, if we pass a plane through any point P and AC, and a plane through P and AB, two complex lines through P are known and hence the polar plane of P:

(ii) *five complex lines which form a twisted pentagon,* for take the vertices $ABCDE$ as points of Σ and the planes BAE, etc. as the corresponding planes in Σ', then the reciprocity is established between Σ and Σ'; also to the plane BAE as belonging to Σ corresponds the intersection of the planes corresponding to B, A and E in Σ', *i.e. the point* A; similarly for the other planes; hence

* The terms null-system, null-plane, etc. are due to Möbius, see *Lehrbuch der Statik*.

the reciprocity is involutory, and in it corresponding points and planes are united; the sides of the pentagon are self-corresponding lines:

(iii) *two pairs of polar lines* (which must form part of the same regulus), the reasoning of Art. 31 applies here also:

(iv) *two polar lines and a complex line*, for as in Art. 31 we can determine a twisted pentagon of complex lines:

(v) *any five complex lines*, a, b, c, d, e; for take the regulus to which a, b, e belong and that to which c, d, e belong, also any line p which meets the first in the (complex) lines a', b' and the second in the (complex) lines c', d', then a', b', c' and d' must all meet the line p' polar to p, hence p' is the other intersector of a', b', c', d', and the pair p, p' together with e determine the complex.

39. Method of Sylvester*. The lines of a linear complex may also be obtained as follows: take corresponding lines of two projective plane pencils which have a common self-corresponding line; all the lines which meet such a pair form a linear complex. For A and A' being the centres of the pencils, any plane π meets the planes of the pencils in two lines p and p' meeting in O, a point of AA'. The pencils determine on p and p' two projective rows of points which are in perspective, with centre P, since O corresponds to itself; hence in the plane π the lines which meet pairs of corresponding lines of the pencils pass through the same point P of π; thus a linear complex is determined.

40. Automorphic Transformations. Linear transformations of the variables for which the expression $\omega(q)$ is unaltered in form, *i.e.* for which $\omega(q) \equiv \omega(q')$, are called *automorphic*; we may then regard $q_1 \dots q_6$ as the coordinates of one line and $q_1' \dots q_6'$ as the coordinates of another line with regard to the same tetrahedron of reference.

The equations

$$q_i' = a_{i1}q_1 + \dots + a_{i6}q_6,$$

then establish a (1, 1) correspondence between the lines q and q'.

Corresponding to the lines q_i which form a plane pencil, *i.e.* for

* See *Comptes Rendus*, T. LII. (1861).

which $q_i = \alpha_i + \lambda\beta_i$, we have lines q_i' also forming a plane pencil; and to lines q for which

$$q_i = \alpha_i + \lambda\beta_i + \mu\gamma_i, \quad \left(\alpha\frac{\partial\omega}{\partial\beta}\right) = \left(\beta\frac{\partial\omega}{\partial\gamma}\right) = \left(\gamma\frac{\partial\omega}{\partial\alpha}\right) = 0,$$

i.e. which either pass through the same point or lie in the same plane, we have lines q' for which

$$q_i' = A_i + \lambda B_i + \mu C_i, \quad \left(A\frac{\partial\omega}{\partial B}\right) = \left(B\frac{\partial\omega}{\partial C}\right) = \left(C\frac{\partial\omega}{\partial A}\right) = 0;$$

for since $\omega(\alpha + \lambda\beta) \equiv \omega(A + \lambda B)$, therefore

$$\left(\alpha\frac{\partial\omega}{\partial\beta}\right) \equiv \left(A\frac{\partial\omega}{\partial B}\right), \text{ etc.},$$

i.e. the lines q' are either concurrent or coplanar. If to one *sheaf* of lines q corresponds one *sheaf* of lines q', then to every sheaf of lines q corresponds a sheaf of lines q'. For let the centres of the given corresponding sheaves be O and O', and let P and P' be centres of any two corresponding pencils; the planes of the latter will not in general go through O and O', hence to two lines of the pencil of centre P and the line PO will correspond two lines of the pencil of centre P' and the line $P'O'$, so that to the sheaf of centre P will correspond the sheaf of centre P'. In like manner if to one sheaf of lines q there corresponds a *plane system* of lines q', then to each sheaf q corresponds a plane system q'. Thus the relationship established between the lines q and q' must arise either from a *collineation* or from a *reciprocity* of space.

Consider for instance the equations

$$p_{12}' = a_{12}p_{12} + a_{34}p_{34} + a_{13}p_{13} + a_{42}p_{42} + a_{14}p_{14} + a_{23}p_{23},$$
$$p_{34}' = b_{12}p_{12} + \ldots\ldots\ldots\ldots\ldots\ldots\ldots\ldots\ldots,$$
$$\ldots\ldots\ldots\ldots\ldots\ldots\ldots\ldots\ldots\ldots\ldots\ldots\ldots,$$
$$p_{23}' = f_{12}\,p_{12} + \ldots\ldots\ldots\ldots\ldots\ldots\ldots + f_{23}p_{23}.$$

The quantities which are vertically underneath each other are the coordinates of the correlatives of the edges of the tetrahedron of reference regarded as belonging to Σ (*e.g.* make all the p's zero except p_{12}): the quantities in the same row are the coordinates of the correlatives of the edges of the tetnahedron of reference regarded as belonging to Σ', *e.g.* for the side A_3A_4, $p_{12}' = 0$ is a special complex formed of the lines which meet A_3A_4; thus the a's being the coordinates of the directrix of this complex are those of the correlative in Σ of A_3A_4 regarded as belonging to Σ'.

41. Ruled surfaces and curves of a linear complex. A *ruled surface* of a linear complex is one whose generators belong to the complex. On each generator a correlation of points is established by means of the planes through it, for each such plane determines a point of contact, and also a pole in the complex. This correlation has two *united points* and at each of them the tangent plane of the surface coincides with the polar plane in the complex.

These points may be determined analytically as follows: if a line y is a tangent to the surface, it meets the generator x and a consecutive generator $x + \frac{dx}{d\theta} d\theta$, where θ is the single parameter of which the coordinates x are functions, thus

$$(yx) = 0, \ \left(y \frac{dx}{d\theta}\right) = 0, \ (y^2) = 0.$$

Now take the lines y which also cut any given line α and belong to the given complex $(ax) = 0$, we then have the additional equations

$$(y\alpha) = 0, \ (ay) = 0.$$

These five equations give two lines y, which, with x, determine the two points.

A curve all of whose tangents belong to a linear complex is called a *curve of the complex*. At any point of such a curve the osculating plane is the polar plane of the point in the complex, since two (consecutive) tangents through the point belong to the complex. The locus of points just discussed is such a complex curve, and in it since the tangent plane of the surface is the osculating plane of the curve, it follows that the curve is a *principal tangent curve* of the surface.

If through any point P a plane be drawn which osculates this curve at Q, PQ is then a complex line, hence the points of contact of the osculating planes which pass through a given point P lie on the polar plane of P. *The degree of the curve is equal to the class of the plane section of the surface*, for if any plane π meets the curve in a point A, the polar plane of A (the tangent plane to the surface at A) passes through F, the pole of π in the complex, and every such line FA is a tangent to the section of the surface by π*.

Curves of a linear complex. If the coordinates of Lie, (Art. 6), be used in the simplest form of the equation of a linear complex (Art. 23), it becomes

$$xdy - ydx = kdz.$$

The complex curves through any point (x, y, z) are those the direction ratios $dx : dy : dz$ of whose tangents satisfy this equation, and we have seen that they all have the same osculating plane at the

* See Picard, *Ann. de l'école normale supér.* Sér. 2, T. VI.; also Lie and Scheffer's *Berührungstransformationen*, S. 235.

point, viz. the polar plane of the point for the complex. Moreover, they all have *the same torsion*, for

$$\frac{ds}{d\tau} = \frac{X^2 + Y^2 + Z^2}{\begin{vmatrix} dx & dy & dz \\ d^2x & d^2y & d^2z \\ d^3x & d^3y & d^3z \end{vmatrix}}, \quad \text{where } X = dyd^2z - dzd^2y, \quad Y = dzd^2x - dxd^2z, \quad Z = dxd^2y - dyd^2x^*.$$

By aid of the equation of the complex this denominator is easily seen to reduce to

$$\frac{1}{k}(dxd^2y - dyd^2x)^3,$$

while

$$X = \frac{y}{k}(dxd^2y - dyd^2x), \quad Y = -\frac{x}{k}(dxd^2y - dyd^2x),$$

hence

$$\frac{ds}{d\tau} = \frac{\frac{x^2 + y^2}{k^2} + 1}{\frac{1}{k}} = \frac{x^2 + y^2 + k^2}{k};$$

a result depending only upon the coordinates of the point, and therefore the same for each complex curve through the point†.

42. It has been seen, (Art. 33), that a linear complex contains two lines of every regulus; *of any ruled surface whose degree is n it contains n generators*; for the equations of any generator of such a surface are, in Cartesian coordinates,

$$x = zf_1(m) + f_2(m),$$
$$y = zf_3(m) + f_4(m);$$

where m is a parameter.

Hence the coordinates p_{ik} of such a generator are given by equations of the form

$$p_{ik} = f_{ik}(m).$$

Since the ruled surface is, by hypothesis, of degree n, the equation

$$a_{12}f_{34}(m) + a_{34}f_{12}(m) + \ldots = 0 \ldots\ldots\ldots\ldots\ldots\ldots\text{(i)},$$

gives n values of m, provided that the a_{ik} are the coordinates of a line, *i.e.*

$$a_{12}a_{34} + a_{13}a_{42} + a_{14}a_{23} = 0.$$

But this restriction as to the a_{ik} will not, in general, affect the number of solutions of equation (i); hence that equation gives n

* See Salmon, *Geometry of Three Dimensions*, 3rd edition, p. 345.

† This theorem is due to Lie; see *Proceedings* of the Society of Sciences at Christiania (1883).

values of m when the a_{ik} are *any quantities whatever*. That is to say, *any given linear complex will contain, in general, n generators of a ruled surface of the nth degree*. So that, for instance, any linear complex contains, in general, three lines of a given ruled cubic.

43. In any rational curve whose points are given by the equations

$$x_1 = f_1(t),\ x_2 = f_2(t),\ x_3 = f_3(t),\ x_4 = f_4(t),$$

where the functions f are of degree n in t, the coordinates of its tangents $x_1dx_2 - x_2dx_1$, &c. are seen to be of degree $2n-2$ in t, hence $2n-2$ of them will intersect any given line. The number of its tangents which intersect any line is known as the *rank* of the curve. It is clearly equal to the degree of the developable of which the curve is the cuspidal edge.

Now any twisted cubic can be so expressed, n being equal to 3, and hence the rank to 4; also a linear complex is determined by any five tangents of the curve, this complex therefore contains five tangents of the developable, and must therefore contain it altogether (since a linear complex which does not contain a ruled surface of degree n can have only n lines in common with it), hence the tangents of any twisted cubic belong entirely to some linear complex.

Every rational quartic curve with two stationary tangents* has the rank 6, hence a linear complex through the two stationary tangents and three other tangents contains seven generators of the developable, *i.e.* contains it altogether.

Every rational quintic with four stationary tangents is seen similarly to belong to a linear complex.

44. The polar surface. If P be any point and π any plane through it, the pole of π for a given linear complex A is some point P' in π, while the polar plane of P is some plane π' through PP'. If now P and π are so related that while P describes a surface S the plane π touches S at P, it follows, neglecting small quantities of the second order, that all positions of P consecutive to P lie in π, hence all planes π' consecutive to π' pass through P', *i.e.* the plane π' touches the surface S' the locus of P', at P'. The surface S' thus obtained may be called the *polar surface* of S for the given linear complex. The tangents of S' are the polar lines for A of the tangents of S.

* See Salmon-Fiedler, *Geometrie des Raumes*, Bd. II. S. 140.

If the coordinates of P are x_i and of π are u_i, x_i' and u_i' denoting corresponding quantities for P' and π', we have

$$\rho \,.\, u_1' = a_{12}x_2 + a_{13}x_3 + a_{14}x_4, \text{ \&c.} \qquad \text{(Art. 16)},$$

hence it follows that the *class* of S', *i.e.* its degree in plane coordinates, is equal to the *degree* of S, and *vice versâ*.

If p and p' are a pair of corresponding tangents of S and S' respectively, we have

$$\rho \,.\, p_{12}' = 2A \,.\, a_{34} - \Omega(a)\, p_{12}, \text{ \&c.}, \qquad \text{(Art. 20)},$$

where $A = \Sigma a_{ik} p_{ik}$, $\frac{1}{2}\Omega(a) = a_{12}a_{34} + a_{13}a_{42} + a_{14}a_{23}$. Now the tangents to a surface form a complex, since they are the ∞^3 lines which satisfy the single condition of touching the surface, hence if this complex for the surface S' is $f(p_{12}', \ldots) = 0$, the corresponding complex for S will be *of the same degree.* But the degree of the "complex equation" of a surface is equal to the class of its plane section, which latter is known as the *rank* of the surface; for the complex cone of any point P being of degree r, any plane π through P meets the cone in r lines, *i.e.* there are r tangents to the section of the surface by π through the point P.

Hence, if any surface be polarized with regard to a linear complex, we obtain a new surface of equal rank, whose degree and class are respectively equal to the class and degree of the original surface.

45. Applying the same process to a curve c we derive a new curve c', the rank of c is equal to that of c', for to each tangent of c which meets any line p will correspond a tangent of c' which meets p' the polar of p, and *vice versâ*.

The *order* of c is the number of points in which it meets any plane π, at such a point of intersection we have two consecutive tangents of c and a line of the pencil (P, π) *concurrent*, corresponding to this we have the three corresponding lines *lying in one plane* which passes through P' the pole of π. Hence the order of c is equal to the number of osculating planes of c' which pass through any point P', *i.e.* the *class* of c'. Thus the order of each curve is equal to the class of the other.

For a twisted cubic the order and class are each equal to three. If the cubic touches four lines and we take one of the ∞^1 linear complexes through these lines, the polar curve of the cubic for this complex is also a twisted cubic, moreover it will touch the four given lines, since each of these lines is its own polar for the complex. Hence, *if a twisted cubic touches four lines there are* ∞^1 *twisted cubics which touch these lines**.

46. Complex equation of the quadric. The polar plane u of a point x with regard to the quadric F, or $\Sigma a_{ik} x_i x_k = 0$, is given by the equations

$$u_i = \sum_m a_{im} x_m,$$

similarly the polar plane v of y is determined by the equations

$$v_i = \sum_m a_{im} y_m.$$

Hence if the line (x, y) is p and its polar line for the quadric is p', the latter line has coordinates π_{ik}', where

$$\pi_{ik}' = u_i v_k - u_k v_i = \Sigma a_{im} x_m \Sigma a_{km} y_m - \Sigma a_{im} y_m \Sigma a_{km} x_m$$
$$= \sum_{h, l} (a_{hi} a_{lk} - a_{hk} a_{li}) p_{hl}.$$

If p intersects p', both p and p' are tangents of the quadric, hence the complex equation of the tangents of F is

$$\Psi \equiv \Sigma \pi_{ik}' p_{ik} \equiv \sum_{i, k} p_{ik} \sum_{h, l} (a_{hi} a_{lk} - a_{hk} a_{li}) p_{hl} = 0,$$

or $$\Psi \equiv \Sigma p_{ik}^2 (a_{ii} a_{kk} - a_{ik}^2) + 2\Sigma (a_{hi} a_{lk} - a_{hk} a_{li}) p_{hl} p_{ik} = 0,$$

where in the second term of the last equation $p_{ik} \neq p_{hl}$.

It follows that

$$\pi_{ik}' = \frac{\partial \Psi}{\partial p_{ik}}.$$

Observe that if u and v are two planes through a tangent line, the equation $\Psi = 0$ may be written

$$\begin{vmatrix} a_{11} & \dots & a_{14} & u_1 & v_1 \\ \dots & \dots & \dots & \dots & \dots \\ a_{41} & \dots & a_{44} & u_4 & v_4 \\ u_1 & \dots & u_4 & 0 & 0 \\ v_1 & \dots & v_4 & 0 & 0 \end{vmatrix} = 0.$$

If the tetrahedron of reference is self-conjugate for the quadric, which has then for its equation $\sum_1^4 a_i x_i^2 = 0$, the equation $\Psi = 0$ becomes

$$a_1 a_2 p_{12}^2 + a_3 a_4 p_{34}^2 + a_1 a_3 p_{13}^2 + a_2 a_4 p_{42}^2 + a_1 a_4 p_{14}^2 + a_2 a_3 p_{23}^2 = 0.$$

* See Voss, "Ueber vier Tangenten einer Raumcurve dritter Ordnung," *Math. Ann.* XIII.; also Dixon, *Quarterly Journal*, vol. XXIV.

47. Simultaneous bilinear equations*. Two special types of bilinear equations will now be considered, viz.

$$f \equiv \Sigma a_{ik} y_i x_k = 0, \text{ in which } a_{ik} = a_{ki};$$

$$A \equiv \Sigma \alpha_{ik} y_i x_k = 0, \text{ in which } \alpha_{ik} = -\alpha_{ki}, \text{ and } \alpha_{ii} = 0.$$

The first assigns to each point x_i its polar plane with regard to the quadric F, or $\Sigma a_{ik} x_i x_k = 0$; the second gives the polar plane of x_i with regard to the linear complex $\Sigma \alpha_{ik} p_{ik} = 0$.

There are in general four points x for each of which the two corresponding planes *coincide*, viz. those determined by the four equations

$$\lambda \sum_k a_{ik} x_k + \sum_k \alpha_{ik} x_k = 0\,;$$

where λ is therefore a root of the biquadratic equation

$$\Delta(\lambda) \equiv |\lambda a_{ik} + \alpha_{ik}| = 0.$$

Since for the linear complex the point x is always united to its corresponding plane, each of these four points must lie on the quadric F. Moreover, since in general a linear complex contains two generators of every regulus, it is clear that these four points are the vertices of a twisted quadrilateral formed by these two pairs of generators of F.

If these generators be taken as four edges of the tetrahedron of reference, the equations of F and A assume the forms

$$2X_1X_4 + 2X_2X_3 = 0, \quad \alpha p_{14} + \beta p_{23} = 0.$$

Taking as the equations of transformation which reduce F and A to this form

$$x_i = \Sigma \beta_{ik} X_k,$$

the quantity $\Delta(\lambda)$ is seen to be an invariant, since if it vanishes a point x_i can be found such that the plane corresponding to x_i in $A + \lambda f$, viz. $\sum_i y_i \sum_k (\lambda a_{ik} + \alpha_{ik}) x_k = 0$, becomes indeterminate. This property does not depend on the coordinate system, hence if D is the determinant of transformation

$$D^2 . \Delta(\lambda) = \begin{vmatrix} 0 & 0 & 0 & \lambda + \alpha \\ 0 & 0 & \lambda + \beta & 0 \\ 0 & \lambda - \beta & 0 & 0 \\ \lambda - \alpha & 0 & 0 & 0 \end{vmatrix} = (\lambda^2 - \alpha^2)(\lambda^2 - \beta^2).$$

* The present and following Articles form a brief account of a lengthy investigation in *Vorlesungen über Geometrie*, Bd. II., Clebsch-Lindemann, S. 343—414.

By direct calculation of $\Delta(\lambda)$ we find

$$\Delta(\lambda) = \lambda^4 A_1 + \lambda^2 \Phi + A'^2,$$

where A_1 is the discriminant of F, $A' = \alpha_{12}\alpha_{34} + \alpha_{13}\alpha_{42} + \alpha_{14}\alpha_{23}$, and $\Phi = \Sigma \alpha_{ik}\alpha_{jh} A_{ik, jh}$, where $A_{ik, jh}$ is the coefficient of $a_{ik}a_{jh}$ in A_1. Thus Φ is the result of substituting the quantities α_{ik} for π_{ik} in Ψ.

It follows that with reference to the tetrahedron of which four edges are the generators of F which belong to A, the equations of F and A can be brought to the form

$$2X_1X_4 + 2X_2X_3 = 0, \quad \lambda_1 p_{14} + \lambda_3 p_{23} = 0,$$

where λ_1, λ_3 are roots of $\Delta(\lambda) = 0$.

In the reciprocity determined by the equation

$$A + \lambda f = 0,$$

the locus of points x which are united to their corresponding planes is clearly $\Sigma a_{ik}x_i x_k = 0$, or F; for the planes u which are united to their corresponding points the following five equations hold,

$$u_i = \sum_k (\alpha_{ik} + \lambda a_{ik}) x_k,$$

$$\Sigma u_i x_i = 0;$$

hence eliminating the x_i we obtain as the envelope of such planes a quadric Λ whose equation is

$$\Sigma \Delta_{ik} u_i u_k = 0,$$

in which the Δ_{ik} are minors of $\Delta(\lambda)$.

The left side of the last equation is clearly a contravariant of $A + \lambda f$, hence

$$D^2 . \Sigma \Delta_{ik} u_i u_k = 2\lambda(\lambda^2 - \lambda_3^2) U_1 U_4 + 2\lambda(\lambda^2 - \lambda_1^2) U_2 U_3.$$

In point-coordinates Λ will therefore have as its equation

$$\lambda^2 (2X_1X_4 + 2X_2X_3) - 2\lambda_3^2 X_2X_3 - 2\lambda_1^2 X_1X_4 = 0.$$

Thus Λ is a member of a "pencil" of quadrics of which one is F, while another is

$$2\lambda_3^2 X_2X_3 + 2\lambda_1^2 X_1X_4 = 0.$$

This latter quadric is the locus of the polars of the generators of F with regard to the linear complex A; for one system of generators of F being $X_1 = \mu X_2$, $\mu X_4 + X_3 = 0$, the polar plane for A of a point X_i on the line μ is

$$\lambda_1 (X_1Y_4 - X_4Y_1) + \lambda_3 (X_2Y_3 - X_3Y_2) = 0,$$

or $$\mu(\lambda_1 X_2 Y_4 + \lambda_3 X_4 Y_2) - \lambda_1 X_4 Y_1 + \lambda_3 X_2 Y_3 = 0;$$

hence the polar of the line 'μ' has the equations

$$\mu\lambda_1 Y_4 + \lambda_3 Y_3 = 0, \quad \mu\lambda_3 Y_2 - \lambda_1 Y_1 = 0;$$

and this line clearly lies on the quadric

$$\lambda_1^2 Y_1 Y_4 + \lambda_3^2 Y_2 Y_3 = 0.$$

Again there is a singly infinite number of linear complexes which have in common four generators of F, one of them being A, another, B, is the locus of the polar lines of A with regard to F; the system of linear complexes through the four given lines is then $A + \mu B = 0$.

If p is a line of A its polar p' with regard to F is determined by the equations $\pi_{ik} = \dfrac{\partial \Psi}{\partial p_{ik}}$, moreover the equation of A being $\Sigma \pi_{ik} \dfrac{\partial A'}{\partial a_{ik}} = 0$, the equation of B will be

$$\Sigma \frac{\partial \Psi}{\partial p_{ik}} \cdot \frac{\partial A'}{\partial a_{ik}} = 0,$$

or, which is the same thing,

$$\Sigma \frac{\partial \Phi}{\partial a_{ik}} \cdot \frac{\partial P}{\partial p_{ik}} = 0,$$

where $P = p_{12}p_{34} + p_{13}p_{42} + p_{14}p_{23}$, and Φ is obtained by substituting a_{ik} for π_{ik} in Ψ.

The vanishing of the invariant Φ is thus seen to be the condition that A and its polar complex B with regard to F should be in involution.

48. Linear transformations which leave a quadric unaltered in form. By means of a linear complex a general class of linear transformations is obtained which leave the form of a quadric surface unaltered. For let the given quadric be

$$F \equiv \Sigma a_{ik} x_i x_k = 0,$$

and the required equations of transformation

$$\xi_i = \sum_k c_{ik} x_k \quad \text{.........................(i).}$$

Now the equations

$$x_i = kt_i + \lambda\tau_i, \quad \xi_i = kt_i - \lambda\tau_i, \quad \text{..............(ii),}$$

lead to $F(x) \equiv F(\xi)$ provided that t and τ are the coordinates of two points which are conjugate with regard to the quadric $F = 0$. If therefore t_i and τ_i can be linearly connected, so that from the equations (ii) the equations (i) can be deduced, a transformation of the required kind will be obtained.

Let u_i be the polar plane of t_i for F, then

$$t_i = \sum_k A_{ik} u_k \quad\text{.......................(iii)},$$

where the quantities A_{ik} are the first minors of the discriminant of F, and the fact that τ_i lies in the plane u_i, *i.e.* that t_i and τ_i are conjugate points, may then be established by linear equations of the form

$$\tau_i = \Sigma a_{ik} u_k \quad\text{..........................(iv)},$$

provided that $a_{ik} = -a_{ki}$ (and hence that $a_{ii} = 0$).

Substituting in equations (ii) from (iii) and (iv) we obtain

$$\left.\begin{aligned} x_i &= \kappa \sum_k A_{ik} u_k + \lambda \sum_k a_{ik} u_k \\ \xi_i &= \kappa \sum_k A_{ik} u_k - \lambda \sum_k a_{ik} u_k \end{aligned}\right\} \quad\text{.................(v)},$$

the elimination of the u_k from these last equations will give the equations (i) which have been sought.

The solution of the first set of equations (v) leads to

$$\Delta(\kappa, \lambda) \,.\, u_i = \sum_k \Delta_{ki} x_k,$$

where $\Delta(\kappa, \lambda)$ is the determinant of the quantities $\kappa A_{ik} + \lambda a_{ik}$ and Δ_{ki} a first minor of $\Delta(\kappa, \lambda)$.

Hence $$\Delta(\kappa, \lambda) \sum_i A_{li} u_i = \sum_k \sum_i \Delta_{ki} A_{li} x_k,$$

also from equations (v)

$$x_l + \xi_l = 2\kappa \sum_i A_{li} u_i,$$

so that $$\Delta(\kappa, \lambda)\, \xi_l = 2\kappa \Sigma \Sigma \Delta_{ki} A_{li} x_k - \Delta(\kappa, \lambda)\, x_l$$

or the required equations of transformation are

$$\xi_l = \Sigma c_{lk} x_k,$$

where $$c_{lk} = \frac{2\kappa \sum_i \Delta_{ki} A_{li}}{\Delta(\kappa, \lambda)}, \text{ when } k \neq l,$$

and $$c_{ll} = \frac{2\kappa \sum_i \Delta_{li} A_{li} - \Delta(\kappa, \lambda)}{\Delta(\kappa, \lambda)}.$$

Equations equivalent to these may be obtained from the solutions of the second set of equations (v) by changing the sign of λ.

The above solution of the problem contains the indeterminate quantity $\frac{\kappa}{\lambda}$, hence *with any given linear complex are associated* ∞^1 *linear transformations which leave the form of any given quadric surface unaltered.*

When F is referred to a self-conjugate tetrahedron its equation may be taken to be

$$F \equiv x_1^2 + x_2^2 + x_3^2 + x_4^2 = 0,$$

and the equations of transformation are then

$$x_i = \kappa u_i + \lambda \Sigma a_{ik} u_k,$$
$$\xi_i = \kappa u_i - \lambda \Sigma a_{ik} u_k.$$

When the four lines which F and the linear complex have (in general) in common are taken as four edges of the tetrahedron of reference,

$$F \equiv 2X_1X_4 + 2X_2X_3, \quad A \equiv \lambda_1 p_{14} + \lambda_3 p_{23}.$$

The equations of transformation are here, (putting $\lambda = 1$),

$$X_1 = (\kappa + \lambda_1)\, U_4,\ X_2 = (\kappa + \lambda_3)\, U_3,\ X_3 = (\kappa - \lambda_3)\, U_2,\ X_4 = (\kappa - \lambda_1)\, U_1;$$
$$\Xi_1 = (\kappa - \lambda_1)\, U_4,\ \Xi_2 = (\kappa - \lambda_3)\, U_3,\ \Xi_3 = (\kappa + \lambda_3)\, U_2,\ \Xi_4 = (\kappa + \lambda_1)\, U_1.$$

Hence

$$X_1 = \frac{\kappa + \lambda_1}{\kappa - \lambda_1}\Xi_1, \quad X_2 = \frac{\kappa + \lambda_3}{\kappa - \lambda_3}\Xi_2, \quad X_3 = \frac{\kappa - \lambda_3}{\kappa + \lambda_3}\Xi_3, \quad X_4 = \frac{\kappa - \lambda_1}{\kappa + \lambda_1}\Xi_4.$$

It is to be observed that this transformation changes not only $X_1X_4 + X_2X_3$ into $\Xi_1\Xi_4 + \Xi_2\Xi_3$, but also $X_1X_4 + \mu X_2X_3$ into $\Xi_1\Xi_4 + \mu\Xi_2\Xi_3$, *i.e.* the pencil of quadrics through the intersection of F and A is unaltered in form by this transformation.

49. Collineations which leave a linear complex unaltered in form. In the most general collineation four points coincide with their corresponding points (Art. 34), hence the edges of the tetrahedron thus determined correspond to themselves.

If the collineation is such that it transforms a given linear complex into the same complex, and if one edge of the preceding tetrahedron does not belong to the complex, its polar line for the given complex must remain unaltered by the collineation, *i.e.* must be the opposite edge of the tetrahedron; the other four edges of the tetrahedron must belong to the given complex. Hence when a linear complex is transformed into itself by a collineation, in general four of its lines remain unaltered in position.

The general collineation is represented by the equations

$$Y_1 = a_1X_1,\ Y_2 = a_2X_2,\ Y_3 = a_3X_3,\ Y_4 = a_4X_4, \quad \text{(Art. 34)},$$

and the given complex is by hypothesis, with regard to this tetrahedron of reference, of the form $ap_{14} + bp_{23} = 0$; in order that the collineation should not alter the form of this latter equation we must have $\alpha_1\alpha_4=\alpha_2\alpha_3$; and if this condition is satisfied every complex $p_{14}+\mu p_{23}=0$ is unaltered by the collineation; also every quadric of the pencil $X_1X_4+\lambda X_2X_3=0$ is then transformed into itself.

Hence the general linear transformation of a linear complex into itself is given by the formulae of Art. 48 by which the surface $\Sigma a_{ik}x_ix_k=0$ is transformed into itself; but in the present case the α_{ik} are given quantities and the a_{ik} undetermined parameters. It is easily seen that every quadric so transformed must be a member of the pencil $X_1X_4+\lambda X_2X_3=0$.

50. Reciprocal transformations. It has just been seen that the general collineation as a rule neither leaves a quadric nor a linear complex unaltered in form, but if by it one complex is thus unaltered so are ∞^1 complexes; it will now be shown that the general *reciprocity* leaves two (and not more than two) linear complexes unaltered in form and also two quadrics.

The general reciprocity is given by the equations

$$u_i=\Sigma\beta_{ik}x_k',$$

in which

$$\beta_{ik}\neq\beta_{ki};$$

the surface F which is the locus of points united to their corresponding planes is

$$\Sigma\beta_{ik}x_i'x_k'=0,$$

and to this surface the reciprocity

$$u_i'=\Sigma\beta_{ki}x_k$$

is similarly related.

Taking $x_i\equiv x_i'$, (Art. 35), and solving for x_i from each equation we obtain

$$Bx_i=\sum_k B_{ki}u_k=\sum_k B_{ik}u_k';$$

hence the planes u_i and u_i' which are united with their corresponding points envelope the quadric

$$\Lambda\equiv\Sigma\Sigma B_{ik}u_iu_k=0.$$

Again writing

$$\beta_{ik}+\beta_{ki}=2a_{ik}=2a_{ki},$$
$$\beta_{ik}-\beta_{ki}=2\alpha_{ik}=-2\alpha_{ki},$$

and taking A as the linear complex $\Sigma\Sigma\alpha_{ik}x_iy_k=0$, v_i as the polar plane of x for F (or $\Sigma a_{ik}x_ix_k=0$), w_i as the polar plane of x for A,

$$v_i+w_i=\Sigma a_{ik}x_k+\Sigma\alpha_{ik}x_k=u_i,$$
$$v_i-w_i=\Sigma a_{ik}x_k-\Sigma\alpha_{ik}x_k=u_i'.$$

Hence, *the most general reciprocity is determined by a quadric and a linear complex.*

The equations determining the general reciprocity may be, by use of a suitable tetrahedron of reference, stated in the form (Art. 35)

$$U_1 = m_1 X_4,\ \ U_2 = m_2 X_3,\ \ U_3 = m_3 X_2,\ \ U_4 = m_4 X_1;$$

hence denoting by p_{ik} the coordinates of the line joining two points X_i, Y_i and by p_{ik}' the coordinates of the intersection of the corresponding planes U_i, V_i we obtain

$$p_{12} = X_1 Y_2 - X_2 Y_1,\ \ p_{12}' = U_3 V_4 - U_4 V_3 = m_3 m_4 p_{21},$$

similarly $\quad p_{34}' = m_1 m_2 p_{43}$, &c.

Hence the coordinates of corresponding lines p and p' are connected by the equations

$$p_{12}' = m_3 m_4 p_{21},\quad p_{34}' = m_1 m_2 p_{43},$$
$$p_{13}' = m_2 m_4 p_{13},\quad p_{42}' = m_1 m_3 p_{42},$$
$$p_{14}' = m_2 m_3 p_{32},\quad p_{23}' = m_1 m_4 p_{41}.$$

It follows that the linear complex $p_{14} + \mu p_{23} = 0$ is transformed into $m_2 m_3 p_{23} + \mu m_1 m_4 p_{14} = 0$, hence for the *two* values of μ given by $\dfrac{m_2 m_3}{\mu m_1 m_4} = \mu$, this complex is unaltered in form by the transformation.

It is clear that no linear complex which is not of the form $ap_{14} + bp_{23} = 0$ can be unaltered by the general reciprocity.

Two quadrics of the pencil $X_1 X_4 + \mu X_2 X_3 = 0$ are unaltered by the reciprocity, for the transformation applied to the last quadric gives

$$U_1 U_4 + \frac{\mu m_1 m_4}{m_2 m_3} U_2 U_3 = 0,$$

and this in point coordinates has the equation

$$X_2 X_3 + \frac{\mu m_1 m_4}{m_2 m_3} X_1 X_4 = 0,$$

which is the same as the original quadric provided that

$$\mu^2 = \frac{m_2 m_3}{m_1 m_4}.$$

CHAPTER IV.

SYSTEMS OF LINEAR COMPLEXES.

51. AMONG the complexes of the system $(ax) + \lambda(bx) = 0$, where λ has all values, there are two which are *special* (Art. 19), the corresponding values of λ being the roots of

$$\Omega(a) + 2\lambda\Omega(a \mid b) + \lambda^2\Omega(b) = 0 \quad \ldots\ldots\ldots\ldots\text{(i)},$$

where
$$\Omega(a \mid b) = \tfrac{1}{2}\Sigma a_i \frac{\partial \Omega}{\partial b_i}.$$

The directrices d_1 and d_2 of these special complexes were seen to be polar lines both in $(ax) = 0$ and in $(bx) = 0$, (Art. 24), (which will be referred to as the complexes A and B). Every line which belongs both to A and to B belongs also to $(ax) + \lambda_1(bx) = 0$ and to $(ax) + \lambda_2(bx) = 0$, and hence intersects d_1 and d_2. Thus the congruence of lines common to two linear complexes consists of all the lines which meet the two lines d_1 and d_2 thus determined; d_1 and d_2 are called the *directrices* of the congruence $(ax) = 0$, $(bx) = 0$. Every line of this congruence belongs to each member of the "*system of two terms*" $(ax) + \lambda(bx) = 0$.

Moreover, d_1 and d_2 are polar in each member of the system, for if a line x belongs to $(ax) + \lambda(bx) = 0$ and also meets d_1, *i.e.* belongs to $(ax) + \lambda_1(bx) = 0$, then it must satisfy both $(ax) = 0$ and $(bx) = 0$ and hence also $(ax) + \lambda_2(bx) = 0$, *i.e.* it meets d_2.

The lines of a linear congruence are said to form a *system of lines of the first order and first class.* For through any point one line of the congruence can be drawn, viz. the line of intersection of the polar planes of the point in the two given complexes; and in any plane there is one line of the congruence, viz. the join of the poles of the plane in the two complexes.

Four lines which do not belong to the same regulus determine a congruence, whose directrices are the (two) common intersectors of the four given lines. There are ∞^1 linear complexes through four such given lines, since the coefficients of the equation of a linear complex through them satisfy four independent equations. Any linear complex contains ∞^4 linear congruences; since a line p may be chosen in ∞^4 ways, its polar line p' for the given complex is then known, and hence a congruence which belongs to the complex.

52. Double ratio of two complexes*. If a_i and β_i are the coordinates of the directrices d_1 and d_2 of the congruence determined by $(ax)=0$ and $(bx)=0$, then any two complexes of the system are

$$(ax)+\lambda(\beta x)=0, \quad (ax)+\mu(\beta x)=0,$$

the coordinates of Klein being here used.

Let any plane cut d_1 and d_2 in two points P, Q and have P_1, Q_1 for its poles in these two complexes; P, Q, P_1, Q_1 are collinear and their double ratio is λ/μ; for let the plane be determined by two intersecting lines y and z which respectively meet d_1 and d_2, then

$$(ay)=0, \quad (\beta z)=0, \quad (yz)=0;$$

join the point yz to the poles of the two complexes by the lines $y+\rho_1 z$, $y+\rho_2 z$, *i.e.* express that these lines belong respectively to $(ax)+\lambda(\beta x)=0$ and to $(ax)+\mu(\beta x)=0$; this gives

$$(ay)+\lambda(\beta y)+\rho_1\{(az)+\lambda(\beta z)\}=0,$$

$$(ay)+\mu(\beta y)+\rho_2\{(az)+\mu(\beta z)\}=0;$$

or

$$\lambda(\beta y)+\rho_1(az)=0,$$

$$\mu(\beta y)+\rho_2(az)=0,$$

hence $\dfrac{\lambda}{\mu}=\dfrac{\rho_1}{\rho_2}=$ double ratio of the lines y, z, $y+\rho_1 z$, $y+\rho_2 z$ (Art. 11).

53. Double ratio of four complexes. If any line common to four complexes of the system $(ax)+\lambda(bx)=0$ be taken, the double ratio of the four polar planes of any point on this line is a constant for these complexes. For if $u=0$ be the equation of the polar plane of this point in A and $v=0$ its polar plane in B, then its polar planes in

$$A+\lambda_1 B, \; A+\lambda_2 B, \; A+\lambda_3 B, \; A+\lambda_4 B$$

are

$$u+\lambda_1 v=0, \quad u+\lambda_2 v=0, \quad u+\lambda_3 v=0, \quad u+\lambda_4 v=0;$$

and the double ratio of these four planes is equal to that formed from the quantities λ_1, λ_2, λ_3, λ_4, and hence is constant for the four given complexes.

54. Special congruence. The roots of equation (i) may be equal, in which case

$$\Omega(a)\,\Omega(b)=\{\Omega(a\,|\,b)\}^2 \text{..................(ii)}.$$

The congruence is here said to be *special*. The directrix z belongs to each complex of the system, for its coordinates are given by

$$\rho\,.\,z_i=\frac{\partial\Omega}{\partial a_i}+\lambda_1\frac{\partial\Omega}{\partial b_i},$$

* Voss, "Zur Theorie der windschiefen Flächen," *Math. Ann.* Bd. VIII.

and

$$\tfrac{1}{2}\rho\Sigma(a_i+\lambda b_i)z_i=\Omega(a)+\lambda_1\Omega(a\,|\,b)+\lambda\{\Omega(a\,|\,b)+\lambda_1\Omega(b)\}=0,$$

since λ_i is the (one) root of

$$\Omega(a)+2\lambda\Omega(a\,|\,b)+\lambda^2\Omega(b)=0.$$

If x be any line common to A and B, and which therefore meets z, each line of the pencil zx belongs to A and to B and therefore to the system $A+\lambda B$, thus when equation (ii) holds each member of the system determines on z the same correlation between its points and planes.

In one case there is an infinite number of special complexes belonging to a system of two terms, namely, when the directrices of the two special complexes intersect; for if $(\alpha x)=0$ and $(\beta x)=0$ are the two special complexes, the system may be written $(\alpha x)+\lambda(\beta x)=0$, where $\Omega(\alpha)=0$, $\Omega(\beta)=0$, and if also α and β intersect, *i.e.* if $\Omega(\alpha\,|\,\beta)=0$, the equation (i) is satisfied for all values of λ. The system consists exclusively of special complexes whose directrices form a plane pencil.

55. Metrical Properties. In the present and following articles we investigate the relations of a complex and a system of two terms to the plane at infinity and the "sphere-circle," *i.e.* the circle in which all spheres meet the plane at infinity.

Denoting by A the pole of the plane at infinity in a given complex, it has been shown (Art. 22) that the diameters of the complex all pass through A; let a be the axis of the complex. *The line a' conjugate to a is the polar of A with regard to the sphere-circle*; for if a complex line meets a in P and a' in Q, then since APQ is a right angle, A and Q are conjugate points with regard to the sphere-circle.

We have seen that any complex line which meets the axis of the complex cuts it at right angles, (and any complex line at right angles to the axis meets it). Also any line which cuts the axis at right angles is a complex line. Hence the "shortest distance" between any line l and a is a complex line and hence meets l' the polar of l.

Again if L and L' are the points in which two polar lines l and l' meet the plane at infinity, the line LL' passes through A, hence the pole B of LL' with regard to the sphere-circle lies on a', but a line which is perpendicular to l and to l' passes through B and therefore meets a', hence the "shortest distance" of l and l' meets a' and therefore meets a, or, *the shortest distance between any two polar lines meets the axis at right angles.*

The complex lines which meet any diameter d and two polar lines l and l' form a regulus consisting of one set of generators of a hyperbolic paraboloid (since they also meet d' the polar of d).

From the (1, 1) correspondence of the points of a complex line and its polar planes it follows that *the double ratio of four points of the line is equal to the double ratio of their polar planes.* Now take a complex line which meets the axis at the point B, then if P and Q are any two points on this line and C its point at infinity, the polar plane of C is that of the axis and the given line; let θ and θ' be the angles which the polar planes of P and Q make with this plane. Then by the theorem just stated, the double ratio of the points $BPQC$ is equal to the double ratio of their polar planes, or,

$$\frac{BP}{BQ} = \frac{\cot\theta}{\cot\theta'},$$

hence

$$BP\tan\theta = BQ\tan\theta'.$$

This is the property of the complex previously proved, (Art. 23).

56. Axes of a system of two terms. Let h be the line in the plane at infinity of the congruence of the system, and H its pole with regard to the sphere-circle; the line k of the congruence which passes through H will therefore cut the directrices of the system at right angles. To any point P of k belong as polar planes for the system the planes of the pencil through k. Any complex of the system has a definite plane of the pencil as polar plane at P and one point on h as its pole in the plane at infinity. Thus a (1, 1) correspondence is established between the pencil of planes and the points of h. Another such correspondence arises between the same plane pencil and the points where the normals at P to the planes meet h. There is thus a correlation established between the points of h which has therefore two "united" points. The two corresponding normals at P are axes of their respective complexes, because each of them passes through the pole in the plane at infinity of its complex and is perpendicular to its polar plane at P.

Hence through each point of k there pass two axes of the system of complexes; and we have a (1, 2) correspondence between the points of h and k, viz. to each point of h one point of k (where the axis of the particular complex meets k), to each point P of k two points of h (where the axes through P meet h).

It follows that the axes of a system of complexes form a ruled cubic in which k is the doubled and h the single directrix, see Chapter V.; the generators are all perpendicular to k; this surface is called a "cylindroid."

57. The cylindroid. The pairs of planes which unite k to the axes through the different points of k give rise to an involution. To this involution belong the pair of planes which touch the sphere-circle; for if T is the point of contact of a tangent from H to the sphere-circle, HT is the normal at H to the plane through k and HT, since it passes through the pole T of HT with regard to the sphere-circle, thus the axes of the system through the point H are the tangents from H to the sphere-circle.

The involution has two "double" planes η_1 and η_2, which are harmonic with any corresponding pair in the involution, and therefore with the tangents to the sphere-circle; hence η_1 and η_2 are at right angles, and therefore bisect the angles between any corresponding pair.

Let r be the distance of any point P of k from O an arbitrary origin on k, and θ the angle which one of the pair of planes for P makes with η_1, we have an equation connecting r and $\tan\theta$ which is linear in r and quadratic in $\tan\theta$, and hence of the form

$$(Ar + B)\tan^2\theta + (Cr + D)\tan\theta + Er + F = 0.$$

Since η_1 bisects the angle between the pair of planes for P we have $C = D = 0$; also for $\theta = 0$ let $r = r_1$ and for $\theta = \frac{\pi}{2}$ let $r = r_2$, this gives

$$Er_1 + F = 0,$$
$$Ar_2 + B = 0.$$

Now take the origin as the mid-point of the distances r_1 and r_2, then $r_1 + r_2 = 0$, and the equation becomes

$$A(r + r_1)\tan^2\theta + E(r - r_1) = 0.$$

Lastly, since 2θ for H is the angle between two planes which touch the sphere-circle, $\tan^2\theta = -1$ if $r = \infty$, hence $A = E$.

Thus the equation of the cylindroid is

$$(r + r_1)\tan^2\theta + r - r_1 = 0$$

or,

$$r = r_1 \cos 2\theta.$$

This equation may be more directly obtained from the property of a linear complex of Art. 55. For if N and N' are the points where the shortest distance of the directrices of the system meets them, then if θ be the angle which an axis through a point P of NN' makes with the line bisecting the angle $2a$ between the directrices we have

$$PN \tan(\theta + a) = PN' \tan(a - \theta),$$

or if $NN' = 2c$, $PO = r$, where O is the mid-point of NN'

$$\frac{c-r}{c+r} = \frac{\tan(a-\theta)}{\tan(a+\theta)}, \quad \text{or} \quad \frac{r}{c} = \frac{\sin 2\theta}{\sin 2a},$$

and writing $\theta + \frac{\pi}{4}$ for θ we obtain the equation of the cylindroid in the form

$$r = \frac{c}{\sin 2a} \cos 2\theta.$$

58. System of three terms. The lines belonging to three linear complexes $(ax) = 0$, $(bx) = 0$, $(cx) = 0$ form a regulus. For the lines which satisfy these three equations and also meet any other line α are two in number, viz. those given by

$$(ax) = (bx) = (cx) = (\alpha x) = (x^2) = 0.$$

Hence since of the lines common to the complexes A, B and C *two* meet any line they must form a regulus. This may also be seen as follows: take the directrices α and β of the special complexes of the system of two terms $(ax) + \lambda(bx) = 0$; to each point P of one of these lines, say α, there is one plane assigned as its polar plane in C, if this meets β in Q we have thus established a (1, 1) correspondence between the points of α and β, whose joins PQ (which belong to A, B and C) must therefore give rise to a regulus.

If two reguli have two lines a, b, in common, then if c and c' are any other lines of the respective reguli, the two common intersectors of a, b, c, c' are directrices of a linear congruence which contains both reguli; the two respective complementary reguli have two lines in common, viz. these directrices, and belong to the congruence whose directrices are a and b. If two linear congruences *have a regulus in common*, their directrices lie upon the complementary regulus, and these four directrices determine one linear complex in which they are two pairs of polar lines; this complex includes both congruences. If $(ax) = 0$ is this complex and $(ax) = (\beta x) = (\gamma x) = 0$ the regulus, the two congruences belong to the system $(ax) = 0$, $(\beta x) + \lambda(\gamma x) = 0$.

From A, B and C we have a "system of three terms"

$$\Sigma(\lambda a_i + \mu b_i + \nu c_i) x_i = 0.$$

The special complexes are given by the values of λ, μ and ν which satisfy the equation

$$\lambda^2 \Omega(a) + 2\lambda\mu\Omega(a|b) + \ldots = 0 \quad \ldots\ldots\ldots\ldots\text{(i)}.$$

Their directrices z whose coordinates are given by

$$\rho z_i = \lambda \frac{\partial \Omega}{\partial a_i} + \mu \frac{\partial \Omega}{\partial b_i} + \nu \frac{\partial \Omega}{\partial c_i}$$

also form a regulus, since *two* of them can be found which meet any line α, viz. by taking the values of λ, μ, ν which satisfy (i) and also the equation $\left(z_i \frac{\partial \omega}{\partial \alpha_i}\right) = 0$. The regulus formed by the lines common to A, B and C and the regulus composed of directrices of special complexes of the system are *complementary*; for each line x which belongs to A, B and C belongs to each complex of the system and hence to the special complexes and therefore meets each directrix z. From three of the lines α, β, γ common to A, B and C a second system of three terms can be formed, viz.

$$\Sigma\,(\lambda' \alpha_i + \mu' \beta_i + \nu' \gamma_i)\, x_i = 0.$$

We have thus obtained two systems of three terms with the property that each member of one system is in involution with each member of the other system; the lines common to one system or *the lines of one system* form a regulus and are the directrices of the special complexes of the other system. Conversely any two systems of three terms in mutual involution are thus related; for, since every member of one system is in involution with every member of the other system, every directrix of a special complex of one system meets every directrix of a special complex of the other system.

59. Expression of the coordinates of a generator of a quadric in terms of one parameter. Select from the first system two special complexes α and γ and the complex β which is in involution with them*.

Then the directrix of any special complex z is given by

$$\rho\,.\,z_i = \lambda \alpha_i + \mu \beta_i + \nu \gamma_i$$

with the condition $\quad \mu^2 (\beta^2) + 2\lambda\nu\,(\alpha\gamma) = 0,$

since by hypothesis

$$(\alpha^2) = (\gamma^2) = (\alpha\beta) = (\beta\gamma) = 0.$$

Moreover we can suppose the coefficients α, β, γ divided by such quantities as will make $(\beta^2) = 1$, $2\,(\alpha\gamma) = -1$; then the relation between λ, μ and ν is $\mu^2 = \lambda\nu$, which is satisfied by putting $\lambda = t^2$, $\mu = t$, $\nu = 1$.

Thus the line z is determined by

$$\rho\,.\,z_i = \alpha_i t^2 + \beta_i t + \gamma_i.$$

* Using Klein coordinates the latter is determined from the equations

$$\Sigma \alpha_i\,(la_i + mb_i + nc_i) = 0,$$
$$\Sigma \gamma_i\,(la_i + mb_i + nc_i) = 0.$$

In a similar manner the coordinates x of the complementary regulus are determined by the equations

$$\sigma \,.\, x_i = \alpha_i' \tau^2 + \beta_i' \tau + \gamma_i'.$$

Since each line x meets each line z we must have

$$(\alpha_i \beta_i') = (\gamma_i \beta_i') = (\alpha_i' \beta_i) = (\gamma_i' \beta_i) = 0;$$

whence the theorem, that if we take any two generators of one system of a quadric and any two generators of the other system, there is one complex of the first system and one complex of the second system which contains all four generators.

In the case when the discriminant of (i) Art. 58 is zero, the quadratic relation between the quantities λ, μ and ν

$$\Omega(\lambda a + \mu b + \nu c) = 0,$$

breaks up into two linear factors, hence we have either

$$p\lambda + q\mu + r\nu = 0 \quad \text{......................(ii)},$$

or

$$p'\lambda + q'\mu + r'\nu = 0 \quad \text{....................(iii)}.$$

Thus z here describes one of two plane pencils which have *one* common line, viz. for the values of λ, μ, ν given by taking (ii) and (iii) simultaneously.

60. Complex equation of a quadric. Let us now select from the first "system of three terms" three complexes which are themselves in mutual involution, this may be done in ∞^3 ways. Similarly from the second system select three complexes themselves in mutual involution. We have then six complexes in mutual involution. Take them for coordinate complexes. One regulus is then determined by $x_1 = x_2 = x_3 = 0$; the other regulus by $x_4 = x_5 = x_6 = 0$. Any tangent line of the quadric to which the reguli belong is one member of a plane pencil of which two are the generators at the point of contact of this tangent, hence if α and β are these generators and y the tangent line we have

$$\alpha_4 = \alpha_5 = \alpha_6 = 0; \quad \beta_1 = \beta_2 = \beta_3 = 0;$$

and

$$y_1 = \lambda\alpha_1, \quad y_2 = \lambda\alpha_2, \quad y_3 = \lambda\alpha_3,$$

$$y_4 = \mu\beta_4, \quad y_5 = \mu\beta_5, \quad y_6 = \mu\beta_6.$$

Since

$$\alpha_1^2 + \alpha_2^2 + \alpha_3^2 = \beta_4^2 + \beta_5^2 + \beta_6^2 = 0,$$

we see that y satisfies the complex equation

$$y_1^2 + y_2^2 + y_3^2 = 0,$$

or, which is the same thing,

$$y_4^2 + y_5^2 + y_6^2 = 0.$$

It is clear that there are ten quadrics associated in this manner with six complexes in mutual involution, viz. the series

$$y_i^2 + y_j^2 + y_k^2 = 0.$$

61. The ten fundamental quadrics. Any six coordinate complexes of Klein in mutual involution involve fifteen independent constants, viz., five for each complex diminished by the fifteen conditions. The complex $x_i = 0$ will be denoted by C_i. The system of two terms of C_i and C_j is $x_i + \lambda x_j = 0$. This is special if $1 + \lambda^2 = 0$. Thus the special complexes are

$$x_i + ix_j = 0, \quad x_i - ix_j = 0.$$

Their directrices being denoted by d_{ij} and d_{ji}, the coordinates of

d_{ij} are $x_i = 1$, $x_j = i$, and the other coordinates zero;

of d_{ji} are $x_i = 1$, $x_j = -i$, " " " " .

There are fifteen congruences C_{ij} and thirty lines d; observe that *each d belongs to the complex not mentioned in its suffix.* In C_{ij} and C_{kl} the directrices form a twisted quadrilateral (for d_{ij} belongs to C_k and C_l and therefore to C_{kl} and hence meets d_{kl} and d_{lk}), similarly for d_{ji}, d_{kl}, d_{lk}. In C_{ij} and C_{ik} the directrices belong to C_l, C_m and C_n and hence to the regulus ρ_{lmn}, so that

$$d_{ij},\ d_{ji},\ d_{ik},\ d_{ki},\ d_{jk},\ d_{kj} \text{ belong to } \rho_{lmn},$$

$$d_{mn},\ d_{nm},\ d_{ln},\ d_{nl},\ d_{lm},\ d_{ml} \text{ belong to } \rho_{ijk}.$$

The reguli ρ_{lmn}, ρ_{ijk} are complementary, the quadric to which they belong is

$$y_l^2 + y_m^2 + y_n^2 = 0.$$

There are ten such quadrics which have been termed "*fundamental*" by Klein.

Two reguli ρ_{ikl}, ρ_{imn} have no common line,

" " ρ_{ijk}, ρ_{ijl} have d_{mn}, d_{nm} in common.

Thus the quadrics

$$(\rho_{ijk},\ \rho_{mnl}),\quad (\rho_{ijl},\ \rho_{mnk})$$

meet in the quadrilateral formed by

$$d_{mn},\ d_{nm},\ d_{ij},\ d_{ji}.$$

In the quadric (ρ_{ikl}, ρ_{jmn}) d_{ij} and d_{ji} are *polar lines*; for d_{ij}

and d_{ji} meet the four lines in the figure, (which lie on the given quadric since they have in common with it the suffixes kl and mn). Hence d_{ij} and d_{ji} are the lines OO_1 and $O'O_1'$, and the polar plane of O passes through O' and O_1', also the polar plane of O_1 passes through O' and O_1', hence OO_1 and $O'O_1'$ are polar lines. We observe also that

$$d_{ij},\ d_{ji},\ d_{lk},\ d_{kl},\ d_{mn},\ d_{nm}$$

form a tetrahedron. [There are fifteen of these tetrahedra, which are the fundamental tetrahedra of Art. 29.] From this result we see that the four quadrics

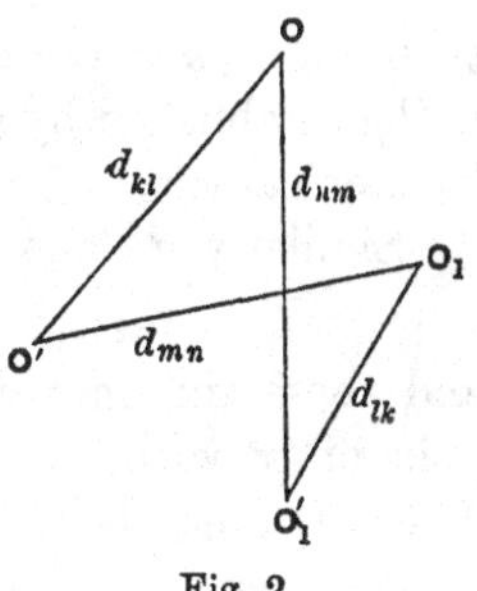

Fig. 2.

$$(\rho_{ikm},\ \rho_{jnl}),\ (\rho_{ilm},\ \rho_{jkn}),\ (\rho_{ikn},\ \rho_{jlm}),\ (\rho_{iln},\ \rho_{jkm})$$

have the tetrahedron $(ij,\ kl,\ mn)$ as a common self-conjugate tetrahedron.

Let π be any plane with poles O_i, O_j, O_k in the complexes C_i, C_j, and C_k. On O_iO_j an involution is determined of poles in C_i and C_j, the double points of this involution being where d_{ij} and d_{ji} meet O_iO_j, (Art. 25). Hence O_i, O_j and these two points are harmonic; similarly for O_iO_k and O_jO_k. But d_{ij}, d_{ji}, d_{ik}, d_{ki}, d_{jk}, d_{kj} lie on ρ_{lmn}, hence the trace of ρ_{lmn} on π has $O_i\,O_j\,O_k$ for a self-conjugate triangle. In the same manner the trace of ρ_{ijk} on π has $O_lO_mO_n$ for a self-conjugate triangle, but these traces are the *same*, hence the six poles O being the vertices of triangles self-conjugate with regard to the same conic lie on a conic, and the sides of the triangles $O_iO_jO_k$, $O_lO_mO_n$ touch a conic.

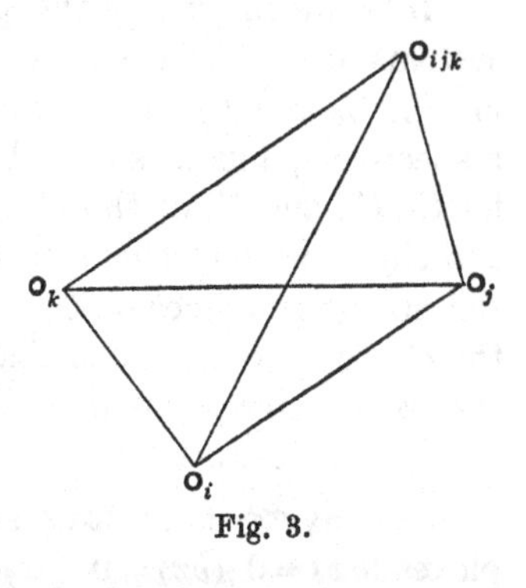

Fig. 3.

62. Closed system of sixteen points and planes*. Since C_i and C_j are in involution there is a plane π_{ij} through O_iO_j which is the polar plane of O_i in C_j and of O_j in C_i; there are clearly fifteen such planes as π_{ij}; take three of them π_{ij}, π_{jk}, π_{ki}, and let O_{ijk} be their common point.

Now O_iO_{ijk} belongs to C_j and to C_k,
O_jO_{ijk} „ C_i „ C_k;

* See Art. 26.

hence $\qquad O_{ijk}$ is the pole of π_{ij} in C_k,
but $\qquad O_i$ " " " π_{ij} " C_j,
therefore O_i and O_{ijk} are corresponding points of the involution on O_iO_{ijk} established by C_j and C_k, thus d_{jk} and d_{kj} divide O_i and O_{ijk} harmonically.

Similarly d_{ij} and d_{ji} divide O_kO_{ijk} harmonically,
$\qquad d_{ik}$ " d_{ki} " O_jO_{ijk} " ;
and these six lines d belong to ρ_{lmn}, hence the point O_{ijk} is the pole of π with reference to the quadric which contains ρ_{lmn}. The point O_{lmn} being in like manner the pole of π with reference to the same quadric is thus seen to be *the same point* as O_{ijk}.

Starting from the plane π we obtain fifteen planes such as π_{ij} and sixteen points, viz., the poles of π in the six fundamental complexes and the poles of π with reference to the ten fundamental quadrics. These points and planes form a closed system, for if we start with another of these planes, *e.g.* π_{ij}, we obviously arrive at the same set of planes and points. Thus by aid of the six fundamental complexes, and their derivatives the ten fundamental quadrics, we can divide the points and planes of space into closed systems of sixteen planes and points*. The point O_{ijk} is the pole for C_i, C_j and C_k of the planes joining O_{ijk} to O_jO_k, O_kO_i, O_iO_j respectively, and it is also (being the same point as O_{lmn}) the pole for C_l, C_m and C_n of the planes joining O_{ijk} to O_mO_n, O_nO_l, O_lO_m, and since the two triangles $O_iO_jO_k$, $O_lO_mO_n$ touch the same conic, the six polar planes of O_{ijk} touch the same quadric cone. Hence the closed system is such that the six points in each plane lie on the same conic† and *the six planes through each point touch the same quadric cone.*

63. Systems of four and of five terms. When four complexes $(ax)=0$, $(bx)=0$, $(cx)=0$, $(dx)=0$ are given, if a complex $(ux)=0$ is in involution with each of them, we have

$$\left(u\frac{\partial\Omega}{\partial a}\right)=0,\quad \left(u\frac{\partial\Omega}{\partial b}\right)=0,\quad \left(u\frac{\partial\Omega}{\partial c}\right)=0,\quad \left(u\frac{\partial\Omega}{\partial d}\right)=0.$$

These four equations are equivalent to the six equations

$$u_i=\sigma p_i+\tau q_i,$$

where σ and τ are indeterminate, and p, q any two particular solutions for u. Thus the system in involution with the system of four terms $\qquad \Sigma(\lambda a_i+\mu b_i+\nu c_i+\rho d_i)\,x_i=0,$

* This system is, of course, identical with that of Arts. 14, 29; the connexion of the system with the 10 fundamental quadrics being here established.

† See Art. 27.

is the system of two terms

$$\Sigma\,(\sigma p_i + \tau q_i)\,x_i = 0.$$

As in the case of systems of three terms, each directrix of a special complex of one system meets each directrix of a special complex of the other system; and the lines common to one system are the directrices of the special complexes of the other system. In general therefore the system of four terms has two lines common to the system, viz., the two directrices of the complementary system of two terms. If the system of two terms has its directrices coincident, that of four terms has only one common line.

One complex can be found which is in involution with five given complexes and hence in involution with the system of five terms

$$\Sigma\,(\lambda a_i + \mu b_i + \nu c_i + \rho d_i + \sigma e_i)\,x_i = 0.$$

If this complex is special the five given complexes have one line in common.

64. Invariants of a system of complexes. A series of invariants is obtained by bordering the discriminant Δ of the equation $\omega(q) = 0$ with the coefficients of one or several complexes; this gives the invariants

$$\Delta_a = \begin{vmatrix} \alpha_{11} \ldots \alpha_{16} & a_1 \\ \cdots\cdots & \\ \alpha_{61} \ldots \alpha_{66} & a_6 \\ a_1 \ldots a_6 & 0 \end{vmatrix}, \quad \Delta_{ab} = \begin{vmatrix} \alpha_{11} \ldots \alpha_{16} & a_1 & b_1 \\ \cdots\cdots & & \\ \alpha_{61} \ldots \alpha_{66} & a_6 & b_6 \\ a_1 \ldots a_6 & 0 & 0 \\ b_1 \ldots b_6. & 0 & 0 \end{vmatrix}, \text{ \&c.}$$

We shall find the geometrical significance of the vanishing of these invariants; it is worth while, in the first place, to notice that they may be expressed in terms of the functions Ω, as follows: understanding by A_{ik} the coefficient of α_{ik} in Δ,

$$\Delta_a = \Omega(a) \qquad \text{(Art. 18)},$$

$$\Delta_{ab}\,\Delta^5 =$$

$$\begin{vmatrix} \alpha_{11} \ldots \alpha_{16} & a_1 & b_1 \\ \cdots\cdots & & \\ \alpha_{61} \ldots \alpha_{66} & a_6 & b_6 \\ a_1 \ldots a_6 & 0 & 0 \\ b_1 \ldots b_6 & 0 & 0 \end{vmatrix} \times \begin{vmatrix} A_{11} \ldots A_{16} & 0 & 0 \\ \cdots\cdots & & \\ A_{61} \ldots A_{66} & 0 & 0 \\ 0 \ldots 0 & 1 & 0 \\ 0 \ldots 0 & 0 & 1 \end{vmatrix} = \begin{vmatrix} \Delta & 0 \ldots\ldots 0 & a_1 & b_1 \\ 0 & \Delta \ldots\ldots & a_2 & b_2 \\ \cdots\cdots & & & \\ 0 & \ldots\ldots \Delta & a_6 & b_6 \\ -\frac{1}{2}\frac{\partial\Omega}{\partial a_1} & \ldots -\frac{1}{2}\frac{\partial\Omega}{\partial a_6} & 0 & 0 \\ -\frac{1}{2}\frac{\partial\Omega}{\partial b_1} & \ldots -\frac{1}{2}\frac{\partial\Omega}{\partial b_6} & 0 & 0 \end{vmatrix}.$$

Now multiply the first column by $-\frac{a_1}{\Delta}$, the second by $-\frac{a_2}{\Delta}$ &c. and add to the seventh column, proceed similarly with the eighth column and we obtain

$$\Delta_{ab}\,\Delta^5 = \Delta^4 \begin{vmatrix} \Omega(a) & \Omega(a|b) \\ \Omega(a|b) & \Omega(b) \end{vmatrix}.$$

This method applies generally and we obtain

$$\Delta^2\,\Delta_{abc} = \begin{vmatrix} \Omega(a) & \Omega(a|b) & \Omega(a|c) \\ \Omega(b|a) & \Omega(b) & \Omega(b|c) \\ \Omega(c|a) & \Omega(c|b) & \Omega(c) \end{vmatrix},$$

$$\Delta^3\,\Delta_{abcd} = \begin{vmatrix} \Omega(a) & \Omega(a|b) & \Omega(a|c) & \Omega(a|d) \\ \Omega(b|a) & \cdots & \cdots & \cdots \\ \Omega(c|a) & \cdots & \cdots & \cdots \\ \Omega(d|a) & \cdots & \cdots & \Omega(d) \end{vmatrix}.$$

It has been seen, (Art. 19), that if $\Omega(a) = 0$ the complex $(ax) = 0$ is special.

If $\Delta_{ab} = 0$ the complexes $(ax) = 0$, $(bx) = 0$ have a *special* congruence, (Art. 54).

If $\Delta_{abc} = 0$ the regulus of the system of three terms formed by $(ax) = 0$, $(bx) = 0$, $(cx) = 0$, degenerates into two plane pencils, (Art. 58).

It will now be shown that if $\Delta_{abcd} = 0$, the two lines common to the system of four terms given by $(ax) = 0$, $(bx) = 0$, $(cx) = 0$, $(dx) = 0$, will coincide.

For the conditions that the *same line* may be the directrix of a special complex both in the system of four terms and in the system of two terms which is in involution with it, are

$$\frac{\partial\Omega(\lambda a_i + \mu b_i + \nu c_i + \rho d_i)}{\partial(\lambda a_i + \mu b_i + \nu c_i + \rho d_i)} = \frac{\partial\Omega(pe_i + qf_i)}{\partial(pe_i + qf_i)}, \ (i = 1, \ldots 6) \ \ldots\ldots(\text{i}),$$

or $$\lambda\frac{\partial\Omega(a)}{\partial a_i} + \mu\frac{\partial\Omega(b)}{\partial b_i} + \nu\frac{\partial\Omega(c)}{\partial c_i} + \rho\frac{\partial\Omega(d)}{\partial d_i} = p\frac{\partial\Omega(e)}{\partial e_i} + q\frac{\partial\Omega(f)}{\partial f_i},$$

together with $$\Omega(\lambda a + \mu b + \nu c + \rho d) = 0,$$

$$\Omega(pe + qf) = 0.$$

Since the two systems are in involution the last two equations follow from the first six, and we have to find the (single) condition for the coexistence of the equations (i).

Multiplying by e_i and by f_i and adding for the six equations we obtain

$$p\Omega(e) + q\Omega(e|f) = 0,$$
$$p\Omega(e|f) + q\Omega(f) = 0;$$

hence

$$\begin{vmatrix} \Omega(e) & \Omega(e|f) \\ \Omega(e|f) & \Omega(f) \end{vmatrix} = 0 \quad\text{.....................(ii).}$$

But this is the condition that the system of two terms should have a special congruence and hence that the system of four terms should have its two common lines *coincident.* Again multiplying the equations (i) by a_i, b_i, c_i, and d_i, adding, and eliminating λ, μ, ν, ρ we obtain

$$\begin{vmatrix} \Omega(a) & \Omega(a|b) & \Omega(a|c) & \Omega(a|d) \\ \dots & \dots & \dots & \dots \\ \dots & \dots & \dots & \dots \\ \Omega(d|a) & \dots & \dots & \Omega(d) \end{vmatrix} = 0, \text{ or } \Delta_{abcd} = 0.$$

Hence either of the equations $\Delta_{ef} = 0$, $\Delta_{abcd} = 0$ involves the coexistence of the equations (i). Thus if $\Delta_{abcd} = 0$, the system of four terms has its two common lines coincident; this line then belongs to each system and is the directrix of the special congruence in the system of two terms.

Complex equation of the quadric determined by a system of three terms. If a, b, c and d are lines which have their two intersectors coincident, each line must touch the quadric containing the regulus to which the other three lines belong.

If a line y touches the quadric of which one regulus consists of lines common to three complexes $(ax) = 0$, $(bx) = 0$, $(cx) = 0$ we have $\Delta_{abcy} = 0$; hence the complex equation of the quadric is, if we use Klein coordinates,

$$\begin{vmatrix} (a^2) & (ab) & (ac) & (ay) \\ (ba) & (b^2) & (bc) & (by) \\ (ca) & (cb) & (c^2) & (cy) \\ (ay) & (by) & (cy) & 0 \end{vmatrix} = 0.$$

If five complexes A, B, C, D, E have one line in common, the complex in involution with them must be special, and we obtain equations similar to (i) which require that $\Delta_{abcde} = 0$.

65. If five linear complexes A, B, C, D, E have two lines in common there exists a linear relation of the form

$$\lambda(ax) + \mu(bx) + \nu(cx) + \rho(dx) + \sigma(ex) \equiv 0;$$

for in this case any one of the complexes belongs to the system of four terms determined by the remaining complexes.

If five linear complexes have only *one* line in common no such linear relation exists; hence if K and F be any other two complexes we can express K in terms of A, B, C, D, E, F or,

$$(kx) \equiv \lambda (ax) + \mu (bx) + \nu (cx) + \rho (dx) + \sigma (ex) + \tau' (fx).$$

If now K passes through the line common to A, B, C, D, E then since F is *any* complex it is necessary in the identity that $\tau = 0$, and we have between six complexes which pass through the same line an identity of the above form. If we suppose the constants λ, μ, &c. to be absorbed within the brackets and call the contents of these brackets $x_1, x_2, \ldots$, it follows that between six linear complexes through the same line an identity exists of the form

$$\sum_1^6 x_i = 0.$$

Now take any five complexes $x_1, \ldots x_5$ which have one line in common, and any other complex X; we will determine the form of the identical relation $\omega(x) = 0$ when these six complexes are taken as coordinate complexes.

Observe, first, that $\omega(x)$ will not contain X^2, since $X = 0$ does not contain the line common to $x_1 = 0, \ldots x_5 = 0$. Hence

$$\omega(x) \equiv \sum_1^5 a_{ii} x_i^2 + 2 \sum_1^5 a_{ij} x_i x_j + 2X \sum_1^5 b_i x_i = 0.$$

Moreover since $\Omega(b)$ clearly vanishes, the coefficient of X equated to zero is a special complex, (Art. 19), and it is seen to be in involution with each of the five complexes $x_1=0, \ldots x_5=0$; hence $\sum_1^5 b_i x_i = 0$ is the special complex whose directrix is the common line of intersection of $x_1, \ldots x_5$. Any four of the complexes $x_1, \ldots x_5$ will have in common both the given line and one other line, *e.g.* for the complexes x_1, x_2, x_3, x_4 this latter line is seen to be

$$x_1 = x_2 = x_3 = x_4 = 0, \quad X = -\frac{a_{55}}{2b_5} x_5.$$

Taking each combination of four of the given five complexes $x_1, \ldots x_5$ we thus obtain five lines, through which passes the linear complex

$$\frac{a_{11}x_1}{b_1} + \frac{a_{22}x_2}{b_2} + \frac{a_{33}x_3}{b_3} + \frac{a_{44}x_4}{b_4} + \frac{a_{55}x_5}{b_5} + 2X = 0.$$

This complex clearly does not contain the common line of intersection. The complex X has so far been taken arbitrarily; let it be *identical with the last complex,* then

$$a_{11} = a_{22} = a_{33} = a_{44} = a_{55} = 0.$$

The form which $\omega(x)$ now assumes is that proper to six complexes of which five pass through the same line and the sixth is the complex passing through the five lines of second intersection of each four of the preceding five complexes.

For convenience the complex X will be called the "residual" of the five complexes $x_1, \ldots x_5$. Any other sixth complex x_6 through the given line being taken, we have the identity $\sum_1^6 x_i \equiv 0$. To each set of five of the complexes $x_1, \ldots x_6$ belongs one residual, and it will now be shown that the *six residuals have one line in common.* For, taking the five complexes x_2, x_3, x_4, x_5, x_6, the five lines which determine the residual are

$x_2 = x_3 = x_4 = x_5 = X = 0$ from the complexes x_2, x_3, x_4, x_5;

$x_3 = x_4 = x_5 = 0,\ x_2 = b_2 - b_1,\ X = a_{12}$ .... x_3, x_4, x_5, x_6;

$x_2 = x_4 = x_5 = 0,\ x_3 = b_3 - b_1,\ X = a_{13}$ .... x_2, x_4, x_5, x_6;

$x_2 = x_3 = x_5 = 0,\ x_4 = b_4 - b_1,\ X = a_{14}$ .... x_2, x_3, x_5, x_6;

$x_2 = x_3 = x_4 = 0,\ x_5 = b_5 - b_1,\ X = a_{15}$ .... x_2, x_3, x_4, x_6.

The complex through these five lines is seen to be

$$\frac{a_{12}}{b_2 - b_1} x_2 + \frac{a_{13}}{b_3 - b_1} x_3 + \frac{a_{14}}{b_4 - b_1} x_4 + \frac{a_{15}}{b_5 - b_1} x_5 - X = 0.$$

By a similar process the equations of the other four residuals are seen to be, if $a_{ik} = a_{ki}$,

$$\frac{a_{21}}{b_1 - b_2} x_1 + \frac{a_{23}}{b_3 - b_2} x_3 + \frac{a_{24}}{b_4 - b_2} x_4 + \frac{a_{25}}{b_5 - b_2} x_5 - X = 0,$$

$$\frac{a_{31}}{b_1 - b_3} x_1 + \frac{a_{32}}{b_2 - b_3} x_2 + \frac{a_{34}}{b_4 - b_3} x_4 + \frac{a_{35}}{b_5 - b_3} x_5 - X = 0,$$

$$\frac{a_{41}}{b_1 - b_4} x_1 + \frac{a_{42}}{b_2 - b_4} x_2 + \frac{a_{43}}{b_3 - b_4} x_3 + \frac{a_{45}}{b_5 - b_4} x_5 - X = 0,$$

$$\frac{a_{51}}{b_1 - b_5} x_1 + \frac{a_{52}}{b_2 - b_5} x_2 + \frac{a_{53}}{b_3 - b_5} x_3 + \frac{a_{54}}{b_4 - b_5} x_4 - X = 0.$$

Now the last five equations and the equation $X = 0$ have one common solution, since their determinant, when $X = 0$, being skew-symmetrical of order five, is *zero*. Moreover these values of $x_1, x_2, x_3, x_4, x_5, X$ are coordinates of a line, since if we multiply the last four equations by $x_2(b_1 - b_2), x_3(b_1 - b_3), x_4(b_1 - b_4), x_5(b_1 - b_5)$ respectively and add, we obtain the result of putting $X = 0$ in $\omega(x)$.

Hence the six residuals have a line in common*.

* This theorem is due to Mr J. H. Grace, see "Circles, spheres and linear complexes," *Camb. Phil. Trans.*, vol. XVI., part III. (1897).

CHAPTER V.

RULED CUBIC AND QUARTIC SURFACES.

66. In the present chapter is given the classification of ruled quartic surfaces due originally to Cremona*, but following the order adopted by R. Sturm†. The surfaces were dealt with in the first place by Cayley‡.

A plane through a generator g of a ruled surface of degree n meets the surface also in a curve of degree $n-1$, and g meets this curve in $n-1$ points, while through each point of the ruled surface and hence of this curve there passes in general only *one* generator; thus as we travel round the curve there is a generator for each point, g is itself the generator for one point, while for each of the remaining $n-2$ intersections of g and the curve there are two generators, viz., g itself and the generator belonging to the point; these $n-2$ points are therefore double points of the surface; on each generator there are $n-2$ double points which lie on a double curve of the surface.

In the case of a ruled cubic surface there is a double curve which must be a straight line, because otherwise the line joining any two points of the double curve would meet the surface in four points.

67. Ruled Cubics. For a ruled cubic the double line d may be either (i) a double directrix, or (ii) a generator and a single directrix, understanding by a *directrix* of a ruled surface a line which meets every generator.

In *the first case* any section of the surface by a plane through a generator g consists of g and a conic c^2; *one* of the points of intersection of g and c^2 is, by the foregoing, a point K of d, the two generators through any point P of d meet c^2 in points X and X' giving rise to an involution on c^2, hence all the lines XX' must intersect in the same point O, (Art. x.); thus the planes of the two

* "Sulle superficie gobbi di quarto grado," *Bologna Accad. Sci. Mem.* VIII. (1868).

† See *Liniengeometrie*, Bd. I., S. 52—61.

‡ *Phil. Trans.*, vol. 159 (1869).

generators through each point of d pass through O, and hence that of K passes through O, which requires that O should lie on g. The plane PXX' cuts the surface in the lines PX, PX' and hence also in another line e which must meet g in O; this line e is the same for each plane PXX' since if two lines e passed through O the line XOX' would meet the surface in four points; thus the planes of the pair of generators through each point of d form a pencil whose axis e is therefore a simple directrix of the surface. These planes are the bitangent planes of the surface. *There is a* (1, 1) *correspondence between the points of e and c^2 made by the generators,* which affords the simplest means of generating the surface.

The second case in which d is a simple directrix is got from the first by the coincidence of d and e; this surface (Cayley's) is obtained by establishing a (1, 1) correspondence between the points of a conic and a line which intersects it.

68. **Ruled Quartics.** The points of any two plane sections of a ruled surface are connected by the generators in a (1, 1) correspondence, hence all plane sections of the surface have the same *deficiency**, this is called the deficiency of the surface. Since the residual cubic curve of a plane section of a quartic surface through a generator is met by each generator, the deficiency of this cubic is therefore that of the surface, *hence the deficiency of the surface is that of this cubic and is therefore unity or zero*; when the deficiency is unity, the plane sections of the surface are quartic curves with two double points, therefore there are two double points of the surface in any plane section; hence the double curve is of the second order, and, since it is met twice by each generator, cannot be a conic, and must consist of two non-intersecting lines d_1 and d_2.

Classes I. and II. correspond to this case; there are two double directrices d_1 and d_2; from each point of d_1 proceed two generators which meet d_2, and from each point of d_2 proceed two generators which meet d_1, hence d_1 and d_2 constitute the envelope of the bitangent planes of the surface; class II. arises when d_1 and d_2 coincide.

* The "deficiency" of a plane curve of degree n is equal to $\alpha-\beta$, where α = maximum number of double points that may be possessed by a curve of degree n, β = number of double points possessed by the given curve.

The word "genus" is also used to designate this number. For proofs of the theorem quoted, viz., that two plane curves in (1, 1) correspondence have the same deficiency, see Clebsch, *Vorlesungen über Geometrie*, Bd. I., S. 458.

69. The surface of zero deficiency has now to be considered, for it the section by any plane is a quartic curve with the maximum number, three, of double points; any plane section, therefore, contains three double points of the surface; the most general classes, viz., III. and IV., arise when the double curve is a twisted cubic; from each point P of the double curve proceed two generators, meeting it again in the points Q and R, thus giving rise to an involution [2] on the double curve, and the ruled quartic is obtained which has already been discussed (Art. xvi.). If this involution [2] is *cubic* we have class IV. The residual cubic in a plane section through a generator g has one double point outside g.

The generators through a point P of the double curve being PQ and PR, the plane PQR is a double tangent plane of the surface, since it contains two generators, and through the point P there pass three such planes, viz., those corresponding to P, Q and R, hence the developable of the bitangent planes is of the third class. In class IV. this bitangent developable is replaced by a pencil of triply tangent planes whose axis is the directrix of the surface. Any bitangent plane cuts out a conic; on two such conics c^2 and c'^2, by means of the generators, a (1, 1) correspondence is established: conversely a (1, 1) correspondence on two conics in different planes gives rise to a ruled quartic; for if A and B be two points in the planes of c^2 and c'^2 respectively, any plane through AB meets c^2 in two points L, L' and c'^2 in two points M, M' and there are related to L, L' by the (1, 1) correspondence two points N, N' on c'^2, the pairs of points N, N' form an involution, hence NN' passes through a fixed point C (Art. x.); also since the lines MM', NN' which pass through B and C respectively are in (1, 1) correspondence, the locus of their intersection is a conic meeting c'^2 in four points, hence four of the lines LN meet AB, or *the locus of the lines joining corresponding points of a* (1, 1) *correspondence on c^2 and c'^2 is a ruled quartic.*

Classes V., VI. arise when the double curve consists of a conic c^2 and a line d which must meet c^2*; if d is a double directrix we have class V.; from each point of d pass two generators which meet the residual conic c'^2 of a section by *any* bitangent plane. Hence an

* For if not, if P is any point of the surface, since the cone (P, c^2) meets d in two points, two generators of this cone would meet the surface in five points, *i.e.*, would be generators of the surface, or through any point P of the surface there would pass two generators, which is impossible.

involution is generated on c'^2 of such pairs of points Q and Q'; therefore QQ' passes through a fixed point O, or *the bitangent planes through the pair of generators from each point of d envelope a cone of vertex O*; this cone is of the second degree since through any line OQQ' only one such bitangent plane can be drawn in addition to the plane of c'^2; the other set of bitangent planes, viz., those which contain the pair of generators through each point of c^2, pass through d, hence *in class V. the envelope of the bitangent planes consists of the double directrix and a quadric cone.*

We have class VI. when d is a simple directrix and also a generator; each plane through the generators from a point of c^2 is triply tangent since d is itself a generator.

Classes VII. and VIII. arise when the double conic breaks up into two intersecting lines d' and e of which one, say e, must meet d; e cannot be a directrix, for if so the surface would consist partly of the plane (d, e), hence e must be a double generator; d and d' are double directrices and the envelope of the bitangent planes is formed by the three lines d, d' and e. Class VIII. is where d and d' come into coincidence.

In classes IX., X., XI., XII. we have a triple line, which is a triple directrix (IX. and X.), a double directrix and simple generator (XI.), a simple directrix and double generator (XII.). In every case one other generator lies in each plane through the triple line d.

In class IX. from each point of d pass three generators, and *the bitangent developable is of the third class*; for any given bitangent plane cuts the surface in a conic c^2 on which a cubic involution is determined by the generators from the different points of d, and, since the Direction Curve, (Art. xv.), of this involution is of the second class, it is clear that through each point of the given plane there pass two bitangent planes besides the given plane. This species is reciprocal to IV. If the generators through each point of d are coplanar, class X. arises; since here from each point of d only *one* triply tangent plane proceeds, the envelope of such planes must be a line.

In class XI., in which d is a double directrix and simple generator, it follows by exactly the same process as in class V. that the envelope of the bitangent planes consists of d and a quadric cone.

In class XII., in which d is a simple directrix and double generator, every plane through d touches the surface three times, viz., once at a point on the generator in the given plane, and also in the points in which d is met by the two generators consecutive to it.

The following Tables show the connexion of the above classification with that given by Cremona and Cayley, and exhibit the special features of the different classes.

Sturm	I	II	III	IV	V	VI	VII	VIII	IX	X	XI	XII
Cremona	11	12	1	7	2	4	5	6	8	9	3	10
Cayley	1	4	10	8	7		2	5	9	3		6

In the next Table the general twisted cubic is denoted by k^3, the developable of the third class by σ^3, the general conic by c^2, the quadric cone by κ^2.

	I	II	III	IV	V	VI	VII
Double Curve	$d+d'$	$d+d$	k^3	k^3	c^2+d	c^2+d	$d+d'+e$
Bitangent Developable	$d+d'$	$d+d$	σ^3	$d+d+d$	κ^2+d	$d+d+d$	$d+d'+e$

	VIII	IX	X	XI	XII
Double Curve	$d+d+e$	$d+d+d$	$d+d+d$	$d+d+d$	$d+d+d$
Bitangent Developable	$d+d+e$	σ^3	$d'+d'+d'$	κ^2+d	$d+d+d$

70. Ruled Surfaces whose deficiency is zero. A good illustration of the use of the Klein coordinates is afforded by Voss's classification by analysis of ruled quartics whose deficiency is zero.

The coordinates of a generator of a ruled quartic belonging to class III., or to the sub-varieties IV.—XII., may be expressed in terms of one parameter, as has been shown by Voss*, of whose method we now proceed to give an account.

If every section of a ruled surface of degree n possesses the maximum number, viz., $\frac{\overline{n-1}\,.\,\overline{n-2}}{2}$, of double points, the coordinates of each point of such a section are expressible as rational integral functions, of degree n, of one variable.

Since by means of the generators a (1, 1) correspondence is established between pairs of points of any two given sections, it follows that the coordinates x_i of any generator are expressible as rational functions of one parameter λ, and since in general a line meets n generators of the surface we may write

$$\rho\,.\,x_i = \phi_i(\lambda),$$

* "Zur Theorie der windschiefen Flächen," *Math. Annalen*, Bd. VIII.

where the functions ϕ are rational, integral and of degree n in λ.

The Rank of the surface (*i.e.* the degree of the tangent cone of any point) is $2(n-1)$; for since any tangent line y of the surface meets two consecutive generators x and $x+dx$ we have $(yx)=0$, $(ydx)=0$, or

$$(y\phi)=0, \qquad \left(y\frac{\partial\phi}{\partial\lambda}\right)=0;$$

the result of eliminating λ between these equations is of degree $2(n-1)$ in y, say $F(y)=0$, and supposing y to pass through a given point P, the equation $F(y)=0$ is the equation of the tangent cone from P to the surface, which is thus seen to be of degree $2(n-1)$. This result shows that the surface must contain a double curve of order $\frac{\overline{n-1}\,.\,\overline{n-2}}{2}$, for if δ is the order of the double curve, since any plane π through P meets $F(y)=0$ in $2(n-1)$ lines, there are $2(n-1)$ tangents to the surface in the pencil (P, π), hence

$$2(n-1)=n(n-1)-2\delta,$$

hence

$$\delta=\frac{\overline{n-1}\,.\,\overline{n-2}}{2}.$$

If the parameters of two generators x and x' which meet (in a point of the double curve) are λ and λ', since

$$(x^2)=0, \quad (x'^2)=0, \quad (xx')=0,$$

it follows that

$$\sum_i \{\phi_i(\lambda)-\phi_i(\lambda')\}^2=0.$$

This last equation may be divided by $(\lambda-\lambda')^2$, and the equation connecting the parameters of two generators meeting on the double curve is of the form

$$\psi(\lambda, \lambda')=0. \quad \text{.......................(I.)}$$

This equation is of degree $n-2$ in λ and λ', thus each generator is met by $n-2$ other generators, and hence meets the double curve in $n-2$ points. The equation (i) will be called the 'equation of the double curve.' If in it we write $\lambda=\lambda'$ we obtain an equation of degree $2(n-2)$ in λ corresponding to generators each of which is intersected by a *consecutive* generator; such a generator is said to be *singular*.

71. Ruled cubics. For a ruled cubic we have

$$\rho . x_i = a_i + \lambda b_i + \lambda^2 c_i + \lambda^3 d_i,$$

and sinee $(x^2) = 0$, it follows that

$$(a^2) = (ab) = (d^2) = (dc) = (b^2) + 2(ac) = (c^2) + 2(bd) = (ad) + (bc) = 0.$$

The equation $\psi(\lambda, \lambda') = 0$ is here

$$(b^2) + 2(bc)(\lambda + \lambda') + (c^2)\lambda\lambda' = 0.$$

There are two singular generators given by the equation

$$(b^2) + 4(bc)\lambda + (c^2)\lambda^2 = 0,$$

if $$4(bc)^2 \neq (b^2)(c^2).$$

The quantity λ may be so chosen that $\lambda = 0$, $\lambda = \infty$ correspond to the two singular generators; in this case we have $(b^2) = (c^2) = 0$, hence $(ac) = (bd) = 0$. The equation $\psi = 0$ reduces to $\lambda + \lambda' = 0$.

From the equations

$$(a^2) = (b^2) = (c^2) = (d^2) = (ac) = (bd) = (ab) = (cd) = 0,$$

it appears that the lines a_i, b_i, c_i, d_i are edges of a tetrahedron in which a, d and b, c are opposite.

Also since $\rho . x_i = a_i + c_i\lambda^2 + \lambda(b_i + d_i\lambda^2)$ with the condition $(ad) + (bc) = 0$, it is clear that the lines $a_i + c_i\lambda^2$, $b_i + d_i\lambda^2$ intersect, therefore x meets the line of intersection γ of the planes (a, c) and (b, d). The generator x' which meets x is obtained by changing the sign of λ (since $\lambda + \lambda' = 0$), hence x and x' intersect upon the line γ, which is therefore the double directrix of the surface; the single directrix being the intersection of the planes (a, b) and (c, d).

If $4(bc)^2 = (b^2)(c^2)$ the singular generators coincide. Taking zero as the value of λ which gives the singular generator we must have $(b^2) = (bc) = 0$; and $\psi = 0$ becomes $\lambda\lambda' = 0$, that is, each generator meets the single generator and no other. This gives Cayley's ruled cubic.

72. Ruled quartics of zero deficiency. For a ruled quartic of this species we have

$$\rho . x_i = a_i + b_i\lambda + c_i\lambda^2 + d_i\lambda^3 + e_i\lambda^4,$$

with the conditions

$$0 = (a^2) = (ab) = (e^2) = (de) = 2(ac) + (b^2) = 2(ce) + (d^2)$$
$$= (ad) + (bc) = (be) + (cd) = 2(ae) + (c^2) + 2(bd).$$

The quartic belongs in general to *one* linear complex $(\gamma x) = 0$ whose coefficients are determined by the equations

$$(\gamma a) = (\gamma b) = (\gamma c) = (\gamma d) = (\gamma e) = 0.$$

The double curve is of the third degree, its *equation* is

$$(ac) + (ad)(\lambda + \lambda') + (ae)(\lambda + \lambda')^2 + (bd)\lambda\lambda' + (be)\lambda\lambda'(\lambda + \lambda') + (ce)\lambda^2\lambda'^2 = 0. \quad \text{......(II.)}$$

There are four singular generators; if λ be so chosen that $\lambda = 0$, $\lambda = \infty$ give two of them, we have the additional equations $(ac) = (ce) = 0$, and therefore $(b^2) = (d^2) = 0$.

If the linear complex $(\gamma x) = 0$ is general, we have class III. The complex is special when, (Art. 64),

$$\begin{vmatrix} (a^2) & (ab) & (ac) & (ad) & (ae) \\ (ab) & (b^2) & (bc) & (bd) & (be) \\ (ac) & (bc) & (c^2) & (cd) & (ce) \\ (ad) & (bd) & (cd) & (d^2) & (de) \\ (ae) & (be) & (ce) & (de) & (e^2) \end{vmatrix} = 0.$$

By aid of the equations connecting the constants, the last equation is seen to reduce to

$$(bd)\{(ad)(be) - (ae)(bd)\}\{2(ad)(be) + (c^2)(ae)\} = 0;$$

while the equation of the double curve breaks into two factors, if

$$(bd)\{(ad)(be) - (ae)(bd)\} = 0.$$

There are thus two cases to be considered:

(i) $(bd) = 0$, here $\lambda + \lambda' = 0$ is a factor of the equation of the double curve, which breaks up into a conic c^2, and a line γ meeting c^2, γ is the directrix of the special complex.

If x and x' are a pair of generators for which $\lambda + \lambda' = 0$, we have

$$\rho \,.\, x_i = a_i + b_i\lambda + c_i\lambda^2 + d_i\lambda^3 + e_i\lambda^4,$$
$$\rho' \,.\, x_i' = a_i - b_i\lambda + c_i\lambda^2 - d_i\lambda^3 + e_i\lambda^4;$$

hence $$\rho \,.\, x_i - \rho' \,.\, x_i' = 2\lambda(b_i + \lambda^2 d_i);$$

therefore a line of the pencil (x, x') is a line of the pencil (b, d), *i.e.*, the plane (x, x') passes through the fixed point (b, d). The figure shows the relative positions of the lines a, b, d, e, γ and it is clear that x, x' meet on γ while their plane passes through (b, d). In this case, therefore, the double curve consists of c^2 and γ, and the bitangent developable consists of γ and a quadric cone whose vertex is the point (b, d); the surface belongs to class V.

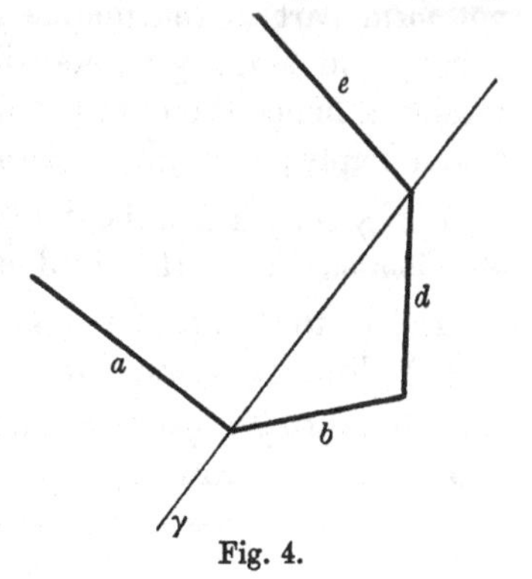

Fig. 4.

(ii) If $(ad)(be)-(ae)(bd)=0$, the equation of the double curve breaks up into two equations of the form

$$\lambda+\lambda'=A, \quad \frac{1}{\lambda}+\frac{1}{\lambda'}=B;$$

where A and B are constants.

In this case, since

$$\rho\,.\,x_i-\rho'\,.\,x_i'=(\lambda-\lambda')\{b_i+c_i(\lambda+\lambda')+d_i(\overline{\lambda+\lambda'}^2-\lambda\lambda')$$
$$+e_i\overline{\lambda+\lambda'}(\overline{\lambda+\lambda'}^2-2\lambda\lambda')\},$$

for the lines x, x' which meet upon the curve corresponding to $\lambda+\lambda'=A$, we have

$$\rho x_i-\rho'\,.\,x_i'=(\lambda-\lambda')(\alpha_i+\lambda\lambda'\beta_i),$$

where the α_i and β_i are constants, and must therefore be the coordinates of two intersecting lines. Similarly, since

$$\rho\,.\,x_i=\lambda^4\{e_i+d_i\mu+c_i\mu^2+b_i\mu^3+a_i\mu^4\},$$

where $\lambda\,.\,\mu=1$, it is clear that

$$\sigma\,.\,x_i-\sigma'\,.\,x_i'=(\mu-\mu')\{d_i+c_i\overline{\mu+\mu'}+b_i(\overline{\mu+\mu'}^2-\mu\mu')$$
$$+a_i\overline{\mu+\mu'}(\overline{\mu+\mu'}^2-2\mu\mu')\},$$

hence for the lines x, x' which meet upon the curve corresponding to $\mu+\mu'=B$, we have as before

$$\sigma\,.\,x_i-\sigma'\,.\,x_i'=(\mu-\mu')(\alpha_i'+\mu\mu'\beta_i');$$

so that, as in the case $(bd)=0$, the plane of the two generators through a point of γ passes through a fixed point, and we have again the class V.

In the next place if $2(ad)(be)+(c^2)(ae)=0$, the complex $(\gamma x)=0$ is special, but the equation of the double curve does not factorize; hence we have the two following cases:

(i) γ is a simple directrix of the surface, and therefore does not form part of the double curve which is the general twisted cubic; and since γ meets all generators, it meets those which intersect, hence the envelope of the bitangent planes is the line γ taken triply; this gives class IV.

(ii) γ constitutes the double curve, and the bitangent developable is a surface of the third class; this gives class IX.

In (i) there are three generators in each plane through γ, in (ii) three generators meet in each point of γ.

If γ is itself a generator, let $\lambda=0$ represent it, and then since γ meets each generator, we have

$$(a^2)=(ab)=(ac)=(ad)=(ae)=0,$$

and by reference to the equation (II.) of the double curve, it is clear that two singular generators now coincide with a (or γ). As before we take $\lambda = \infty$ to represent one of the two other singular generators, in which case we have $(ce) = 0$.

The equation of the double curve is here

$$(bd) + (be)(\lambda + \lambda') = 0.$$

There are two cases: (i) the generators which meet in a point of the double curve lie in a plane with a, the bitangent developable consists of the line a taken triply, and the double curve breaks up into a and a conic c^2, which gives class VI.; or, (ii) the generators meet in a point of a which constitutes the double curve and the bitangent developable consists of a together with a quadric cone, which is class XI.

It has been seen that the surface belongs in general to one linear complex; if however, there exist six linear relations of the form

$$Aa_i + Bb_i + Cc_i + Dd_i + Ee_i = 0,$$

the surface will belong to an *infinite number* of linear complexes.

The double curve consists of the two lines γ, γ', which are given by

$$(\gamma^2) = (\gamma a) = (\gamma b) = (\gamma c) = (\gamma d) = 0.$$

In the present case the surface may have a *double generator*, viz. when two distinct values of λ give rise to the same generator; as before we may take the values $\lambda = 0$ and $\lambda = \infty$ as relating to it; the generator x is then given by

$$\rho \,.\, x_i = a_i + b_i\lambda + c_i\lambda^2 + d_i\lambda^3 + a_i\lambda^4,$$

and the linear relation is then $a_i = e_i$.

From the identity $(x^2) \equiv 0$, we have the system of equations

$$(a^2) = (ab) = (ad) = (cd) = (bc) = (b^2) + 2(ac) = (c^2) + 2(bd)$$
$$= (d^2) + 2(ac) = 0.$$

The equation of the double curve, which factorizes, is

$$(ac) + (bd)\lambda\lambda' + (ac)\lambda^2\lambda'^2 = 0.$$

This gives class VII.

There are two subvarieties in which the lines γ, γ' coincide; one arises when $(bd)^2 = 4(ac)^2$, in which case the double curve becomes

$$\lambda\lambda' = 1,$$

hence for two generators x, x' which intersect, we have

$$\rho . x_i = a_i + b_i\lambda + c_i\lambda^2 + d_i\lambda^3 + a_i\lambda^4,$$

$$\rho' . x_i' = a_i\lambda^4 + b_i\lambda^3 + c_i\lambda^2 + d_i\lambda + a_i;$$

hence $$\rho . x_i - \rho' . x_i' = (\lambda - \lambda^3)(b_i - d_i),$$

the generators x, x', therefore, intersect upon the line $b_i - d_i$, and lie in the same plane with it. The double curve consists of the double generator a, and two indefinitely near lines $b_i - d_i$. This gives class VIII.

The second subvariety occurs when $(ac) = 0$, we then have

$$(a^2) = (ab) = (ac) = (ad) = (b^2) = (d^2) = (cd) = (bc) = 0.$$

The lines a, b, d all belong to the complex $(cx) = 0$, while a meets both b and d; to the same complex belong the lines γ, γ', which must therefore both coincide with a. This is the class which has a double generator in coincidence with two indefinitely near directrices, *i.e.* class XII.

Finally, the surface may belong to an infinite number of linear complexes and not possess a double generator; of the two lines γ, γ' one is a triple and the other a single directrix; this gives class X.

CHAPTER VI.

THE QUADRATIC COMPLEX.

73. WHEN the equation of the complex, $f(x)=0$, is of the second degree in the coordinates, the complex is called quadratic. By the same reasoning as that employed in Chapter II., it is seen that the lines of the complex through any point form a quadric cone, and that the lines of the complex in any plane envelope a curve of the second class.

The general quadratic expression in six variables involves 21 terms, but, by means of the identical relation $\omega(x)=0$, it may be deprived of one of its terms without loss of generality; so that the quadratic complex is seen to involve 19 constants.

It will be shown in Chapter XI. that $f(x)$ and $\omega(x)$ can, in general, be brought by the same linear transformation to the forms

$$\lambda_1 x_1^2 + \lambda_2 x_2^2 + \lambda_3 x_3^2 + \lambda_4 x_4^2 + \lambda_5 x_5^2 + \lambda_6 x_6^2,$$

and $$x_1^2 + x_2^2 + x_3^2 + x_4^2 + x_5^2 + x_6^2, \text{ respectively*}.$$

This *canonical* form of the equation of the quadratic complex will be generally used in the present chapter. With this form of equation it is clear that if x belongs to the complex, so also do the 31 lines associated with x, (Art. 14); hence the polars of the lines of a quadratic complex $(\lambda x^2)=0$, for a fundamental linear complex, belong to $(\lambda x^2)=0$.

This is a *characteristic* property of the fundamental complexes of $f(x)$; for if L, or $(lx)=0$, is a linear complex such that the polar line x', with regard to L, of any line x of the quadratic complex $f(x)=0$, also belongs to $f(x)=0$, then, taking $\omega(x)\equiv(x^2)$,

$$\rho \,.\, x_i' = x_i + \sigma l_i, \text{ where } \sigma = -\frac{2\,(lx)}{(l^2)}, \quad \text{(Art. 20)},$$

* The complexes x_i which appear in its canonical form may be called the fundamental complexes of $f(x)=0$.

if $f(x') = 0$, we have

$$f(x) + \sigma \Sigma x_i \frac{\partial f(l)}{\partial l_i} + \sigma^2 f(l) = 0;$$

and, since $f(x) = 0$,

$$\Sigma x_i \frac{\partial f(l)}{\partial l_i} - 2 \frac{(lx)}{(l^2)} \cdot f(l) = 0.$$

If the last equation holds for *every* line x of $f(x) = 0$, it follows that

$$\frac{\partial f}{\partial l_i} = \kappa \, . \, l_i, \quad (i = 1, 2, \ldots 6).$$

These equations determine six values of κ, and hence *six* linear complexes which have the required property. The equations serve, therefore, to determine the fundamental complexes of any quadratic complex $f(x) = 0$, $(x^2) = 0$, where $f(x)$ is *general* in form.

Any given quadratic complex C^2, or $f(x) = 0$, *possesses two lines in every plane pencil* (A, α) *of space*, viz. the intersection of the plane α with the complex cone of A (or the tangents from A to the complex conic of α). In any regulus A, B, C there are four lines of C^2, viz. those determined by the equations

$$(ax) = 0, \ (bx) = 0, \ (cx) = 0, \ f(x) = 0, \ \omega(x) = 0.$$

74. The tangent linear complex. The linear complex

$$\Sigma y_i \left(\mu \frac{\partial f}{\partial x_i} + \nu x_i \right) = 0 \quad \ldots\ldots\ldots\ldots\ldots\ldots\text{(i)},$$

is called a *polar* complex of C^2 for the line x*; if x belongs to C^2 it is seen that this complex contains x, and every line $x + dx$ of C^2 consecutive to x, since

$$\left(x \frac{\partial f}{\partial x} \right) = 0, \ (x dx) = 0, \ \left(\frac{\partial f}{\partial x} dx \right) = 0:$$

the complex is then called a *tangent* linear complex of C^2.

On any line x of C^2 a correlation is established between its points X and its planes u, such that u is the plane whose complex curve has X for point of contact with x, while u is at the same time the tangent plane along x to the complex cone of X. Hence the pencils (X, u) belong to every tangent linear complex of x, so that each of these ∞^1 complexes determines on x the foregoing correlation.

* $\omega(x)$ is here taken as (x^2); for the general form of $\omega(x)$ the polar complex is

$$\Sigma y_i \left(\mu \frac{\partial f}{\partial x_i} + \nu \frac{\partial \omega}{\partial x_i} \right) = 0.$$

The *special* complexes of (i) are given by the equation

$$\omega\left(\mu\frac{\partial f}{\partial x}+\nu x\right)=0,$$

which, since x belongs to C^2, reduces to

$$\mu^2\omega\left(\frac{\partial f}{\partial x}\right)=0,$$

hence in general the two directrices coincide with x.

75. Singular points and planes of the complex. It will now be shown that the complex cones of the points of a certain surface Φ_1 break up into pairs of planes, the complex lines of such a point therefore consisting of two plane pencils; while the complex conics of the tangent planes of a surface Φ_2 break up into pairs of points, so that the complex lines in such a plane form two pencils.

Let x be any line of C^2, it will be shown that the complex cones of *four* points on x break up into pairs of planes; for if $x+\mu a$ is one line of a pencil which belongs wholly to C^2, the following conditions must be satisfied, viz.

$$f(x+\mu a)=0, \text{ for all values of } \mu, \text{ while } \Sigma a_i\frac{\partial\omega}{\partial x_i}=0.$$

This gives the following four equations:

$$f(x)=0,\ \Sigma a_i\frac{\partial f}{\partial x_i}=0,\ f(a)=0,\ \Sigma a_i\frac{\partial\omega}{\partial x_i}=0.$$

If now a be that line of the pencil which meets any given line b, $\Sigma a_i\dfrac{\partial\omega}{\partial b_i}=0$, and the coordinates of a satisfy the four equations

$$\Sigma a_i\frac{\partial\omega}{\partial b_i}=0,\ \Sigma a_i\frac{\partial f}{\partial x_i}=0,\ \Sigma a_i\frac{\partial\omega}{\partial x_i}=0,\ f(a)=0.$$

Hence the lines a are those common to a regulus, and the complex C^2, and are therefore *four* in number; we shall denote them by $a^{\rm I}$, $a^{\rm II}$, $a^{\rm III}$, $a^{\rm IV}$. It has therefore been shown that on every line x of C^2 there are four points A_1, A_2, A_3, A_4 called *singular points*, whose complex cones break up into a pair of planes; since this is true in general for each of the ∞^3 lines of C^2, the locus of such points is a surface of the *fourth degree* which we may denote by Φ_1.

Again in each of these four planes (x, a^i) since there exists one pencil of complex lines there must also be a second, *i.e.* the complex conic in each plane consists of a pair of pencils; hence, through any line x of C^2 we can draw four planes β_1, β_2, β_3, β_4, called *singular planes*, for each of which the complex conic breaks up into two pencils; the envelope of these planes is therefore a surface of the *fourth class* which we may denote by Φ_2. It will shortly be shown that the surfaces Φ_1 and Φ_2 are identical.

76. Singular Lines. No two of the four lines a^i will in general intersect, since they belong to the same regulus; if they do intersect, so that for instance A_1 and A_2 coincide or β_1 and β_2 coincide, the regulus must break up into two plane pencils, for which the condition is that the discriminant of

$$\Omega\left(\lambda\frac{\partial\omega}{\partial b_i}+\mu\frac{\partial f}{\partial x_i}+\nu\frac{\partial\omega}{\partial x_i}\right)$$

should be zero (Art. 59).

If $\omega(x)\equiv\Sigma x^2$, this condition becomes that the discriminant of

$$\Sigma\left(\lambda b_i+\mu\frac{\partial f}{\partial x_i}+\nu x_i\right)^2$$

should vanish or that

$$\Sigma\left(\frac{\partial f}{\partial x_i}\right)^2=0,\quad \text{since } (bx)\neq 0^*.$$

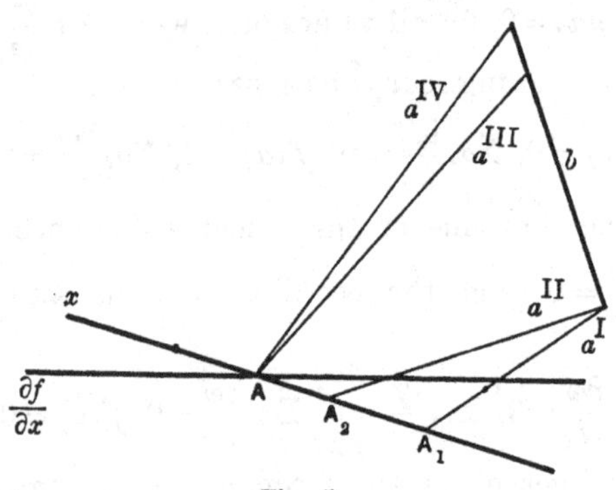

Fig. 5.

In this case the lines a which satisfy the equations $(ax)=0$, $\left(a\frac{\partial f}{\partial x}\right)=0$, $(ab)=0$ consist of the two pencils

(i) that which has $\left(x, \frac{\partial f}{\partial x}\right)$ for its plane, and its centre on b;

(ii) that which has $\left(x, \frac{\partial f}{\partial x}\right)$ for its centre, and its plane through b.

* For the general form of $\omega(x)$ the condition is that $\Omega\left(\frac{\partial f}{\partial x}\right)=0$.

The lines a^{I}, a^{II}; a^{III}, a^{IV} are the complex lines of these respective pencils. Thus two of the singular points A, say A_3, A_4, come into coincidence and *also* two of the planes β, *e.g.* β_1 and β_2; so that we have on x the singular points A, A_1, A_2 and through x the singular planes β, β_3, β_4.

A line x of C^2 for which $\frac{\partial f}{\partial x}$ is a line, thus touches Φ_1 at the point A, or $\left(x, \frac{\partial f}{\partial x}\right)$, and is the intersection of the plane-pair of A; it meets Φ_1 in two other points A_1, A_2 which form the point-pair for the tangent plane β, or $\left(x; \frac{\partial f}{\partial x}\right)$ of Φ_2. Such a line is called a *singular line* of C^2.

Taking $\Sigma\lambda_i x_i^2 = 0$ as the equation of C^2, the singular lines are given by the equations

$$\Sigma\lambda_i x_i^2 = 0, \quad \Sigma\lambda_i^2 x_i^2 = 0.$$

77. Singular points and planes of any line. On any line l let its points of intersection with Φ_1 be A_1, A_2, A_3, A_4, and let the tangent planes through it to Φ_2 be β_1, β_2, β_3, β_4; let the point-pair of β_1 be B_1, B_1' and their join b_1, &c.; let the plane-pair of A_1 be α_1, α_1' and their intersection a_1, &c. Then it is clear that the eight singular planes α meet the four lines b in the points B, *e.g.* the plane α_1 meets the plane (l, b_1) in a line through A_1 which belongs to C^2 and hence must pass through B_1 or B_1', since all lines of C^2 in (l, b_1) pass through one of these points.

Thus through each point B_i there pass four planes α and through B_i' the other four planes α; similarly in each plane α there lie four points B; so that if we take three pairs of planes α_1, α_1'; α_2, α_2'; α_3, α_3' they exactly determine the eight points B; let the notation of these points be determined as follows:—

$$B_1 = (\alpha_1'\alpha_2\alpha_3), \quad B_2 = (\alpha_1\alpha_2'\alpha_3), \quad B_3 = (\alpha_1\alpha_2\alpha_3'), \quad B_4 = (\alpha_1\alpha_2\alpha_3),$$

$$B_1' = (\alpha_1\alpha_2'\alpha_3'), \quad B_2' = (\alpha_1'\alpha_2\alpha_3'), \quad B_3' = (\alpha_1'\alpha_2'\alpha_3), \quad B_4' = (\alpha_1'\alpha_2'\alpha_3').$$

Then it is easily seen that the only possible remaining arrangement by fours of the points B, so that no three of them are in any of these six planes $\alpha_1, \ldots \alpha_3'$, is

$$(B_1B_2B_3B_4'), \quad (B_1'B_2'B_3'B_4).$$

Calling these planes α_4 and α_4', we have the following arrangement of the points B and planes α:

$$\alpha_1 = (B_1', B_2, B_3, B_4),\quad \alpha_1' = (B_1, B_2', B_3', B_4'),$$
$$\alpha_2 = (B_1, B_2', B_3, B_4),\quad \alpha_2' = (B_1', B_2, B_3', B_4'),$$
$$\alpha_3 = (B_1, B_2, B_3', B_4),\quad \alpha_3' = (B_1', B_2', B_3, B_4'),$$
$$\alpha_4 = (B_1, B_2, B_3, B_4'),\quad \alpha_4' = (B_1', B_2', B_3', B_4).$$

Now consider the tetrahedron whose vertices are $B_1B_2B_3'B_4'$; the planes joining l to its vertices are respectively $\beta_1, \beta_2, \beta_3, \beta_4$; and

the plane through $B_2B_3'B_4'$ is α_2', which meets l in A_2;
„ „ $B_1B_3'B_4'$ „ α_1', „ „ A_1;
„ „ $B_1B_2B_4'$ „ α_4, „ „ A_4;
„ „ $B_1B_2B_3'$ „ α_3, „ „ A_3;

hence, by von Staudt's theorem, (Art. 12),

$$(\beta_1, \beta_2, \beta_3, \beta_4) = (A_2, A_1, A_4, A_3) = (A_1, A_2, A_3, A_4).$$

From this result the identity of the surfaces Φ_1 and Φ_2 follows immediately, for if l touches Φ_1, *i.e.* if two of the points A coincide two of the planes β must also coincide, *i.e.* l touches Φ_2, and conversely; hence Φ_1 and Φ_2 have the same tangent lines and must therefore be identical. The surface Φ, with which Φ_1 and Φ_2 coincide, is known as the Singular Surface of the complex.

78. The identity of the surfaces Φ_1 and Φ_2 also follows from the fact, that if x is a singular line of C^2, the plane $(x, \lambda x)$, or π, touches at P the locus of the point $(x, \lambda x)$, or P; for take any point P' consecutive to P, the singular line corresponding to P' being $x+dx$, then since $\Sigma\lambda_i x_i dx_i=0$, it follows that the four lines x, λx, $x+dx$, $\lambda(x+dx)$ form a twisted quadrilateral, so that if the distance PP' is of the first order of small quantities, the distance of P' from π is of the second order, *i.e.* π touches the locus of P at P*.

79. Polar Lines. If P be any point on a line l, π any plane through l, and u the fourth harmonic to l and the two complex lines of the pencil (P, π), these complex lines must be

$$l + \mu u \text{ and } l - \mu u;$$

expressing that they belong to C^2, whose equation we take in the form $(\lambda x^2) = 0$, we have

$$(\lambda l^2) + \mu^2(\lambda u^2) = 0,\quad (\lambda lu) = 0,\quad (lu) = 0.$$

* The identity of the surfaces Φ_1 and Φ_2 was shown by Pasch, *Ueber die Brennflächen der Strahlensysteme und die Singularitätenflächen der Complexe*; Crelle, Bd. 76, S. 156.

The first equation gives the values of μ for which $l+\mu u$ and $l-\mu u$ are complex lines, the second equation states that u belongs to the complex $(\lambda l x)=0$, hence since u meets l it must also meet l', the polar of l for the last complex (Art. 24), that is, it meets $l_i-\kappa\lambda_i l_i$, where $\kappa=\dfrac{2(\lambda l^2)}{(\lambda^2 l^2)}$.

But for different positions of P, all the lines u in π pass through the pole of l for the complex conic of π, so that l' must pass through this pole.

Hence we have the following theorem, *the locus of the poles of l with reference to the complex conics of planes through l is a straight line l', called the polar of l with respect to C^2.*

The coordinates of l' were seen to be $l_i-\kappa\lambda_i l_i$, where

$$\kappa=\frac{2(\lambda l^2)}{(\lambda^2 l^2)};$$

so that if l belongs to C^2, $\kappa=0$, and l' coincides with l. If both (λl^2) and $(\lambda^2 l^2)$ are zero, any line of the pencil $(l, \lambda l)$ is polar to l.

If l is situated in the plane at infinity, the planes through l are parallel to each other and we deduce the result that the locus of the centres of the complex conics in a system of parallel planes is a straight line, which is called a *diameter* of the complex.

Considering again the system of singular points and planes connected with any line l (Art. 77), each of the lines a and b is singular and satisfies the equations

$$(\lambda x^2)=0,\quad (\lambda^2 x^2)=0,\quad (lx)=0,\quad (l\lambda x)=0;$$

from the last two equations we deduce that

$$\Sigma x_i(l_i-\kappa\lambda_i l_i)=0$$

is an equation satisfied by each of the lines a and b, hence the eight lines a and b meet both l and its polar l' for C^2.

Referring to the table of Art. 77, it is seen that B_1' and B_2 lie on both α_1 and α_2', hence $B_1'B_2$ (which meets b_1 and b_2), meets a_1 and a_2; similarly B_1B_2' meets a_1 and a_2, so that the lines a_1, a_2, b_1, b_2 being met by l, l', $B_1'B_2$ and B_1B_2', *lie on the same regulus.* Consider the linear complex K determined by a_1, b_1; a_2, b_2 as pairs of polar lines (Art. 31); then in K the *polar of a_3 is b_3*, for α_3' contains the lines $B_2'B_3$, $B_1'B_3$ which meet a_2, b_2; a_1, b_1 respectively, hence its pole is B_3, and α_3 contains B_1B_3', B_2B_3', hence its pole is B_3', *i.e.* a_3 and b_3 are polar lines; similarly a_4 and b_4 are polar lines in K.

It is clear, therefore, that for K the polar plane of A_1 is β_1, of A_2 is β_2, of A_3 is β_3 and of A_4 is β_4, whence we again see that

$$(A_1, A_2, A_3, A_4) = (\beta_1, \beta_2, \beta_3, \beta_4).$$

There are three other such linear complexes K', K'', K''';

in K' the pairs of lines a_1, b_2; a_2, b_1; a_3, b_4; a_4, b_3 are polar

... K'' .... a_1, b_3; a_3, b_1; a_2, b_4; a_4, b_2 ....

... K''' .... a_1, b_4; a_4, b_1; a_2, b_3; a_3, b_2 ....

The points A_1, A_2, A_3, A_4 have for polar planes in these four complexes the planes

$$\beta_1, \beta_2, \beta_3, \beta_4;\ \beta_2, \beta_1, \beta_4, \beta_3;\ \beta_3, \beta_4, \beta_1, \beta_2;\ \beta_4, \beta_3, \beta_2, \beta_1,$$

respectively, which shows that the four complexes are mutually in involution.

80. The singular lines of the complex of the second and third orders. It has been shown that the lines which satisfy the equations

$$f(x) = 0, \qquad \Sigma\left(\frac{\partial f}{\partial x_i}\right)^2 = 0, \qquad \Sigma x^2 = 0,$$

belong to C^2 and touch the surface Φ, the *singular surface* of C^2.

The point at which x touches Φ was seen to be the intersection of the lines x and $\frac{\partial f}{\partial x}$, the plane of these lines being the tangent plane at the point. Taking for f its canonical form these equations are

$$(\lambda x^2) = 0, \quad (\lambda^2 x^2) = 0.$$

The singular lines thus form a congruence of the fourth *class* and fourth *order*, *i.e.* through any point there pass four lines of the congruence and any plane contains four of its lines, which are touched by the complex conic of the plane.

The complex conics in the pencil of planes through a singular line x touch x in its point of contact with Φ. For a tangent linear complex T of a singular line, being $\Sigma y_i\left(\frac{\partial f}{\partial x_i} + \mu x_i\right) = 0$, is *special*, and its directrix belongs to the pencil $\left(x, \frac{\partial f}{\partial x}\right)$, so that the point $\left(x, \frac{\partial f}{\partial x}\right)$ is the pole in T for any plane through x; but the pole of such a plane is the intersection of x and $x + dx$, the latter line being common to T and C^2, *i.e.* the point of contact of x with the complex curve in the plane, hence all these points of contact coincide with $\left(x, \frac{\partial f}{\partial x}\right)$.

If in addition to the equations of Arts. 75, 76 which give the singular lines, viz.

$$f(a)=0, \quad (ax)=0, \quad \left(a\frac{\partial f}{\partial x}\right)=0, \quad \Sigma\left(\frac{\partial f}{\partial x_i}\right)^2=0, \quad f(x)=0,$$

we have also $f\left(\frac{\partial f}{\partial x}\right)=0$, it is clear that one value of a is $\frac{\partial f}{\partial x}+\mu x$, so that one of the points A_1, A_2 must coincide with A, or one of the pencils of complex lines in β is (A, β), and the plane β_3 coincides with β; so that we have two points A on x, viz. A and A_1; and two planes β through x, viz. β and β_4; while the lines a^{II}, a^{III} coincide. Thus x meets Φ in three consecutive points, *i.e.* is a principal tangent to Φ, and all the lines (A, β) are complex lines. A curve on Φ is thus obtained whose tangents are singular lines of C^2 and principal tangents of Φ; at each point of this curve all the tangent lines of Φ belong to C^2.

In any plane the complex conic touches the four singular lines of the plane and has apart from them $2.4.3-2.4=16$ common tangents with the section of Φ by the plane. Each of these 16 lines touches Φ and is not a singular line of C^2, hence its point of contact is a point of the above curve; thus any plane meets this curve in 16 points, *i.e.* the curve is of the order 16.

Lines whose coordinates satisfy the equations

$$(\lambda x^2)=0, \quad (\lambda^2 x^2)=0, \quad (\lambda^3 x^2)=0,$$

are termed *singular lines of the second order* (Segre).

If in addition to the foregoing conditions we have

$$\Sigma\left(\frac{\partial f(y)}{\partial y_i}\right)^2=0,$$

where y is $\frac{\partial f}{\partial x}$, it is easy to see that $\Sigma\left(\frac{\partial f(x+\rho y)}{\partial(x+\rho y)}\right)^2=0$, or, *every line of (A, β) is singular*; for since

$$\frac{\partial f(x+\rho y)}{\partial(x+\rho y)}=\frac{\partial f(x)}{\partial x}+\rho\frac{\partial f(y)}{\partial y},$$

it follows that

$$\Sigma\left\{\frac{\partial f(x+\rho y)}{\partial(x_i+\rho y_i)}\right\}^2=\Sigma\left(\frac{\partial f}{\partial x_i}\right)^2+2\rho\Sigma\frac{\partial f}{\partial x_i}\cdot\frac{\partial f}{\partial y_i}+\rho^2\Sigma\left(\frac{\partial f}{\partial y_i}\right)^2$$

$$=0, \text{ if } y_i=\frac{\partial f}{\partial x_i}.$$

It is clear that $\frac{\partial f}{\partial y}$ is a line which meets y, and also x, since $\left(x \frac{\partial f}{\partial y}\right) \equiv \left(y \frac{\partial f}{\partial x}\right) \equiv 2(y^2) = 0$. Therefore the lines $x, y, \frac{\partial f}{\partial y}$ are either *coplanar* or *concurrent*; in the former case, any line of the complex which lies in the plane being of the form $x + \rho y + \sigma \frac{\partial f}{\partial y}$, we have

$$f\left(x + \rho y + \sigma \frac{\partial f}{\partial y}\right) = 0, \quad i.e. \quad \sigma^2 f\left(\frac{\partial f}{\partial y}\right) = 0, \quad \text{hence} \quad \sigma = 0.$$

It follows that *all the complex lines in the plane* (x, y) *belong to the pencil* (x, y); hence A_1 coincides with A, and x meets Φ in four consecutive points.

If z is any line of the pencil (x, y), since z is a singular line, it touches Φ at its point of intersection with $\frac{\partial f}{\partial z}$; hence z touches Φ at two distinct points, viz. (x, y) and $\left(z, \frac{\partial f}{\partial z}\right)$; this being true for each line of the pencil (x, y), it follows that the plane (x, y) is a *singular tangent plane* of Φ, *i.e.* it touches Φ along a conic.

In the case for which $x, y, \frac{\partial f}{\partial y}$ are concurrent, it is similarly seen that their point of concurrence is a *double point* of Φ, and all the complex lines through this point lie in the plane (x, y), while the four tangent planes to Φ through x come into coincidence.

It will be shown (Art. 82), that Φ possesses 16 singular tangent planes and 16 double points.

Taking the equation of the complex as being $(\lambda x^2) = 0$, the lines given by the equations

$$(\lambda x^2) = 0, \quad (\lambda^2 x^2) = 0, \quad (\lambda^3 x^2) = 0,$$

are the tangents of the principal tangent curve previously determined; if λx is itself a singular line, we have $(\lambda^4 x^2) = 0$; these four equations determine 32 lines, so that 16 of them are tangents to the (conic) sections of Φ by its singular tangent planes; if x be one of these lines in such a plane σ, the lines $x, \lambda x, \lambda^2 x$ lie in σ, and all the lines of C^2 in σ consist of the pencil $(x, \lambda x)$; 16 of these lines are generators of the tangent cones at double points D of Φ; at such a point D the lines $x, \lambda x, \lambda^2 x$ concur, and all the complex lines through D consist of the pencil $(x, \lambda x)$.

Lines of this kind are called *singular lines of the third order* (Segre).

81. The complex in Plücker coordinates. If in the canonical form of the equation of the complex, we transform to coordinates p_{ik} by using one of the 15 transformations, (Art. 29),

$$x_1 = p_{12} + p_{34}, \quad ix_2 = p_{12} - p_{34} \text{ \&c.},$$

it takes the form

$$k_{12}(p_{12}^2 + p_{34}^2) + k_{13}(p_{13}^2 + p_{42}^2) + k_{14}(p_{14}^2 + p_{23}^2) \\ + 2dp_{12}p_{34} + 2ep_{13}p_{42} + 2fp_{14}p_{23} = 0,$$

and a quadratic complex may be brought to this form in 15 ways.

A quadratic complex whose equation is

$$a_{12}p_{12}^2 + a_{34}p_{34}^2 + a_{13}p_{13}^2 + a_{42}p_{42}^2 + a_{14}p_{14}^2 + a_{23}p_{23}^2 \\ + 2dp_{12}p_{34} + 2e\,p_{13}p_{42} + 2fp_{14}p_{23} = 0,$$

may be brought to the preceding form, for since the coordinates of any point are, as usual, the ratios of the perpendiculars from the point on the faces of the tetrahedron of reference to the perpendiculars from a fixed point E on those faces, if we take a point E' for which

$$\frac{\perp \text{ from } E' \text{ on face } a_1}{\perp \text{ from } E \ldots\ldots\ldots\ldots a_1} = m_1, \text{ \&c.};$$

it is easy to see that a line whose Plücker coordinates are p_{ik} with reference to the former system, are p'_{ik} with reference to the latter, where $p_{ik} = m_i m_k p'_{ik}$. The quantities m may now be chosen so that

$$a_{12}m_1^2m_2^2 = a_{34}m_3^2m_4^2, \quad a_{13}m_1^2m_3^2 = a_{42}m_2^2m_4^2, \quad a_{14}m_1^2m_4^2 = a_{23}m_2^2m_3^2;$$

and on substituting for p_{ik} the resulting equation is of the required form.

82. The singular surface. It has been shown that Φ is the locus of points of intersection of x and λx, where x is a singular line of C^2; let y be the coordinates of such a point, α the point at infinity on x, and β the point at infinity on λx, we have then

$$y_1\lambda_1\alpha_2 - y_2\lambda_1\alpha_1 + y_3\lambda_1\alpha_4 - y_4\lambda_1\alpha_3 \equiv \rho\,(y_1\beta_2 - y_2\beta_1 + y_3\beta_4 - y_4\beta_3),$$

and five similar equations, ρ being a factor of proportionality which is the same for the six equations.

Also
$$a_1\alpha_1 + a_2\alpha_2 + a_3\alpha_3 + a_4\alpha_4 = 0,$$
$$a_1\beta_1 + a_2\beta_2 + a_3\beta_3 + a_4\beta_4 = 0,$$

where the quantities a are connected with the tetrahedron of reference. Eliminating the quantities α and $\rho\beta$ from these equations we obtain as the equation of Φ,

$$\begin{vmatrix} -y_2\lambda_1 & y_1\lambda_1 & -y_4\lambda_1 & y_3\lambda_1 & -y_2 & y_1 & -y_4 & y_3 \\ -y_2\lambda_2 & y_1\lambda_2 & y_4\lambda_2 & -y_3\lambda_2 & -y_2 & y_1 & y_4 & -y_3 \\ -y_3\lambda_3 & y_4\lambda_3 & y_1\lambda_3 & -y_2\lambda_3 & -y_3 & y_4 & y_1 & -y_2 \\ -y_3\lambda_4 & -y_4\lambda_4 & y_1\lambda_4 & y_2\lambda_4 & -y_3 & -y_4 & y_1 & y_2 \\ -y_4\lambda_5 & -y_3\lambda_5 & y_2\lambda_5 & y_1\lambda_5 & -y_4 & -y_3 & y_2 & y_1 \\ -y_4\lambda_6 & y_3\lambda_6 & -y_2\lambda_6 & y_1\lambda_6 & -y_4 & y_3 & -y_2 & y_1 \\ \alpha_1 & \alpha_2 & \alpha_3 & \alpha_4 & 0 & 0 & 0 & 0 \\ 0 & 0 & 0 & 0 & \alpha_1 & \alpha_2 & \alpha_3 & \alpha_4 \end{vmatrix} = 0.$$

This determinant is divisible by $(\alpha_1 y_1 + \alpha_2 y_2 + \alpha_3 y_3 + \alpha_4 y_4)^2$, after division by it is effected, the expanded form of the equation is

$$\begin{aligned} &(\lambda_2\lambda_4\lambda_6 - \lambda_1\lambda_3\lambda_5)(\Sigma y^2)^2 + (\lambda_1\lambda_3\lambda_6 - \lambda_2\lambda_4\lambda_5)(y_1^2 - y_2^2 - y_3^2 + y_4^2)^2 \\ &+ (\lambda_1\lambda_4\lambda_5 - \lambda_2\lambda_3\lambda_6)(y_1^2 - y_2^2 + y_3^2 - y_4^2)^2 \\ &\qquad\qquad + (\lambda_2\lambda_3\lambda_5 - \lambda_1\lambda_4\lambda_6)(y_1^2 + y_2^2 - y_3^2 - y_4^2)^2 \\ &+ 4(\lambda_1\lambda_2\lambda_3 - \lambda_4\lambda_5\lambda_6)(y_1y_3 - y_2y_4)^2 + 4(\lambda_1\lambda_5\lambda_6 - \lambda_2\lambda_3\lambda_4)(y_1y_2 - y_3y_4)^2 \\ &+ 4(\lambda_3\lambda_4\lambda_5 - \lambda_1\lambda_2\lambda_6)(y_1y_4 - y_2y_3)^2 + 4(\lambda_1\lambda_2\lambda_5 - \lambda_3\lambda_4\lambda_6)(y_2y_3 + y_1y_4)^2 \\ &+ 4(\lambda_3\lambda_5\lambda_6 - \lambda_1\lambda_2\lambda_4)(y_2y_4 + y_1y_3)^2 + 4(\lambda_1\lambda_3\lambda_4 - \lambda_2\lambda_5\lambda_6)(y_1y_2 + y_3y_4)^2. \end{aligned}$$

Finally writing

$$\begin{aligned} \lambda_1 - \lambda_2 &= a_1, & \lambda_1 + \lambda_2 &= a_2, \\ \lambda_3 - \lambda_4 &= b_1, & \lambda_3 + \lambda_4 &= b_2, \\ \lambda_5 - \lambda_6 &= c_1, & \lambda_5 + \lambda_6 &= c_2, \end{aligned}$$

$$\begin{aligned} &a_1b_1c_1 = A, \\ &a_1\{b_1^2 + c_1^2 - (b_2 - c_2)^2\} = 2B, \\ &b_1\{c_1^2 + a_1^2 - (c_2 - a_2)^2\} = 2C, \\ &c_1\{a_1^2 + b_1^2 - (a_2 - b_2)^2\} = 2D, \\ &a_1^2(b_2 - c_2) + b_1^2(c_2 - a_2) + c_1^2(a_2 - b_2) + (b_2 - c_2)(c_2 - a_2)(a_2 - b_2) = -2E; \end{aligned}$$

the equation of the singular surface assumes the form

$$\begin{aligned} A\Sigma y_i^4 + 2B(y_1^2y_2^2 + y_3^2y_4^2) + 2C(y_1^2y_3^2 + y_2^2y_4^2) + 2D(y_1^2y_4^2 + y_2^2y_3^2) \\ + 4Ey_1y_2y_3y_4 = 0 \ldots\ldots\ldots\ldots\ldots\ldots\ldots\text{(I)}. \end{aligned}$$

It is easily verified, that between the coefficients of this equation the following relation exists:

$$A(A^2 + E^2 - B^2 - C^2 - D^2) + 2BCD = 0 \ldots\ldots\ldots\text{(II)}.$$

No other relation holds between A, B, C, D, E.

This surface may be shown to possess 16 double points; for, the equations to be satisfied by the coordinates of such a point are

$$\left.\begin{aligned} Ay_1^3 + y_1(By_2^2 + Cy_3^2 + Dy_4^2) + Ey_2y_3y_4 = 0, \\ Ay_2^3 + y_2(By_1^2 + Cy_4^2 + Dy_3^2) + Ey_1y_3y_4 = 0, \\ Ay_3^3 + y_3(By_4^2 + Cy_1^2 + Dy_2^2) + Ey_1y_2y_4 = 0, \\ Ay_4^3 + y_4(By_3^2 + Cy_2^2 + Dy_1^2) + Ey_1y_2y_3 = 0. \end{aligned}\right\} \ldots\ldots\text{(III)}.$$

We notice that if there is *one* solution $(\alpha_1\alpha_2\alpha_3\alpha_4)$, of these equations, there are 15 others, viz. $(\alpha_2\alpha_1\alpha_4\alpha_3)$, $(\alpha_3\alpha_4\alpha_1\alpha_2)$, $(\alpha_4\alpha_3\alpha_2\alpha_1)$, together with the solutions got from any one of these four by taking any two of the coordinates negatively. (See Art. 14.)

To find the condition that a point can be found whose coordinates y_i satisfy the four equations (III), we observe that, writing $\rho = -E y_1 y_2 y_3 y_4$, we obtain from them

$$(A+B)(y_1^2+y_2^2)+(C+D)(y_3^2+y_4^2)=\rho\frac{y_1^2+y_2^2}{y_1^2y_2^2},$$

$$(C+D)(y_1^2+y_2^2)+(A+B)(y_3^2+y_4^2)=\rho\frac{y_3^2+y_4^2}{y_3^2y_4^2},$$

whence

$$(A+B)^2+E^2-(C+D)^2=\rho(A+B)\left(\frac{1}{y_1^2y_2^2}+\frac{1}{y_3^2y_4^2}\right);$$

similarly

$$(A-B)^2+E^2-(C-D)^2=-\rho(A-B)\left(\frac{1}{y_1^2y_2^2}+\frac{1}{y_3^2y_4^2}\right);$$

or

$$\frac{(A+B)^2+E^2-(C+D)^2}{(A-B)^2+E^2-(C-D)^2}+\frac{A+B}{A-B}=0;$$

which gives

$$A(A^2+E^2-B^2-C^2-D^2)+2BCD=0;$$

and this condition has been seen to be satisfied.

Moreover if we write

$$B=A\cos\theta_1,\quad C=A\cos\theta_2,\quad D=A\cos\theta_3$$

the identical relation between the constants becomes

$$-\left(\frac{E}{A}\right)^2=1+2\cos\theta_1\cos\theta_2\cos\theta_3-\cos^2\theta_1-\cos^2\theta_2-\cos^2\theta_3,$$

and it is easily verified that one solution of equations (III) is

$$a_1^2:a_2^2:a_3^2:a_4^2=-\sin s:\sin(s-\theta_1):\sin(s-\theta_2):\sin(s-\theta_3),$$

where $2s=\theta_1+\theta_2+\theta_3$.

The equation (I) of the singular surface can also be obtained, by expressing the equation $(\lambda x^2)=0$ in terms of Plücker coordinates, (Art. 81), and then finding the condition that the complex cone of the point y should be a pair of planes.

The complex equation of Art. 81 when expressed in terms of the coordinates π_{ik} (Art. 4) has precisely the same form, viz.,

$$k_{12}(\pi_{34}^2+\pi_{12}^2)+k_{13}(\pi_{42}^2+\pi_{13}^2)+k_{14}(\pi_{23}^2+\pi_{14}^2)$$
$$+2d\pi_{34}\pi_{12}+2e\pi_{42}\pi_{13}+2f\pi_{23}\pi_{14}=0.$$

From this it follows that the singular surface, being the envelope of planes whose complex-conics break up into point-pairs, will have exactly the same form of equation in plane- as in point-coordinates, so that Φ must have 16 singular tangent planes. The quartic surface which has 16 double points and 16 singular tangent planes is known as Kummer's surface. It should be noticed that the coordinates of the 16 double points are the same as those of the 16 singular tangent planes and form a system already discussed, (Art. 14), so that through each double point there pass six singular tangent planes and in each singular tangent plane there are six double points which lie on the (conic) section of the surface by the plane.

The polar line π'_{ik} of any line p_{ik}, with regard to the quadric $\Sigma a_i \xi_i^2=0$, is given by the equations

$$\pi'_{ik}=a_i a_k p_{ik}, \qquad \text{(Art. 46)}.$$

Hence it follows that if p_{ik} belongs to the complex C^2, or

$$a(p_{12}^2+p_{34}^2)+b(p_{13}^2+p_{42}^2)+c(p_{14}^2+p_{23}^2)+2dp_{12}p_{34}+2ep_{13}p_{42}+2fp_{14}p_{23}=0,$$

its polar with regard to any one of the quadrics

$$\begin{aligned}\xi_1^2+\xi_2^2+\xi_3^2+\xi_4^2&=0,\\ \xi_1^2-\xi_2^2+\xi_3^2-\xi_4^2&=0,\\ \xi_1^2-\xi_2^2-\xi_3^2+\xi_4^2&=0,\\ \xi_1^2+\xi_2^2-\xi_3^2-\xi_4^2&=0,\end{aligned}$$

also belongs to C^2.

The polar π'_{ik} of a line p_{ik}, for the quadric

$$\alpha\xi_1\xi_2+\beta\xi_3\xi_4=0,$$

is given by the equations,

$$\begin{aligned}&\pi'_{12}=\alpha^2 p_{12},\ \pi'_{13}=\alpha\beta p_{42},\ \pi'_{14}=\alpha\beta p_{32},\\ &\pi'_{34}=\beta^2 p_{34},\ \pi'_{42}=\alpha\beta p_{13},\ \pi'_{23}=\alpha\beta p_{41};\end{aligned}$$

hence it is clear that if p_{ik} belongs to C^2, its polar line with regard to any one of the quadrics

$$\begin{aligned}&\xi_1\xi_2+\xi_3\xi_4=0,\ \xi_1\xi_3+\xi_2\xi_4=0,\ \xi_1\xi_4+\xi_2\xi_3=0,\\ &\xi_1\xi_2-\xi_3\xi_4=0,\ \xi_1\xi_3-\xi_2\xi_4=0,\ \xi_1\xi_4-\xi_2\xi_3=0,\end{aligned}$$

also belongs to C^2.

These ten quadrics, which occur in the equation of the singular surface of C^2, are the fundamental quadrics connected with the fundamental linear complexes, (Art. 61).

83. Double tangents. The line y whose coordinates are $\lambda_i x_i+\mu x_i$ where x is a singular line, is a tangent to Φ at the point $(x, \lambda x)$; and similarly each of the 32 lines $\pm(\lambda_i x_i+\mu x_i)$ is such a tangent to Φ. If one of these 32 lines y belongs to a fundamental complex so do all the others. If (P, π) is one of these 32 pencils, 15 centres of pencils and 16 planes of pencils belong to the closed system determined by P, 15 of the planes and 16 of the centres belong to the system determined by π. Join P to the poles $P_1 \ldots P_6$ of the fundamental complexes in π, these poles belong to the closed system determined by π; the line P_iP, since it belongs to the complex $x_i=0$ is common to the pencil centre P and the pencil centre P_i, thus P_iP touches Φ at P_i, and therefore is a double tangent to Φ. Hence through any point of Φ we can draw six double tangents to Φ, each of which belongs to a fundamental complex.

This may also be seen analytically; for, if the pencils $(x, \lambda x)$, $(x', \lambda x')$ have a common line, we have

$$\lambda_i x_i + \mu x_i = \rho (\lambda_i x_i' + \nu x_i'),$$

whence in general x and x' are the same line, unless $\mu = \nu = -\lambda_i$, and $x_k = x'_k$, except for $k=i$, when $x_i = -x_i'$, and the common line of the two pencils belongs to the fundamental complex C_i; this bitangent line therefore belongs to the two complexes

$$y_i = 0, \quad \sum_k \frac{y_k^2}{\lambda_k - \lambda_i} = 0, \quad (k \neq i);$$

for since

$$y_i = 0; \; y_k = (\lambda_k - \lambda_i)\, x_k,$$

therefore

$$\sum_k \frac{y_k^2}{\lambda_k - \lambda_i} = \Sigma (\lambda_k - \lambda_i)\, x_k^2 = \sum_1^6 \lambda_i x_i^2 - \lambda_i \sum_1^6 x_i^2 = 0.$$

There are thus six congruences to which bitangent lines belong, viz. those obtained by giving to i the successive values 1, ... 6. Hence, *the double tangents form six congruences of the second order and class,* a fact also deducible from consideration of the 28 bitangents of an arbitrary plane section of Φ, which consist of the 16 intersections of this plane with the 16 singular tangent planes, and 12 others which pass in pairs through the poles of the plane with regard to the fundamental complexes. Hence *the double tangents of a Kummer's surface form six congruences of the second order and class, and each congruence belongs to one fundamental complex.*

84. A Kummer's Surface and one singular line determine one C^2. In a closed system of 16 points and planes, two of the planes intersect in a line containing two points of the system; if two such planes are α and β and two such points A and B, we have six planes of the system through A and six planes through B, (including in each case α and β); this leaves six planes of the system; hence there are six points on AB through which three planes of the system pass.

Now taking any tangent line of Φ as a singular line of the complex to be determined, we know at the same time 31 other singular lines, the singular point and plane for each singular line being also determined, and we have two closed systems of 16 points and planes.

If P be the point of intersection of three planes of one system, we know six complex lines for P, viz. the joins of P to the points A_1 and A_2, the centres of the pencils in each of the three planes; hence the complex cone of P is determined. On the intersection of two planes of one of these closed systems there are six such points P, therefore the complex conic of any plane through the line is determined and therefore the complex cone of any point of the line.

Lastly, any arbitrary plane meets all the 120 lines of intersection of the 16 planes of either system, and hence we know 240 tangents in this plane, and therefore the complex conic of this arbitrary plane.

It follows that a Kummer surface is the singular surface for ∞^1 quadratic complexes, viz. those thus determined by the pencil of tangent lines at any point; hence *a Kummer surface contains* 18 *constants.*

There are two quadratic complexes which contain a given line touching Φ at a point P, viz. the complexes determined by the *principal* tangents at P (Art. 80). If the Kummer surface Φ and one line x be given, we can construct four complexes which contain the line and have Φ for singular surface; for draw through x one of the four tangent planes a to Φ, and let x meet Φ in the points $A_1A_2A_3A_4$; *one* of these points is a singular point for a, so that if O is the point of contact of this plane with Φ, the singular line through O must pass through one of the points A; thus taking in succession OA_1, OA_2, OA_3, OA_4 as singular lines, we can by the preceding method construct four complexes which contain x.

85. The singular surface is a general Kummer Surface. The general Kummer surface* is the most general quartic surface which possesses 16 nodes, and it will now be shown that the equation of such a surface is reducible to the form (I) of Art. 82.

The enveloping cone of a surface with 16 double points whose vertex is a double point, is of the sixth degree; if one of the double points S_1 be joined to another S_1', two tangent planes of the tangent cone to the surface at S_1' pass through S_1S_1', hence S_1S_1' is a double edge of the enveloping cone whose vertex is S_1, so that this enveloping cone has 15 double edges; but an irreducible cone of the sixth degree can have only $\frac{5 \times 4}{2}$, *i.e.* 10, double edges, hence each of the enveloping cones whose vertices are double points of the surface breaks up into six planes; each such plane touches the surface along a curve which is necessarily a conic. Therefore through each point S there pass six singular tangent planes and through each pair of points S pass two singular tangent planes. A singular plane σ through a point S is met by the five other planes σ through S in five lines on each of which a second point S lies, therefore in each plane σ there are six points S,

* This surface was investigated by Kummer in papers published in the *Monatsberichte der Akademie zu Berlin* (1864), and the *Abhandlungen der Akademie* (1866).

(which lie on a conic). Through each of the 15 joins of the points S in a plane σ passes one other plane σ, hence there are 16 planes σ.

A consequence of the above arrangement is that on the intersection of any two planes σ there lie two points S.

Two planes σ contain 10 points S of which two are common to them, and through one or other of two points S pass 10 planes σ; let x_1 and x_2 be two singular planes and x_3 one of the six planes which do not pass through the points S on the intersection of x_1 and x_2; take x_1, x_2, x_3 as coordinate planes, then on each of the edges A_4A_1, A_4A_2, A_4A_3 lie two pairs of points S and taking in each of these planes one other point of their conic of contact with the surface we obtain nine points through which can be described one quadric Ψ which contains these three conics.

It follows that the equation of the surface must have the form

$$\Psi^2 - 16Kx_1x_2x_3x_4 = 0 \quad \text{.....................(I.)},$$

and of which, therefore, x_4 is a plane σ.

Taking for Ψ the most general quadric, we may write,

$$\Psi \equiv x_1^2 + x_2^2 + x_3^2 + x_4^2 + 2a_{12}x_1x_2 + 2a_{34}x_3x_4$$
$$+ 2a_{13}x_1x_3 + 2a_{24}x_2x_4 + 2a_{14}x_1x_4 + 2a_{23}x_2x_3;$$

in each edge of the tetrahedron of reference there lie a pair of points S, for instance, in the edge A_3A_4 there are the points S_{34}, S'_{34} whose coordinates are $(0, 0, a_{34}, 1)$, $(0, 0, a'_{34}, 1)$ respectively, where a_{34}, a'_{34} are the values of x_3/x_4 determined from the equation

$$x_3^2 + x_4^2 + 2a_{34}x_3x_4 = 0;$$

hence $\qquad a_{34} \cdot a'_{34} = 1.$

Through the line $S_{34}S_{23}$ there passes a singular plane σ in addition to the plane a_1, and the line of intersection of σ and a_3 contains two singular points, which therefore lie in A_1A_2 and A_1A_4 respectively, say the points S'_{12}, S_{14}, hence

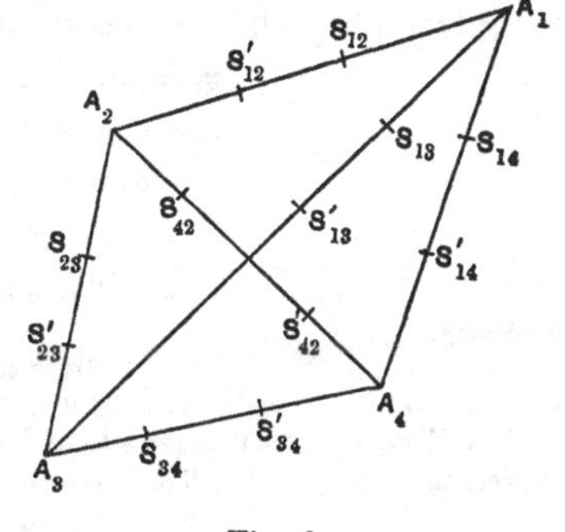

Fig. 6.

$$\begin{vmatrix} a'_{12} & 1 & 0 & 0 \\ 0 & a_{23} & 1 & 0 \\ 0 & 0 & a_{34} & 1 \\ a_{14} & 0 & 0 & 1 \end{vmatrix} = 0, \text{ i.e., } a'_{12} \cdot a_{23} \cdot a_{34} = a_{14} \quad \text{......(i).}$$

The singular plane through $S_{34}S'_{23}$, (distinct from α_1), must pass through S'_{12}*, therefore

$$\alpha'_{12} . \alpha'_{23} . \alpha_{34} = \alpha'_{14} \text{..........................(ii).}$$

It is clear that neither singular plane through $S_{34}S_{12}$ can meet the edges A_2A_3, A_1A_4 in singular points, (for the singular plane through $S_{34}S_{23}$ passes through S'_{12}, &c.); expressing that such a plane through $S_{34}S_{12}$ meets A_1A_3, A_4A_2 in singular points, which may be designated S'_{13}, S_{42}, we have again,

$$\alpha'_{13} . \alpha_{34} . \alpha_{42} = \alpha_{12} \text{..........................(iii).}$$

From (i), (ii), and (iii) it follows that

either $\alpha_{12} = \alpha_{34}, \ \alpha_{13} = \alpha_{42}, \ \alpha_{14} = \alpha_{23};$

or, $\alpha_{12} + \alpha_{34} = \alpha_{13} + \alpha_{42} = \alpha_{14} + \alpha_{23} = 0.$

The first set of equations require that

$$a_{12} = a_{34}, \ a_{13} = a_{24}, \ a_{14} = a_{23},$$

the second that $a_{12} + a_{34} = a_{13} + a_{24} = a_{14} + a_{23} = 0$,

but, by changing the sign of x_1, it is seen that there is no real difference in form of the equation of the surface in the two cases. The equation of the surface is therefore seen to be

$$\{x_1^2 + x_2^2 + x_3^2 + x_4^2 + 2a_{12}(x_1x_2 + x_3x_4) + 2a_{13}(x_1x_3 + x_2x_4)$$
$$+ 2a_{14}(x_1x_4 + x_2x_3)\}^2 = 16Kx_1x_2x_3x_4\text{†...(II.).}$$

By changing to a new tetrahedron of reference, the equation of the surface now arrived at, can be brought to the form (I) of Art. 82; for introducing the new coordinate system given by the equations

$$x_1 = \alpha_1y_1 + \alpha_2y_2 + \alpha_3y_3 + \alpha_4y_4,$$
$$x_2 = \alpha_2y_1 + \alpha_1y_2 + \alpha_4y_3 + \alpha_3y_4,$$
$$x_3 = \alpha_3y_1 + \alpha_4y_2 + \alpha_1y_3 + \alpha_2y_4,$$
$$x_4 = \alpha_4y_1 + \alpha_3y_2 + \alpha_2y_3 + \alpha_1y_4;$$

* For if it passed through S_{12}, (and therefore through S'_{14}), we should have, similarly,

$$a_{12} . a'_{23} . a_{34} = a'_{14};$$

whence from (i) $a_{34}^2 = 1$, since $a_{12} . a'_{12} = a_{23} . a'_{23} = a_{14} . a'_{14} = 1$.

But if $a_{34}^2 = 1$, then $a_{34} = \pm 1$, and the points S_{34}, S'_{34} would coincide; hence expressing that S_{34}, S'_{23}, S'_{12}, S'_{14} are coplanar, we have

$$a'_{12} . a'_{23} . a_{34} = a'_{14}.$$

† To determine K we observe that the equations which give the double points are

$$\Psi(x_1 + a_{12}x_2 + a_{13}x_3 + a_{14}x_4) - 4Kx_2x_3x_4 = 0,$$

together with three similar equations; and, excluding the twelve double points which lie upon the coordinate planes, the coordinates of the remaining four double points satisfy the equations

$$x_1^2 + x_1(a_{12}x_2 + a_{13}x_3 + a_{14}x_4) \pm \sqrt{Kx_1x_2x_3x_4} = 0, \ \&c.$$

But these equations become identical with equations (III) of Art. 82, if we replace x_i by y_i^2; hence from the result there given it follows that

$$K = a_{12}^2 + a_{13}^2 + a_{14}^2 - 2a_{12}a_{13}a_{14} - 1.$$

it is easily seen that each of the expressions

$$x_1^2+x_2^2+x_3^2+x_4^2,\quad x_1x_2+x_3x_4,\quad x_1x_3+x_2x_4,\quad x_1x_4+x_2x_3$$

is linearly expressible in terms of four functions of the y_i of this form; hence

$$\Psi \equiv y_1^2+y_2^2+y_3^2+y_4^2+2A_1(y_1y_2+y_3y_4)$$
$$+2B_1(y_1y_3+y_2y_4)+2C_1(y_1y_4+y_2y_3).$$

The equation of the surface now assumes the form

$$\Psi^2-16K'(\alpha_1y_1+\alpha_2y_2+\alpha_3y_3+\alpha_4y_4)(\alpha_2y_1+\ldots)(\alpha_3y_1+\ldots)(\alpha_4y_1+\ldots)$$
$$=0\ldots\ldots\ldots\ldots\ldots\text{(III.)}.$$

Lastly take $(\alpha_1, \alpha_2, \alpha_3, \alpha_4)$ to be the coordinates of a double point of the surface which does not lie on $\Psi=0$, then writing

$$L=(\alpha_1\alpha_2+\alpha_3\alpha_4)(\alpha_1\alpha_3+\alpha_2\alpha_4)(\alpha_1\alpha_4+\alpha_2\alpha_3),$$

we obtain $\{\Psi(\alpha)\}^2-16K'.\Sigma\alpha_i^2.8L=0$; and expressing that the four equations to determine the double points of the surface are satisfied by the point α_i, we have

$$4\Psi(\alpha)(\alpha_1+A_1\alpha_2+B_1\alpha_3+C_1\alpha_4)-16K'.8L\left\{\alpha_1+\alpha_2\frac{\Sigma\alpha_i^2}{2(\alpha_1\alpha_2+\alpha_3\alpha_4)}\right.$$
$$\left.+\alpha_3\frac{\Sigma\alpha_i^2}{2(\alpha_1\alpha_3+\alpha_2\alpha_4)}+\alpha_4\frac{\Sigma\alpha_i^2}{2(\alpha_1\alpha_4+\alpha_2\alpha_3)}\right\}=0,$$

with three similar equations; these four linear equations to determine A_1, B_1, C_1, and $\dfrac{K'L}{\Psi(\alpha)}$, are seen to be satisfied by

$$A_1=\frac{\Sigma\alpha^2}{2(\alpha_1\alpha_2+\alpha_3\alpha_4)},\quad B_1=\frac{\Sigma\alpha^2}{2(\alpha_1\alpha_3+\alpha_2\alpha_4)},\quad C_1=\frac{\Sigma\alpha^2}{2(\alpha_1\alpha_4+\alpha_2\alpha_3)},$$
$$\Psi(\alpha)=4\Sigma\alpha_i^2=32K'.L.$$

From equation (III) all such terms as $y_1^3y_2$, $y_1^2y_2y_3$ disappear, since the coefficient of $y_1^3y_2=4A_1-\dfrac{16K'L}{\alpha_1\alpha_2+\alpha_3\alpha_4}=0$; and the coefficient of

$$y_1^2y_2y_3=4C_1+8A_1B_1-16K'\left\{\frac{L}{\alpha_1\alpha_4+\alpha_2\alpha_3}+(\alpha_1\alpha_4+\alpha_2\alpha_3)\Sigma\alpha_i^2\right\}=0.$$

Hence the equation of the Kummer surface, when referred to this system of coordinates, has the form

$$A(y_1^4+y_2^4+y_3^4+y_4^4)+2B(y_1^2y_2^2+y_3^2y_4^2)+2C(y_1^2y_3^2+y_2^2y_4^2)$$
$$+2D(y_1^2y_4^2+y_2^2y_3^2)+4Ey_1y_2y_3y_4=0.$$

86. The Complex Surfaces of Plücker. The lines of C^2 which lie in any plane envelope a conic; taking all the planes of a pencil through any line l these conics form a surface, the

Complex Surface of Plücker*. This surface may be regarded either as the locus of points of intersection of consecutive tangents in these planes, *i.e. the locus of points whose complex cones touch l*, or as the envelope of the tangent planes of the complex cones of the points of l. If P is the point of contact of a tangent line of the surface which meets l in Q, the tangent plane at P to the surface is the tangent plane along PQ of the complex cone of Q.

If A and B are two given points on l, and u, v the lines joining a point P of the surface to A and B respectively, on substituting $u + \mu v$ for x in $(\lambda x^2) = 0$ the quadratic in μ must have equal roots, whence we derive the equation of the Plücker surface as being

$$(\lambda u^2)(\lambda v^2) - (\lambda uv)^2 = 0;$$

this is of the fourth degree in the coordinates of P. In like manner if α and β are two given planes through l, and u, v the coordinates of the lines of intersection of a tangent plane π with α and β respectively, the equation of the surface is similarly seen to be of the same form and is therefore of the fourth class.

The equation $(\lambda u^2) = 0$ represents the complex cone of A, $(\lambda v^2) = 0$ the complex cone of B, while $(\lambda uv) = 0$ gives the locus of points P such that the lines PA, PB are harmonic to the complex lines of the pencil (P, APB); for the lines harmonic to u and v being $u + \mu v$, $u - \mu v$, if the latter are complex lines we have

$$\Sigma\lambda_i (u_i + \mu v_i)^2 = 0, \quad \Sigma\lambda_i (u_i - \mu v_i)^2 = 0,$$

whence $(\lambda uv) = 0$ is the required locus.

The complex cone of any point on AB is

$$(\lambda u^2) + 2\mu (\lambda uv) + \mu^2 (\lambda v^2) = 0;$$

it is clear that, for all values of μ, this cone passes through the eight points of intersection of the quadrics

$$(\lambda u^2) = 0, \quad (\lambda uv) = 0, \quad (\lambda v^2) = 0.$$

These points are double points of the complex surface and consist of the eight singular points B of Art. 77. Reciprocally, the complex conics of the planes through l touch the same eight planes (the planes α of Art. 77).

Since any line which meets l intersects the surface in two other points only, it follows that l is a double line of the complex surface; the tangent planes of the surface at any point Q of l are the two tangent planes through l to the complex cone of Q.

* *Neue Geometrie des Raumes*, Bd. I. S. 163. These surfaces are called *meridian* surfaces by Plücker.

The lines of C^2 which meet any two lines l, l', determine upon them a (2, 2) correspondence. If l' meets the complex surface of l in P', then P' is seen to be a branch point for the points of l', hence if l' meets the complex surface of l in P_1', P_2', P_3', P_4' and l meets the complex surface of l' in P_1, P_2, P_3, P_4 we have (Art. xvii),

$$(P_1, P_2, P_3, P_4) = (P_1', P_2', P_3', P_4').$$

Singularities of the surface. The eight singular points of C^2 which lie in the four singular planes through l are double points of the surface; they are given by the equations $(\lambda u^2) = 0, (\lambda uv) = 0, (\lambda v^2) = 0$. If x is the singular line of C^2 which joins a corresponding pair of these singular points it is a simple line of the surface; for if u and v are the lines joining any point P of x to A and B we have

$$\sigma . x_i = u_i + \rho . v_i,$$

therefore,

$$\sigma(\lambda xu) = (\lambda u^2) + \rho(\lambda uv), \quad \sigma(\lambda xv) = (\lambda uv) + \rho(\lambda v^2),$$

and since the line λx meets both u and v, $(\lambda xu) = 0$, $(\lambda xv) = 0$, we have

$$(\lambda u^2) + \rho(\lambda uv) = (\lambda uv) + \rho(\lambda v^2) = 0,$$

and therefore, $(\lambda u^2)(\lambda v^2) - (\lambda uv)^2 = 0$, *i.e.*, P lies on the surface.

The plane (l, x) is a *stationary* tangent plane along x; for any line in the plane touches the surface where it meets x.

The relative positions of the double points of the surface have been already seen (Art. 77).

The complex surface of *any* line l passes through the double points of the singular surface of C^2 and touches its singular tangent planes; for if x is the singular line of the *third order* (Art. 80), which passes through a double point D, then the lines $x, \lambda x, \lambda^2 x$ intersect in D, and the lines AD and BD will be given by the equations

$$u_i = \rho x_i + \mu\lambda_i x_i + \nu\lambda_i^2 x_i,$$
$$v_i = \sigma x_i + \mu'\lambda_i x_i + \nu'\lambda_i^2 x_i,$$

whence

$$(\lambda u^2) = \nu^2(\lambda^5 x^2), \quad (\lambda v^2) = \nu'^2(\lambda^5 x^2), \quad \lambda uv = \nu\nu'(\lambda^5 x^2);$$

therefore $(\lambda u^2)(\lambda v^2) - (\lambda uv)^2 = 0$, hence D lies on the surface; the second result is similarly proved.

Special complex surfaces. Modifications of the form of the complex surface will arise when l has special relations to C^2.

(i) *If l belongs to C^2*, it touches every complex conic in planes through l, which thus becomes a cuspidal line of the surface; since the pairs of tangent planes through l to the complex cones of the points of l here coalesce. There are only four double points of the surface outside l.

(ii) *If l touches Φ*, let A be the point of contact of l with Φ and a_1 the singular line of C^2 which touches Φ at A; then if v is any line through B in the plane (l, a_1), since $(\lambda a_1 v) = 0$, it follows that the surface $(\lambda uv) = 0$ contains the line a_1; moreover $(\lambda u^2) = 0$, representing the complex cone of A, consists of two planes α, α' intersecting in a_1; hence the complex surface is of the form

$$(\lambda v^2)\, \alpha\alpha' = (\alpha\xi + \alpha'\eta)^2,$$

which shows that a_1 is a double line of the surface.

Two of the tangent planes to Φ through l coincide with (l, a_1), hence there are four double points of the surface outside l; denote them by B_1, B_1'; B_2, B_2', where B_1B_1' and B_2B_2' meet l; B_1A and $B_1'A$ are lines of C^2, hence the planes (B_1, a_1), (B_1', a_1) are the plane-pair for A; similarly (B_2, a_1), (B_2', a_1) are also the plane-pair for A, hence the joins of two pairs of points B meet a_1; thus $B_1B_1'B_2B_2'$ form a twisted quadrilateral of which one pair of opposite sides meet l and one pair meet a_1.

The tangent plane along each side of this quadrilateral is stationary.

(iii) *If l is a double tangent to Φ*, the singular lines a_1 and a_2 at its points of contact are each double lines of the surface, which has thus three double lines. The surface is *ruled*, for any plane through a_1 meets the surface in a conic having for a double point its point of intersection with a_2; *i.e.*, the conic is a pair of lines. There are no double points outside the double lines. The surface, having two double directrices and a double generator, l, belongs to class VII. of ruled quartics.

(iv) *If l lies in a singular tangent plane α of Φ*, all complex lines in α pass through the same point O of the section of Φ by α (Art. 80), and all are singular lines; if A and B are the points where l meets this section, then denoting OA and OB by b and c, λb meets b in A and λc meets c in B, hence

$$(\lambda lb) = 0, \quad (\lambda lc) = 0;$$

and expressing that all lines of the pencil (O, α) are singular lines we have

$$(\lambda b^2) = (\lambda bc) = (\lambda c^2) = (\lambda^2 b^2) = (\lambda^2 bc) = (\lambda^2 c^2) = 0.$$

Let d and d' be any two lines through A and B respectively, then

$$(\lambda bd) = 0, \quad (\lambda cd') = 0,$$

and we have for u and v coordinates given by

$$u_i = \rho l_i + \kappa b_i + \nu d_i,$$
$$v_i = \rho' l_i + \kappa' c_i + \nu' d_i',$$

where $\rho = 0$ is the equation of the plane (bd) &c. (Art. 13), thus $\nu = 0$ and $\nu' = 0$ are each the equation of the plane α.

We obtain, rejecting terms which vanish by aid of the above relations,

$$(\lambda u^2) = \rho^2 (\lambda l^2) + \nu^2 (\lambda d^2) + 2\rho\nu (\lambda ld),$$
$$(\lambda v^2) = \rho'^2 (\lambda l^2) + \nu'^2 (\lambda d'^2) + 2\rho'\nu' (\lambda ld'),$$
$$(\lambda uv) = \rho\rho' (\lambda l^2) + \nu\nu' (\lambda dd') + \kappa\nu' (\lambda bd')$$
$$+ \kappa'\nu (\lambda cd) + \rho\nu' (\lambda ld') + \rho'\nu (\lambda ld).$$

Thus from the equation $(\lambda u^2)(\lambda v^2) - (\lambda uv)^2 = 0$, the factor ν may be removed, and the resulting cubic surface is seen to have l, b, and c for simple lines. There are no double lines.

87. Normal form of the equation of a quadratic complex. The equation of a quadratic complex may be written in the form

$$F(p) \equiv a_{12,\,12} p_{12}^2 + 2a_{12,\,34} p_{12} p_{34} + 2a_{12,\,13} p_{12} p_{13} + \ldots = 0;$$

and since the sign of p_{ik} is changed if its suffixes are interchanged, it follows that if in a coefficient $a_{ik,\,jl}$ two suffixes of a pair are interchanged, the sign of the coefficient is changed; if the change is effected in both pairs of suffixes the sign is unaltered, *i.e.*

$$a_{ik,\,jl} = a_{ki,\,lj}.$$

We may therefore write $F(p) = 0$ in the symbolic form

$$(\alpha_{12} p_{12} + \alpha_{34} p_{34} + \alpha_{13} p_{13} + \alpha_{42} p_{42} + \alpha_{14} p_{14} + \alpha_{23} p_{23})^2 = 0 \quad \ldots\text{(i)},$$

where $\alpha_{ik} \cdot \alpha_{jl} = a_{ik,\,jl}$, it being understood that each symbolic coefficient α_{ik} changes its sign if its suffixes are interchanged.

The complex $\sum_1^6 \lambda_i x_i^2 = 0$ becomes in terms of the p_{ik}

$$(\lambda_1 - \lambda_2)(p_{12}^2 + p_{34}^2) + \ldots + 2(\lambda_1 + \lambda_2) p_{12} p_{34} + \ldots = 0;$$

now the given complex is also represented by $\sum_1^6 \lambda_i x_i^2 + \mu \sum_1^6 x_i^2 = 0$, but this indeterminateness of the λ_i may be removed by assuming that $\sum_1^6 \lambda_i = 0$, and on this understanding the symbolic equation (i) is the square of a *special* linear complex, *i.e.* one for which

$$\alpha_{12} \cdot \alpha_{34} + \alpha_{13} \cdot \alpha_{42} + \alpha_{14} \cdot \alpha_{23} = 0,$$

since
$$\alpha_{12} \cdot \alpha_{34} = \lambda_1 + \lambda_2, \text{ \&c.*}$$

This is termed by Clebsch† the Normal form of the equation of the complex.

The symbolic expressions α_{ik} may be replaced by others

$$a_1, a_2, a_3, a_4; \; b_1, b_2, b_3, b_4,$$

such that
$$\alpha_{ik} = a_i b_k - a_k b_i,$$

since, by this substitution,

$$\alpha_{12}\alpha_{34} + \alpha_{13}\alpha_{42} + \alpha_{14}\alpha_{23} \equiv (a_1 b_2 - a_2 b_1)(a_3 b_4 - a_4 b_3) + \ldots + \ldots \equiv 0,$$

while, also, the condition of the α_{ik} changing sign with interchange of their suffixes is observed.

The equation of the complex now assumes the form

$$\{\Sigma (a_i b_k - a_k b_i) p_{ik}\}^2 = 0,$$

or, if
$$p_{ik} = x_i y_k - x_k y_i,$$

$$(a_x b_y - a_y b_x)^2 = 0.$$

Regarding the symbolic quantities a_i, b_i as coordinates of two planes we may introduce new symbolic quantities α_i, β_i regarded as coordinates of two points on the line of intersection of these planes, so that

$$a_1 b_2 - a_2 b_1 = \rho (\alpha_3 \beta_4 - \alpha_4 \beta_3), \text{ \&c.}$$

and the equation of the complex becomes

$$0 = \{(\alpha_1\beta_2 - \alpha_2\beta_1)(x_3 y_4 - x_4 y_3) + \ldots\}^2 \equiv \begin{vmatrix} \alpha_1 & \beta_1 & x_1 & y_1 \\ \alpha_2 & \beta_2 & x_2 & y_2 \\ \alpha_3 & \beta_3 & x_3 & y_3 \\ \alpha_4 & \beta_4 & x_4 & y_4 \end{vmatrix}^2$$

or, as it is usually written,

$$(\alpha\beta x y)^2 = 0.$$

88. Complex equation of a quadric. The quadratic complex which consists of the tangents of a quadric, or $\Psi = 0$, (Art. 46), is a special case, for which the symbolic form of

* A corresponding result holds for a complex of any degree, see Chapter XVII.

† "Ueber die Plückerschen Complexe," *Math. Ann.* II. See also Waelsch, "Zur Invariantentheorie der Liniengeometrie," *Math. Ann.* XXXVII.

equation may easily be found directly; for if $(\Sigma a_i \xi_i)^2 = 0$, or as it is usually written $a_\xi^2 = 0$, is the symbolic form of the equation of the quadric

$$a_{11}\xi_1^2 + 2a_{12}\xi_1\xi_2 + \ldots = 0,$$

so that
$$a_{ik} = a_i \cdot a_k;$$

the points in which the line joining any two points X, Y meets the quadric are determined by substituting $X_i + \lambda Y_i$ for ξ_i in the equation $a_\xi^2 = 0$, giving

$$a_X^2 + 2\lambda a_X a_Y + \lambda^2 a_Y^2 = 0.$$

If these values of λ are equal we have

$$a_X^2 \cdot a_Y^2 - (a_X \cdot a_Y)^2 = 0;$$

if now we employ a second set of symbols a_i', so that $a_\xi^2 \equiv a_\xi'^2$, the last equation is seen to be equivalent to

$$a_X^2 \cdot a_Y'^2 + a_Y^2 \cdot a_X'^2 - 2(a_X \cdot a_Y)(a_X' \cdot a_Y') = 0,$$

or,
$$0 = (a_X \cdot a_Y' - a_Y \cdot a_X')^2 = \{\Sigma (a_i a_k' - a_k a_i') p_{ik}\}^2;$$

which is the required symbolic form of the equation $\Psi = 0$. If U and V are two planes through the line (X, Y), the last equation is, writing π_{12} for p_{34}, &c.,

$$(aa' \, UV)^2 = 0.$$

89. Harmonic Complex. The assemblage of lines which meet any two given quadrics $a_\xi^2 = 0$, $b_\xi^2 = 0$ in points which form a harmonic range is a complex, usually called the Harmonic complex, whose equation is readily found by the symbolic method; for the condition that the roots of

$$a_X^2 + 2\lambda a_X \cdot a_Y + \lambda^2 a_Y^2 = 0,$$
$$b_X^2 + 2\mu b_X \cdot b_Y + \mu^2 b_Y^2 = 0,$$

should be harmonic is known to be

$$a_X^2 \cdot b_Y^2 + a_Y^2 \cdot b_X^2 - 2(a_X \cdot a_Y)(b_X \cdot b_Y) = 0;$$

i.e.
$$(a_X \cdot b_Y - a_Y \cdot b_X)^2 = 0, \text{ or, } \{\Sigma (a_i b_k - a_k b_i) p_{ik}\}^2 = 0.$$

If the complex equation of $a_\xi^2 = 0$ be $\Psi = 0$, it is at once seen that the harmonic complex has the form

$$\Sigma b_{ik} \frac{\partial}{\partial a_{ik}} \Psi = 0;$$

for $\Psi = 0$ has been seen to be

$$\Sigma (a_i a_k' - a_k a_i')^2 p_{ik}^2 + 2\Sigma (a_i a_k' - a_k a_i')(a_h a_l' - a_l a_h') p_{ik} p_{hl} = 0,$$

or,
$$\Sigma (a_{ii} a_{kk} - a_{ik}^2) p_{ik}^2 + 2\Sigma (a_{ih} a_{kl} - a_{kh} a_{il}) p_{ik} p_{hl} = 0,$$

while the equation of the harmonic complex is

$$\Sigma\,(a_{ii}b_{kk}+a_{kk}b_{ii}-2a_{ik}b_{ik})\,p_{ik}^2$$
$$+2\Sigma\,(a_{ih}b_{kl}+a_{kl}b_{ih}-a_{il}b_{kh}-a_{kh}b_{il})\,p_{ik}\,p_{hl}=0.$$

When the quadrics are referred to their common self-conjugate tetrahedron the equation of the harmonic complex is

$$\Sigma\,(a_{ii}b_{kk}+a_{kk}b_{ii})\,p_{ik}^2=0.$$

90. Symbolic form of the equation of a quadric in plane coordinates. The symbolic form of the discriminant of the conic

$$a_{11}x_1^2+2a_{12}x_1x_2+\ldots=0, \text{ is known to be } (aa'a'')^2\,*;$$

and the coordinates of the points of the section of the quadric $a_x^2=0$ by the plane u_i are obtained by substituting

$$x_i=K_1y_i+K_2z_i+K_3w_i \text{ in } a_x^2=0,$$

and giving to K_1, K_2, K_3 all values consistent with the equation

$$(K_1a_y+K_2a_z+K_3a_w)^2=0\ldots\ldots\ldots\ldots\ldots\ldots(\text{i}),$$

provided that y, z and w are three points on the plane u_i.

The quantities K_1, K_2, K_3 may therefore be considered as coordinates of the points of a conic which are connected with those of the section of $a_x^2=0$ by u_i in a (1, 1) correspondence.

Hence if this section has a double point the discriminant of (i) is zero and

* For writing $\qquad a_{ik}=a_i\,.\,a_k=a_i'\,.\,a_k'=a_i''\,.\,a_k''$

and substituting respectively in the three rows of

$$\begin{vmatrix} a_{11} & a_{12} & a_{13}\\ a_{21} & a_{22} & a_{23}\\ a_{31} & a_{32} & a_{33}\end{vmatrix}$$

we obtain

$$a_1a_2'a_3''\begin{vmatrix} a_1 & a_2 & a_3\\ a_1' & a_2' & a_3'\\ a_1'' & a_2'' & a_3''\end{vmatrix},$$

but since it is a matter of indifference in what order the substitutions are applied we obtain also

$$a_1'a_2a_3''\begin{vmatrix} a_1' & a_2' & a_3'\\ a_1 & a_2 & a_3\\ a_1'' & a_2'' & a_3''\end{vmatrix}=-a_1'a_2a_3''\begin{vmatrix} a_1 & a_2 & a_3\\ a_1' & a_2' & a_3'\\ a_1'' & a_2'' & a_3''\end{vmatrix};$$

proceeding similarly it is clear that the discriminant has as its symbolic form

$$\tfrac{1}{6}\begin{vmatrix} a_1 & a_2 & a_3\\ a_1' & a_2' & a_3'\\ a_1'' & a_2'' & a_3''\end{vmatrix}^2.$$

See Clebsch, *Vorlesungen über Geometrie*, I. S. 268.

$$\begin{vmatrix} a_y & a_z & a_w \\ a_y' & a_z' & a_w' \\ a_y'' & a_z'' & a_w'' \end{vmatrix}^2 = 0;$$

it is easy to see that the last equation is, by the definition of the points y, z, w, equivalent to the following:

$$\begin{vmatrix} a_1 & a_2 & a_3 & a_4 \\ a_1' & a_2' & a_3' & a_4' \\ a_1'' & a_2'' & a_3'' & a_4'' \\ u_1 & u_2 & u_3 & u_4 \end{vmatrix}^2 = 0, \text{ or, } (a, a', a'', u)^2 = 0.$$

Since the section of the surface by any of its tangent planes has a double point, the last equation is that of the surface in plane coordinates.

91. Plücker surfaces and singular surface of the complex. Referring to the symbolic form of the equation of the complex, viz. $(a_x b_y - a_y b_x)^2 = 0$, if the point x be given, this equation represents the complex cone of the point x; it may be represented by $\gamma_y^2 = 0$, where $\gamma_i = a_x b_i - b_x a_i$, and hence $\gamma_x \equiv 0$.

It was seen that if the line of intersection of the planes U, V touches this cone, $(\gamma, \gamma', U, V)^2 = 0$; hence the equation of the Plücker surface for this line is

$$(a_x b_i - a_i b_x,\ a_x' b_i' - a_i' b_x',\ U,\ V)^2 = 0.$$

Now the equation $(a, a', a'', u)^2 = 0$ gives the planes whose sections of $a_x^2 = 0$ have a double point; applying this to the cone $\gamma_y^2 = 0$ the equation $(\gamma, \gamma', \gamma'', u)^2 = 0$ gives such planes for the cone; for points not on the singular surface the only planes of this description are those through the point x, *i.e.* those for which $u_x = 0$; hence $(\gamma, \gamma', \gamma'', u)^2 \equiv M \cdot u_x^2$, where M cannot contain u. If x is a point on the singular surface the section of the cone by *every* plane has a double point, since the cone consists of a pair of planes; so that in the case of all such points x we have $M = 0$, which is therefore the equation of the singular surface.

To determine the form of M we observe that

$$(\gamma, \gamma', \gamma'', u) = (a_x b - b_x a, \gamma', \gamma'', u) = a_x (b, \gamma', \gamma'', u) - b_x (a, \gamma', \gamma'', u).$$

But it is easy to verify that

$$a_x (b, \gamma', \gamma'', u) - b_x (a, \gamma', \gamma'', u)$$
$$\equiv \gamma_x'' (a, b, \gamma', u) - \gamma_x' (a, b, \gamma'', u) - u_x (a, b, \gamma', \gamma'');$$

and since $\gamma_x' = 0$, $\gamma_x'' = 0$, we have

$$a_x (b, \gamma', \gamma'', u) - b_x (a, \gamma', \gamma'', u) \equiv -u_x (a, b, \gamma', \gamma''),$$

hence
$$(\gamma, \gamma', \gamma'', u)^2 = (a, b, \gamma', \gamma'')^2 \cdot u_x^2.$$

The equation of the singular surface is therefore

$$(a, b, a_x' \cdot b' - a' \cdot b_x', a_x'' \cdot b'' - b_x'' \cdot a'')^2 = 0.$$

CHAPTER VII.

SPECIAL VARIETIES OF THE QUADRATIC COMPLEX.

92. The Tetrahedral Complex. In illustration of the theory of the quadratic complex, some special varieties of this complex will now be discussed. We begin with one of peculiar interest, the Tetrahedral Complex*; it is composed of the lines which meet the faces of a given tetrahedron in four points whose Double Ratio is constant; this Double Ratio is then, by the Theorem of von Staudt, (Art. 12), equal to that of the four planes through the line and the corresponding vertices of the tetrahedron. To this complex belong the lines of the four sheaves whose centres are the vertices of the tetrahedron; for since three of the points in which a line meets the faces of the tetrahedron come into coincidence at a vertex the fourth point of the given D.R. may have *any position* in the face opposite to this vertex; for a similar reason, any line in the face of the tetrahedron belongs to the complex, and it follows that the complex cone of any point passes through the vertices of the tetrahedron, and the complex curve in every plane touches each face of the tetrahedron; again, the lines of the complex in a given plane, being those which meet four given lines in a constant D.R., envelope a conic, and in any plane pencil there are *two* lines of the complex, which must therefore be *quadratic*.

The *singular planes* of the complex consist of the sheaves of planes through the vertices of the tetrahedron; for in a plane π through a vertex A_i the section of the sheaf of lines (A_i) by π

* This complex was first investigated by Binet from a dynamical standpoint, "Mémoire sur la théorie des axes conjugées et des momens d'inertie des corps," *Journal de l'École polytech.*, T. IX. (1813). For an elegant application of the properties of this complex see Schönflies, *Geometrische Bewegung*. Reye, *Geometrie der Lage*, arranged and extended the various known properties of the complex, which is usually connected with his name; to him is due the method of formation by points in two collinear spaces, (second method).

A historical account of this complex will be found in Lie and Scheffers, *Berührungstransformationen*, Bd. I. S. 320.

belongs to the complex, hence the remaining complex lines in π must form a plane pencil.

The *singular points* of the complex consist of the points of the faces of the tetrahedron; the complex cone of a point P in the face α_i opposite to A_i consists partly of the pencil (P, α_i) and partly of another pencil of lines whose plane must pass through A_i. There are therefore three sets of ∞^2 plane pencils belonging to the complex, viz. those which have either (i) their centres at a vertex of the tetrahedron, or (ii) their planes a face of the tetrahedron, or (iii) their centres in a face of the tetrahedron and their planes through the opposite vertex of the tetrahedron; these pencils include all the lines of the complex.

A tetrahedral complex is determined by its "fundamental" tetrahedron and by one of its lines or by the Double Ratio of the complex; since the fundamental tetrahedron may be chosen in ∞^{12} ways, there are ∞^{13} tetrahedral complexes.

93. Equation of the tetrahedral complex. From Art. 12* it is seen that the equation of a tetrahedral complex in Plücker coordinates is $\frac{p_{12} \cdot p_{34}}{p_{14} \cdot p_{23}} = \text{constant}$, or,

$$Ap_{12}p_{34} + Bp_{13}p_{42} + Cp_{14}p_{23} = 0.$$

The form of this equation suggests a method of construction of the complex, which may be easily verified; viz. *the complex is the locus of lines which meet a pair of corresponding lines of two projective plane pencils.*

For take the centres A and B of the given pencils as two vertices of the tetrahedron of reference, and the united points C, D of the projective rows determined by the pencils on the line of intersection of their planes, as the two other vertices of this tetrahedron; then if AD and AC have coordinates a_{ik} and b_{ik}, any line of one of the given pencils is $a + \mu b$, similarly if c_{ik} and d_{ik} are coordinates of BC and BD, any line of the other pencil is $c + \nu d$; and from the correspondence of the lines of the pencils there must exist an equation of the form

$$\mu\nu + A\mu + B\nu - k = 0.$$

* Or directly as follows:—If the line joining the points ξ_i and η_i meets the planes $a_x=0$, $b_x=0$, $c_x=0$, $d_x=0$ in points whose double ratio is given, we have

$$\left(\frac{a_\xi}{a_\eta} - \frac{b_\xi}{b_\eta}\right)\left(\frac{d_\xi}{d_\eta} - \frac{c_\xi}{c_\eta}\right) \Big/ \left(\frac{b_\xi}{b_\eta} - \frac{c_\xi}{c_\eta}\right)\left(\frac{a_\xi}{a_\eta} - \frac{d_\xi}{d_\eta}\right) = \text{constant}.$$

Taking the given planes as coordinate planes the equation of the complex follows at once.

This relation simplifies, since $\mu = 0, \nu = \infty$, gives a pair of corresponding lines; hence $B = 0$; similarly $A = 0$, and the equation becomes

$$\mu\nu = k.$$

Now if a line whose coordinates are p_{ik} meets each of these two corresponding lines

$$\omega(p \mid a)^* + \mu\omega(p \mid b) = 0,$$

$$\omega(p \mid c) + \nu\omega(p \mid d) = 0,$$

or,
$$\omega(p \mid a)\,\omega(p \mid c) = k\omega(p \mid b)\,\omega(p \mid d);$$

but in $\omega(p \mid a)$ the only term is p_{23} since all the coordinates a_{ik} are zero except a_{14}, which is unity; similarly $\omega(p \mid c) = p_{14}$, &c., and we have as the equation of the locus of p

$$p_{23}\,p_{14} = kp_{42}\,p_{13},$$

i.e. a tetrahedral complex.

This result is shown by Hirst and Sturm† as follows:—let (A, β) and (B, α) be the two given pencils, C and D being the points given as above; then if two corresponding lines of the pencils meet CD in X_1 and X_2, and if x is a line which meets this pair of lines,

$$x(ABCD) = (X_1X_2\,CD).$$

In the same manner if y is any other line which meets a pair of corresponding lines of the two pencils,

$$y(ABCD) = (Y_1\,Y_2\,CD)$$

but
$$(X_1X_2\,CD) = (Y_1\,Y_2\,CD), \qquad \text{(Introd. iii)},$$

therefore
$$x(ABCD) = y(ABCD).$$

Since any two vertices of the fundamental tetrahedron may be taken as the centres of the projective pencils, the complex may be thus generated in six ways.

94. Reguli of the complex. The complex

$$Ap_{12}p_{34} + Bp_{13}p_{42} + Cp_{14}p_{23} = 0$$

contains the regulus whose equations are

$$p_{23} = \rho \,.\, p_{12} - \frac{B+\mu}{C+\mu}\,\sigma \,.\, p_{13},$$

$$p_{34} = \tau \,.\, p_{13} - \frac{C+\mu}{A+\mu}\,\rho \,.\, p_{14},$$

$$p_{42} = \sigma \,.\, p_{14} - \frac{A+\mu}{B+\mu}\,\tau \,.\, p_{12};$$

* $\omega(p|a)$ is a frequently used abbreviation for $\sum\limits_i p_i \dfrac{\partial\omega(a)}{\partial a_i}$.

† Hirst, "On the complexes generated by two correlative planes," *Proc. Lond. Math. Soc.*, vol. x.

for any given values of ρ, σ, τ and μ; since, eliminating these quantities, we obtain the tetrahedral complex. This regulus is seen to pass through each vertex of the tetrahedron of reference, since the three equations are satisfied by taking $p_{23}=p_{34}=p_{42}=0$; which shows that the regulus passes through the vertex A_1, and so for the other vertices; hence, *a tetrahedral complex contains* ∞^4 *reguli which pass through each vertex of its tetrahedron.*

Similarly, by writing π_{ik} in place of p_{ik} in these equations, we obtain ∞^4 reguli of the complex which touch the coordinate planes.

Since the equation of the complex may also, by subtraction of $C(p_{12}p_{34}+p_{13}p_{42}+p_{14}p_{23})$, be written in the form

$$(A-C)\,p_{12}p_{34}+(B-C)\,p_{13}p_{42}=0,$$

it is clear that the regulus whose equations are

$$p_{12}=\rho p_{13},$$
$$\rho(A-C)\,p_{34}+(B-C)\,p_{42}=0,$$
$$p_{14}=\sigma p_{23}+\tau p_{13}+\mu p_{42},$$

is contained in the complex, for any given values of ρ, σ, τ and μ.

These ∞^4 reguli pass through the vertices A_1 and A_4 and touch the coordinate planes α_1 and α_4. There are five other similar sets of ∞^4 reguli, hence, *a tetrahedral complex contains six sets of* ∞^4 *reguli, where every regulus of a set passes through two vertices of the tetrahedron of the complex and touches the opposite faces.*

95. If any two lines, p and p', of the complex, meet the fundamental tetrahedron in points L, M, N, R; L', M', N', R', then on p and p' are determined by these points two projective rows of points, whose joins, therefore, form a regulus ρ, (Introd. viii); each line of the complementary regulus ρ' meets the lines LL', MM' in points L'', M'', &c., so that $(L''M''N''R'')=(LMNR)=$ constant double ratio of the complex; hence each line of ρ' belongs to the complex; there are ∞^6 possible combinations of the lines p and p', but ∞^2 of these belong to the same regulus, so that there are ∞^4 such reguli ρ'; since the line LL' of ρ lies in the plane BCD, this plane contains a line of ρ'.

Similarly we derive ∞^4 reguli ρ'' which are obtained as the complementary reguli of the loci of intersection of projective pencils of planes with p and p' as axes. The reguli ρ' are those previously obtained which touch the faces of the tetrahedron, the reguli ρ'' those which pass through its vertices.

If three lines of the complex p, p', p'' be taken, and the planes (p, A) (p', A) (p'', A); (p, B) (p', B) (p'', B); (p, C) (p', C) (p'', C) be made to correspond by threes, the correspondence of the three pencils of planes of which the axes are p, p', p'' is determined, (Introd. xii), and the locus of points of intersection of three corresponding planes is a twisted cubic, which passes through D, since p, p', p'' belong to the tetrahedral complex. If p''' be any chord of this cubic, then since $p'''(ABCD)=p(ABCD)$, (Introd. xii), p''' belongs to the complex.

Since there are ∞^9 combinations of lines p, p', p'', and since three chords of a twisted cubic may be selected in ∞^6 ways, there are ∞^3 twisted cubics all of whose chords belong to the complex.

Similarly by taking the three projective rows of points determined on three lines of the complex we obtain ∞^3 developables of the third class, for each of which the intersection of two tangent planes belongs to the complex.

Conversely if any twisted cubic is given, and four points A, B, C, D be taken on it, if p is any chord of the cubic, $p(ABCD)$ is constant, and hence all the chords of the cubic belong to a tetrahedral complex whose fundamental tetrahedron is $ABCD$. Also if x is any generator of a regulus which passes through $ABCD$ we have $x(ABCD)$ constant for this regulus (Introd. viii), hence all the generators of one system belong to one tetrahedral complex, while the generators of the other system belong to a second tetrahedral complex. Since a quadric is determined by nine conditions and two intersecting complex lines may be chosen in ∞^5 ways, there are ∞^3 quadrics which pass through four given points and of which the generators of one system belong to a given tetrahedral complex, and those of the other system to another given tetrahedral complex which has the same fundamental tetrahedron.

96. Second method of formation of the complex. The collineation of two spaces Σ and Σ' gives rise to a tetrahedral complex; for, if the united points of Σ and Σ' be taken as vertices of the tetrahedron of reference the equations connecting corresponding points are then, (Art. 34),

$$\mu x_i' = a_i x_i,$$

hence if p_{ik} is the line joining x and x',

$$\mu p_{ik} = \mu (a_k - a_i) x_i x_k$$

or,

$$\frac{p_{12} \cdot p_{34}}{p_{14} \cdot p_{23}} = \frac{(a_2 - a_1)(a_4 - a_3)}{(a_4 - a_1)(a_3 - a_2)} = \text{constant}.$$

If u and u' are a pair of corresponding planes

$$\nu \,.\, u_i = a_i u_i',$$

hence, if π_{ik} is the line of intersection of u and u',

$$\frac{\pi_{12} \cdot \pi_{34}}{\pi_{14} \cdot \pi_{23}} = \frac{(a_2 - a_1)(a_4 - a_3)}{(a_4 - a_1)(a_3 - a_2)};$$

that is, *the locus of intersection of corresponding planes is the same tetrahedral complex.*

Finally if p_{ik} is the line joining two points x_i, y_i and p_{ik}' the line joining their corresponding points x_i', y_i',

$$\mu^2 \cdot p_{ik}' = a_i a_k p_{ik}.$$

Hence, *the locus of lines p_{ik} which intersect their corresponding lines* is

$$(a_1 a_2 + a_3 a_4)\, p_{12} p_{34} + (a_1 a_3 + a_4 a_2)\, p_{13} p_{42} + (a_1 a_4 + a_2 a_3)\, p_{14} p_{23} = 0,$$

which is easily seen to be the same tetrahedral complex.

Let $ABCD$ be the united points of the spaces Σ and Σ', and P, P' a pair of corresponding points, then in the sheaves (P) and (P') certain corresponding lines intersect, the locus of such points of intersection being a twisted cubic, which passes through A, B, C, D, P and P' (Introd. xii). There are ∞^3 such cubics obtained by taking for P all positions in Σ; two such cubics have a regulus of chords in common; for if Q and Q' are any other pair of corresponding points, since PQ, $P'Q'$ are corresponding lines we have two projective pencils of planes, with axes PQ, $P'Q'$ respectively, the intersections of whose corresponding planes form a regulus, (Introd. viii); any generator of this regulus is a chord of the cubic of both P and Q, for if a and a' are two corresponding planes of the pencils which meet in p, the pencils (P, a), (P', a') determine two projective rows on p having two united points, hence p is a chord of the cubic of P, and similarly is seen to be a chord of the cubic of Q; it follows therefore that if x is any chord of any one of these ∞^3 cubics

$$x\,(ABCD) = \text{constant}.$$

Hence, as just shown analytically, the joins of corresponding points P and P' of Σ and Σ' form a tetrahedral complex.

97. Third method of formation of the complex. A third method of formation is the following: *having given a (1, 1) correspondence between the points P of any plane a, and the lines p of any sheaf (A), the lines of the pencils (P, p) constitute a tetrahedral complex.* For the lines of the sheaf give rise to a second point-system in a, collinear to the given one; let BCD be the united points of this correspondence (Introd. xiii), then if to P in the first system corresponds P' in the second, and if to P in the second corresponds P'' in the first, to the line PP' in the second system will correspond the line PP'' in the first system. Now take any line PT of the pencil (P, p) meeting AP' in T, and any point S in the plane (PT, PP''), then the pencils (S, PT), (A, PP') are projective and determine two projective rows on PT; hence, corresponding lines of the pencils meet *twice* on PT. But the sheaves which project the two collinear plane systems from S and A have, as locus of intersection of corresponding lines, a twisted cubic (Introd. xii) which passes through A, B, C, D, and of which PT is therefore seen to be a chord. Hence $PT\,(ABCD)$ is the double ratio of the chords of this twisted cubic; by taking different positions for S we obtain ∞^3 cubics, and since *any two*

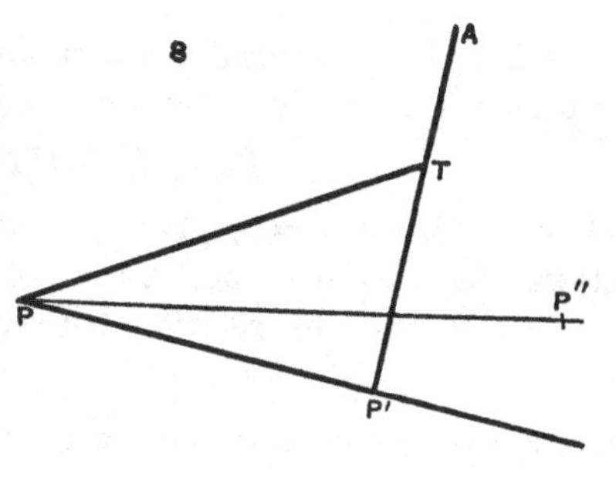

Fig. 7.

of them, as will be shown immediately, have a regulus of chords in common, the double ratio of the chords is the same for each of these cubics, or, $PT\,(ABCD)$ is a constant, which proves the result.

That any two of the twisted cubics have a regulus of chords in common may be seen as follows: take any two points S and S', then any plane π through SS' meets a in a line PP'' to which the corresponding line PP' is given by the correlation, and hence P, and therefore P', is determined; T is the point where AP' meets π, and PT is thus determined as a chord of both cubics: again the locus of P for the pencil of planes π through SS' is a *conic which passes through* O the point of intersection of SS' and a; for PP'' passes through this fixed point O and hence the corresponding line PP' must also pass through a fixed point, hence the locus of P being that formed by the intersection of corresponding lines of two pencils is a *conic*. Thus the chords common to the two cubics for S and S' form a regulus of which SS' is a directrix.

It follows from the method of proof just given that the cubics for *coplanar* points S have one chord in common, for *collinear* points S a regulus in common.

98. The lines which meet every corresponding pair of lines of two projective pencils form, in general, a tetrahedral complex. If certain special connexions exist between the two pencils the complex is modified in character.

Firstly, when the plane of one pencil passes through the centre of the other, we may take the pencils as being (A_1, α_2), (A_3, α_1) and suppose that to A_1A_3 of the first pencil A_3A_2 of the second pencil corresponds, while to the line of intersection A_3A_4 of the planes of the pencils the line A_1A_4 corresponds.

	p_{12}	p_{13}	p_{14}	p_{23}	p_{34}	p_{42}
Now any line of (A_1, α_2) has coordinates	0	1	λ	0	0	0,
" " (A_3, α_1) " "	0	0	0	1	μ	0,

where λ and μ are variable quantities, which, because the pencils are projective, are connected by an equation of the form

$$A\lambda\mu + B\lambda + C\mu + D = 0 \quad \text{.................(I)};$$

and since $\lambda = 0$, $\mu = 0$, gives a corresponding pair of lines, it follows that $D = 0$; since $\lambda = \infty$, $\mu = \infty$, gives a corresponding pair, $A = 0$, and the relation between λ and μ is

$$B\lambda + C\mu = 0.$$

Any line p_{ik} which intersects a pair of corresponding lines must satisfy the equations

$$p_{42} + \lambda p_{23} = 0, \quad p_{14} + \mu p_{12} = 0 \quad \text{..............(II)};$$

hence eliminating $\frac{\lambda}{\mu}$, we obtain as the equation of the complex

$$Cp_{14}\,p_{23} + Bp_{12}\,p_{42} = 0.$$

Secondly, if the plane of one pencil passes through the centre of the other, and if also to the join of the centres corresponds the intersection of the planes of the pencils, then A_1A_3 corresponds to A_3A_4, and we may take A_1A_4 and A_3A_2 as corresponding lines.

In the foregoing relation (I) between λ and μ, since $\lambda=0$, $\mu=\infty$, gives a pair of corresponding lines and also $\lambda=\infty$, $\mu=0$, it follows that $B=C=0$, and eliminating λ and μ from equations (II) we obtain as the equation of the complex

$$Ap_{14}p_{42}+Dp_{12}p_{23}=0.$$

Thirdly, if the plane of each pencil passes through the centre of the other, we take the pencils to be (A_3, α_1), (A_4, α_2), and to A_3A_4 let A_4A_1 correspond in one pencil and A_3A_2 in the other pencil, then

	p_{12}	p_{13}	p_{14}	p_{23}	p_{34}	p_{42}
any line of (A_4, α_2) has coordinates	0	0	1	0	λ	0,
„ „ (A_3, α_1) „ „	0	0	0	1	μ	0;

and the relation between λ, μ being $A\lambda\mu+B\lambda+C\mu+D=0$, since $\lambda=0$, $\mu=\infty$; $\lambda=\infty$, $\mu=0$, give corresponding pairs of lines, it follows that $B=C=0$. A line p_{ik} which meets a pair of corresponding lines satisfies the equations

$$p_{23}+\lambda p_{12}=0, \quad p_{14}+\mu p_{12}=0;$$

giving as the equation of the complex

$$Ap_{14}p_{23}+Dp_{12}^2=0.$$

Fourthly, if the pencils have the same centre, O, the complex consists of the lines of the planes which pass through every pair of corresponding lines of the two pencils.

The two projective pencils determine upon any plane α which does not pass through O, two projective rows of points whose joins envelope a conic (Introduction, vi), therefore the planes which contain the lines of the complex meet α in the tangents to a conic, *i.e.* they touch a quadric cone and the complex consists of the tangents of this cone.

Fifthly, if the pencils have the same plane, the intersections of corresponding pairs of lines lie upon a conic and the complex consists of the lines which meet this conic.

99. Complexes determined by two bilinear equations. The preceding complexes may also be arrived at by means of two equations which are bilinear in two sets of coordinates. For in the equations

$$\left.\begin{aligned} x_1\Sigma a_{k1}x_k'+x_2\Sigma a_{k2}x_k'+x_3\Sigma a_{k3}x_k'+x_4\Sigma a_{k4}x_k'=0\\ x_1\Sigma b_{k1}x_k'+x_2\Sigma b_{k2}x_k'+x_3\Sigma b_{k3}x_k'+x_4\Sigma b_{k4}x_k'=0\end{aligned}\right\}\ \ldots\ldots \text{(i)},$$

regarding the x_i as the coordinates of a point P in a space Σ, and the x_i' as the coordinates of a point P' in a space Σ'; if P' is given, a line p of Σ is determined, and if P is given a line p' of Σ' is determined. Moreover, regarding the x_i' as parameters, the equations (i) establish a (1, 1) correspondence between the planes of Σ, hence the locus of p is, in general, a tetrahedral complex (Art. 96), similarly for p'.

This may be seen otherwise, as follows:—such values of x_i' as make the two planes of Σ *identical* give a singular plane of the locus of p, *i.e.* a plane all of whose lines belong to the complex; these values of x_i' are given by the equations

$$\Sigma a_{k1} x_k' - \rho \Sigma b_{k1} x_k' = 0;\ \Sigma a_{k2} x_k' - \rho \Sigma b_{k2} x_k',\ \Sigma a_{k3} x_k' - \rho \Sigma b_{k3} x_k',$$
$$\Sigma a_{k4} x_k' - \rho \Sigma b_{k4} x_k' = 0.$$

By elimination of the x_k' we find the quartic equation for ρ

$$\begin{vmatrix} a_{11} - \rho b_{11} & a_{21} - \rho b_{21} & a_{31} - \rho b_{31} & a_{41} - \rho b_{41} \\ a_{12} - \rho b_{12} & a_{22} - \rho b_{22} & a_{32} - \rho b_{32} & a_{42} - \rho b_{42} \\ a_{13} - \rho b_{13} & a_{23} - \rho b_{23} & a_{33} - \rho b_{33} & a_{43} - \rho b_{43} \\ a_{14} - \rho b_{14} & a_{24} - \rho b_{24} & a_{34} - \rho b_{34} & a_{44} - \rho b_{44} \end{vmatrix} = 0 \ldots\ldots \text{(ii)}.$$

If this equation has four different roots, there are four points P' which make the planes of (i) identical; and if Σ' be referred to a tetrahedron whose vertices are these four points, and Σ to a tetrahedron whose sides are the corresponding planes, the equations (i) assume the form

$$x_1 x_1' + x_2 x_2' + x_3 x_3' + x_4 x_4' = 0,$$
$$m_1 x_1 x_1' + m_2 x_2 x_2' + m_3 x_3 x_3' + m_4 x_4 x_4' = 0.$$

If x_i, y_i are two points on the line p determined by x_i', we have on elimination of the x_i'

$$\begin{vmatrix} x_1 & x_2 & x_3 & x_4 \\ y_1 & y_2 & y_3 & y_4 \\ m_1 x_1 & m_2 x_2 & m_3 x_3 & m_4 x_4 \\ m_1 y_1 & m_2 y_2 & m_3 y_3 & m_4 y_4 \end{vmatrix} = 0,$$

or,

$$(m_1 m_2 + m_3 m_4)\, p_{12}\, p_{34} + (m_1 m_3 + m_4 m_2)\, p_{13}\, p_{42} + (m_1 m_4 + m_2 m_3)\, p_{14}\, p_{23} = 0$$

as the locus of p; the locus of p' is clearly of the same form.

100. We obtain the variations from the tetrahedral complex which are the first, second and third of those recently given, when the equation (ii) has a pair of equal roots, three equal roots, or two pairs of equal roots.

For, if (ii) has two different roots, each of them determines a singular plane of the locus of p; take these planes as the coordinate planes α_1 and α_2, for Σ; and take the points P' which respectively correspond to them as the vertices A_1', A_2', of the tetrahedron of reference for Σ'; the equations (i) then assume the form

$$x_1'x_1+x_3'\Sigma a_{3k}x_k+x_4'\Sigma a_{4k}x_k=0,$$
$$x_2'x_2+x_3'\Sigma b_{3k}x_k+x_4'\Sigma b_{4k}x_k=0;$$

and of the values of ρ which identify the planes, one is seen to be zero and the other infinite. The last equations may be written, for convenience, in the form

$$\left.\begin{aligned}x_1'x_1+x_3'x_3+x_4'\xi_4=0\\ x_2'x_2+x_4'x_4+x_3'\xi_3=0\end{aligned}\right\}\quad\text{.........................(iii)},$$

where ξ_3, ξ_4 are linear functions of x_1, x_2, x_3, x_4; which latter are taken as new coordinates of P.

If x_i and y_i are the points P and Q in which the line corresponding to a point P' meets the coordinate planes α_1 and α_2, we have

$$\left.\begin{aligned}x_3'x_3+x_4'\xi_4=0\\ x_4'y_4+x_3'\eta_3=0\end{aligned}\right\}\quad\text{............................(iv)},$$

where η_3 is the result of substituting y_i for x_i in ξ_3.

This shows that the line PQ makes the pencils, whose centres are

$$(\alpha_1,\ \alpha_3,\ \xi_4=0),\quad (\alpha_2,\ \alpha_4,\ \xi_3=0)$$

and planes α_1, α_2 respectively, *projective* to each other.

If now the plane α_2 passes through the centre of the pencil in α_1 we have concurrence of the planes $(\alpha_1, \alpha_2, \alpha_3)$ and $\xi_4=0$, hence

$$\xi_4\equiv u_1x_1+u_2x_2+u_3x_3.$$

Also writing
$$\xi_3\equiv v_1x_1+v_2x_2+v_3x_3+v_4x_4,$$

the solutions of the equation which corresponds to (ii) are in this case zero, infinity, and the values of ρ given by the equations

$$\left.\begin{aligned}x_1'+u_1x_4'&=\rho\, x_3'v_1\\ u_2x_4'&=\rho\,(x_2'+v_2x_3')\\ x_3'+u_3x_4'&=\rho\, x_3'v_3\\ 0&=\rho\,(x_4'+v_4x_3')\end{aligned}\right\}\quad\text{............................(v)}.$$

One of the two values of ρ given by these equations is zero, hence (ii) must have a pair of equal roots.

The remaining value of ρ given by (v) is easily seen to be $\dfrac{1-v_4u_3}{v_3}$; this is infinite, *i.e.* (ii) has two pairs of equal roots, if $v_3=0$; which, with the condition already satisfied, viz. $u_4=0$, makes the plane of each pencil pass through the centre of the other.

If, in the two pencils (iv), the line joining the centres corresponds to the line (α_1, α_2), then, for some value of the ratio x_3'/x_4', the first equation reduces to $u_1x_1+u_2x_2=0$, while the corresponding line of the second pencil passes through $(\alpha_1, \alpha_2, \alpha_3)$; this requires that

$$x_3'+x_4'u_3=0,\quad x_4'+x_3'v_4=0,$$

i.e. $1-v_4u_3=0$; but this is the condition that (iii) may have three roots equal.

The centres of the two pencils will coincide if the six planes

$$x_1=0,\ x_2=0,\ x_3=0,\ x_4=0,\ \xi_3=0,\ \xi_4=0$$

concur, in which case the eight original planes

$$\Sigma a_{ik}x_k=0,\quad \Sigma b_{ik}x_k=0,$$

must concur.

101. It has been seen that there are in general four points P' which make the two planes in equations (i) coincide, giving four different planes: in the following special case there are ∞^1 points P' which make the planes (i) coincide with the *same* plane; for, when the coefficients in (i) are such that two points P' can be found, which identify the two planes (i) with the *same plane* α_1, then any point on the line joining these two points P' will make the two planes (i) identical with α; for take this supposed pair of points P' as the vertices $(0, 0, 1, 0)$, $(0, 0, 0, 1)$ of the tetrahedron of reference for Σ', then the equations

$$x_3'x_1+x_1'x_3+x_2'\xi_4=0,$$
$$x_4'x_1+x_1'x_4+x_2'\xi_3=0,$$

give rise to a connexion of the supposed kind, since it is clear that any point P' on the line $x_1'=0$, $x_2'=0$, makes the preceding planes identical with $x_1=0$.

Every line p meets the conic $x_1=0$, $x_3\xi_3-x_4\xi_4=0$, hence the complex in Σ consists of the lines which meet this conic c^2.

To find the complex formed by the lines p', we may for convenience write

$$\xi_3\equiv x_2,\quad \xi_4\equiv \alpha_1x_1+\alpha_2x_2+\alpha_3x_3+\alpha_4x_4;$$

then the complex is obtained from the equation

$$\begin{vmatrix} x_3'+\alpha_1x_2' & \alpha_2x_2' & x_1'+\alpha_3x_2' & \alpha_4x_2' \\ y_3'+\alpha_1y_2' & \alpha_2y_2' & y_1'+\alpha_3y_2' & \alpha_4y_2' \\ x_4' & x_2' & 0 & x_1' \\ y_4' & y_2' & 0 & y_1' \end{vmatrix}=0;$$

i.e. $p'_{12}\{p'_{31}+\alpha_1p'_{21}+\alpha_3p'_{32}+\alpha_2p'_{14}+\alpha_4p'_{42}\}=0$; and consists, therefore, of two linear complexes.

Hence, to the points of c^2 correspond in Σ' the lines which meet $A_3'A_4'$; to each of the other points of Σ corresponds a line of the complex

$$-\alpha_1p'_{12}+\alpha_2p'_{14}-\alpha_3p'_{23}+\alpha_4p'_{42}-p'_{13}=0.$$

An important special case arises when the planes $x_4 = 0$, $\xi_4 = 0$ are taken to be identical, *i.e.* when $\alpha_1 = \alpha_2 = \alpha_3 = 0$, we may then take $\alpha_4 = -1$; in which case c^2 has as its equations

$$x_1 = 0,\ x_2x_3 + x_4^2 = 0,$$

and the complex in Σ' reduces to $p'_{42} + p'_{13} = 0$; the relationship between Σ and Σ' is then determined by the equations

$$x_3'x_1 + x_1'x_3 - x_2'x_4 = 0,$$
$$x_4'x_1 + x_1'x_4 + x_2'x_2 = 0,$$

and is such that to each point of Σ, except those of

$$x_1 = 0,\ x_2x_3 + x_4^2 = 0,\ (\text{or } c^2),$$

there corresponds a line of the complex $p'_{42} + p'_{13} = 0$; to each point of Σ' corresponds a line which meets c^2.

Finally, writing

$$x_2 = x - yi,\quad x_2' = -x',\quad x_1 = x_1' = 1;$$
$$x_3 = x + yi,\quad x_3' = z',$$
$$x_4 = z,\quad x_4' = y',$$

the spaces Σ and Σ' are each referred to Cartesian coordinates, and the bilinear equations become

$$x + iy + zx' + z' = 0\text{*},$$
$$x'(x - iy) - z - y' = 0;$$

while, with reference to the new coordinates, the Σ complex is that formed by lines which meet the trace on the plane at infinity of $x^2 + y^2 + z^2 = 0$, or the *sphere-circle*; the Σ' complex is that formed by the lines of the complex $p'_{12} = p'_{34}$. So that to each point of Σ' corresponds a line which meets the sphere-circle, *i.e.* a *minimal line*; to each point of Σ corresponds a line of the complex $p'_{12} - p'_{34} = 0$; the only exception being that to the points of the sphere-circle of Σ correspond all the lines of a plane parallel to the plane $x' = 0$.

102. Reye's Complex of Axes. Reye† denotes by an *axis* of a quadric, a line which is perpendicular to its polar line for the quadric; such a line is an axis of a plane section of the quadric; for if p and p' are a pair of such lines, and π' the plane through p which is parallel to p', the pole of the section of the quadric by π' is the point at infinity on p'; hence, since the pole

* See Lie and Scheffers, *Berührungstr.* Bd. I. S. 445.

† See Reye, *Geometrie der Lage*, II.

of p for this section is at infinity in a direction perpendicular to p, it follows that p is an axis of the section.

Taking $\frac{x^2}{a}+\frac{y^2}{b}+\frac{z^2}{c}=1$ as the equation of the quadric, if the direction cosines of p are l, m, n, and any point on it is (xyz), then $l'm'n'$ being the direction cosines of p', we have,

$$ll'+mm'+nn'=0;$$

also the polar planes of (xyz), and of the point at infinity on p, being respectively

$$\frac{x\xi}{a}+\frac{y\eta}{b}+\frac{z\zeta}{c}=1,$$

$$\frac{l\xi}{a}+\frac{m\eta}{b}+\frac{n\zeta}{c}=0;$$

therefore $l':m':n'=a(yn-zm):b(zl-xn):c(xm-yl)$,

hence $al(yn-zm)+bm(zl-xn)+cn(xm-yl)=0;$

while for a tetrahedron of reference of which one face is the plane at infinity we have $l=p_{14}$ &c., $yn-zm=p_{23}$ &c.; hence the axes form the complex

$$ap_{14}p_{23}+bp_{13}p_{42}+cp_{12}p_{34}=0.$$

To this complex also belong the normals of the quadrics confocal to the given quadric, and of those similar, similarly situated and concentric to it; for if the line through the point (xyz) having l, m, n for its direction cosines, is normal at (xyz) to the quadric $\frac{x^2}{a}+\frac{y^2}{b}+\frac{z^2}{c}=1$, we must have

$$\frac{\frac{a}{x}}{l}=\frac{\frac{b}{y}}{m}=\frac{\frac{c}{z}}{n},$$

whence $a\left(\frac{y}{m}-\frac{z}{n}\right)+b\left(\frac{z}{n}-\frac{x}{l}\right)+c\left(\frac{x}{l}-\frac{y}{m}\right)=0.$

The quadric being one of a series of similar, similarly situated and concentric quadrics, the ∞^3 normals of the quadrics form a complex, which, from the last equation, is seen to be tetrahedral. Again if $a=A^2+\lambda$, $b=B^2+\lambda$, $c=C^2+\lambda$, *i.e.* if the quadric is one of a series of *confocal* quadrics, the normals again form a complex which is seen to be the *same* tetrahedral complex*.

* From this property the complex is sometimes called the Normal Complex. We easily find that any line of this complex meets the quadric $\frac{x^2}{a}+\frac{y^2}{b}+\frac{z^2}{c}=1$ in points at which the normals to this quadric intersect each other.

103. Differential Equation of the Complex. The tetrahedral complex has been seen to possess an equation of the form

$$ap_{14}p_{23} + bp_{13}p_{42} + cp_{12}p_{34} = 0.$$

Any projective transformation will change the given tetrahedral complex T^2 into another such complex; since, by a projective transformation, the double ratio of four points on a line is equal to the double ratio of the four corresponding points on the corresponding line. A projective transformation which leaves the fundamental tetrahedron unaltered, will interchange the lines of T^2, but leave the complex as a whole unaltered; such a transformation is given by equations of the form $\mu x_i' = a_i x_i$, and of such transformations there are ∞^3, which may also be observed from the fact, that when the united points of the collineation are given, the projection (or collineation) is determined (Introduction, xiv) by connecting a given point P with any point of space Q, and Q may be chosen in ∞^3 ways.

If T^2 be projectively transformed into a complex which has the plane at infinity as one face of its fundamental tetrahedron, since we have now $p_{12} = xdy - ydx, \ldots p_{41} = dx, \ldots$ (Art. 6), the equation of the new complex is

$$adx\,(ydz - zdy) + bdy\,(zdx - xdz) + cdz\,(xdy - ydx) = 0.$$

The complex represented by this equation, (that of the last Article), is therefore the projection of any T^2, one of whose double ratios is $\dfrac{a-b}{c-b}$; its complex curves are parabolas (since they touch the plane at infinity). If P, Q, R be the points in which any complex line meets the coordinate planes of x, y, z respectively, since $(PQR\infty) = \text{constant}$, we have $\dfrac{PQ}{QR} = \text{constant}$.

104. The line element. If with any given point of space a definite direction be associated, we obtain the idea, due to Sophus Lie, of a *line element.* Connected with any point there are ∞^2 line elements (corresponding to the different directions through the point), and in space there are altogether ∞^5 line elements.

By any differential equation of Monge, of the form

$$f(x, y, z, dx, dy, dz) = 0,$$

homogeneous in dx, dy, dz, ∞^4 line elements are selected from the

∞^5 line elements of space. It was seen (Art. 6), that an equation of the form

$$f(ydz - zdy,\ zdx - xdz,\ xdy - ydx,\ dx,\ dy,\ dz) = 0$$

represents a line complex; a line element

$$(x, y, z;\ dx : dy : dz)$$

which satisfies the equation $f = 0$, will be said to *belong* to the complex represented by $f = 0$. In Chapter XVIII. Lie's investigations of this differential equation will be considered.

105. Curves of the Tetrahedral Complex. Any point P of a curve and the tangent p at P determine a line element *of the curve*, the number of such line elements being ∞^1. A *curve of a complex* is one whose line elements belong to the complex. Along any given curve of a tetrahedral complex T^2 the coordinates of its points are functions of a single variable t, and this parameter t may clearly be chosen so that

$$\frac{dx}{x} : \frac{dy}{y} = \frac{1}{a+t} : \frac{1}{b+t}\ *,$$

and since the equation of T^2 may be written

$$(b-c)\frac{dy}{y} \cdot \frac{dz}{z} + (c-a)\frac{dz}{z} \cdot \frac{dx}{x} + (a-b)\frac{dx}{x} \cdot \frac{dy}{y} = 0;$$

it is easily seen that for a curve of T^2 we have

$$\frac{dx}{x} : \frac{dy}{y} : \frac{dz}{z} = \frac{1}{a+t} : \frac{1}{b+t} : \frac{1}{c+t},$$

whence the curves of T^2 are seen to be

$$\log x = \int \frac{F(t)}{a+t}\,dt + \lambda, \quad \log y = \int \frac{F(t)}{b+t}\,dt + \mu, \quad \log z = \int \frac{F(t)}{c+t}\,dt + \nu,$$

the form of F being arbitrary and λ, μ, ν being arbitrary constants.

The following cases are of special interest:

(i) $F(t) = 1,\ x = \lambda(a+t),\ y = \mu(b+t),\ z = \nu(c+t)$;

this gives the complex lines;

(ii) $F(t) = 2,\ x = \lambda(a+t)^2,\ y = \mu(b+t)^2,\ z = \nu(c+t)^2$;

this gives the (parabolic) complex conics;

(iii) $F(t) = -1,\ x = \dfrac{\lambda}{a+t},\ y = \dfrac{\mu}{b+t},\ z = \dfrac{\nu}{c+t}$;

* See Lie, *Berührungstr.* S. 327. An exception occurs if $\frac{dx}{x} : \frac{dy}{y} : \frac{dz}{z} = \frac{1}{\alpha} : \frac{1}{\beta} : \frac{1}{\gamma}$ where α, β and γ are constants. In that case $x^\alpha : y^\beta : z^\gamma = A : B : C$.

this gives the ∞^3 twisted cubics through the vertices of the fundamental tetrahedron the chords of which form T^2.

106. Non-Projective Transformations of the Complex. Any transformation of the form $x_1 = \lambda x^m$, $y_1 = \mu y^m$, $z_1 = \nu z^m$ applied to a line element of T^2 transforms it into a line element which also belongs to T^2, since here

$$m.\frac{dx}{x} = \frac{dx_1}{x_1}, \quad m.\frac{dy}{y} = \frac{dy_1}{y_1}, \quad m.\frac{dz}{z} = \frac{dz_1}{z_1},$$

and substituting in the differential equation of the complex the form of the equation is not changed. Complex curves are therefore by this transformation changed into complex curves; the cases $m = 2$, $m = -1$ are of special importance. For $m = 2$, any *line*

$$Ax_1 + By_1 + Cz_1 + D = 0, \quad A'x_1 + B'y_1 + C'z_1 + D' = 0,$$

becomes, for $\lambda = \mu = \nu = 1$, the twisted quartic

$$Ax^2 + By^2 + Cz^2 + D = 0, \quad A'x^2 + B'y^2 + C'z^2 + D' = 0.$$

Hence the line elements of this twisted quartic belong to the same T^2, and by projection it follows that any tangent to the curve of intersection of two quadrics meets their common self-conjugate tetrahedron in four points of constant double ratio.

To a line in the space (xyz), *e.g.*,

$$x = \alpha + lr, \quad y = \beta + mr, \quad z = \gamma + nr,$$

there corresponds

$$x_1 = \lambda(\alpha + lr)^2, \quad y_1 = \mu(\beta + mr)^2, \quad z_1 = \nu(\gamma + nr)^2;$$

i.e., a complex conic.

For $m = -1$ the transformation is Involutory; the complex line

$$x = \alpha + lr, \quad y = \beta + mr, \quad z = \gamma + nr$$

becomes the twisted cubic

$$x_1 = \frac{\lambda}{\alpha + lr}, \quad y_1 = \frac{\mu}{\beta + mr}, \quad z_1 = \frac{\nu}{\gamma + nr}.$$

The proposition already established (Art. 95), that the chords of these ∞^3 cubics form T^2, is shown by Lie in the following manner:—the transformation $x_1 = \frac{\lambda}{x}$, $y_1 = \frac{\mu}{y}$, $z_1 = \frac{\nu}{z}$ is such that by suitably choosing λ, μ and ν any two points of space may be interchanged, *i.e.* so that to P and Q of the space (xyz) there correspond Q and P of the space $(x_1y_1z_1)$; now take any two points P and Q on one of these twisted cubics, then to the cubic will correspond a *complex line* which passes through P and Q, *i.e.* the chord PQ is a complex line.

107. The Special Quadratic Complex. One species of quadratic complex consists of the tangents to a quadric surface. This complex is said to be *special*. On each line of the complex there is one point for which the complex cone becomes two (coincident) planes through it, viz., its point of contact with the quadric. Hence every line of the complex is singular, so that for every line of the complex $F(x)=0$ we must have

$$\Sigma\left(\frac{\partial F}{\partial x_i}\right)^2 \equiv \alpha F(x)+\beta\Sigma x_i^2,$$

where α and β are constants.

If in place of $F(x)=0$, the equation

$$f(x)\equiv F(x)-\frac{\alpha}{8}\Sigma(x^2)=0,$$

which represents the same complex, be taken, we have

$$\Sigma\left(\frac{\partial f}{\partial x_i}\right)^2 \equiv \Sigma\left(\frac{\partial F}{\partial x_i}\right)^2-\frac{\alpha}{2}\Sigma x_i\frac{\partial F}{\partial x_i}+\frac{\alpha^2}{16}\Sigma x_i^2$$

$$\equiv \alpha F(x)+\beta\Sigma x_i^2-\alpha F(x)+\frac{\alpha^2}{16}\Sigma x_i^2$$

$$\equiv\left(\beta+\frac{\alpha^2}{16}\right)\Sigma x_i^2.$$

Thus the equation of a special quadratic complex being given in the form $F(x)=0$, by the addition of a determinate multiple of (x^2), this equation may be replaced by $f(x)=0$, where

$$\Sigma\left(\frac{\partial f}{\partial x}\right)^2\equiv 4\Delta\,.\,(x^2).$$

The equation of the complex being expressed in the form

$$f(x)\equiv a_{11}x_1^2+\ldots+2a_{ik}x_ix_k+\ldots=0,$$

the result last obtained requires that

$$\sum_i a_{ik}^2=\Delta,\quad \sum_i a_{ij}a_{ik}=0,$$

Δ being the same for all values of k. It follows also that if the quantities x_i are the coordinates of a line so also are $\frac{\partial f}{\partial x_i}$, moreover *this line is the polar of* x *with regard to the quadric*; for the polar line of x with regard to the complex is

$$\lambda x_i+\mu\frac{\partial f}{\partial x_i},\quad \text{whence } \lambda\,.\,\mu=0,\quad \text{and therefore } \lambda=0,$$

(since a line coincides with its polar only when it is a complex line), and this polar line, being the locus of poles for x of the sections of the quadric through x, is therefore the polar line of x for the quadric.

Taking as the equation of the quadric

$$a_1\xi_1^2 + a_2\xi_2^2 + a_3\xi_3^2 + a_4\xi_4^2 = 0,$$

its complex equation is (Art. 46),

$$\Psi \equiv a_1a_2p_{12}^2 + a_3a_4p_{34}^2 + a_1a_3p_{13}^2 + a_4a_2p_{42}^2 + a_1a_4p_{14}^2 + a_2a_3p_{23}^2 = 0.$$

On substituting for p_{12}, as usual, $\frac{1}{2}(x_1 + ix_2)$, &c., and multiplying by two, we obtain for $f(x)$ the form

$$f \equiv \tfrac{1}{2}\{a_1a_2\overline{x_1 + ix_2}^2 + a_3a_4\overline{x_1 - ix_2}^2 + \ldots\ldots\} = 0,$$

which gives $$\Sigma\left(\frac{\partial f}{\partial x}\right)^2 \equiv 4a_1a_2a_3a_4 \,.\, (x^2);$$

showing that Δ is equal to the discriminant of the given quadric.

The determinant of the coefficients of f has the value $-\Delta^3$; for from the equation connecting the coefficients of f we see that $|a_{ik}|^2 = \Delta^6$, and taking a_1, a_2, a_3, a_4 as being each equal to unity

$$f \equiv x_1^2 + x_3^2 + x_5^2 - x_2^2 - x_4^2 - x_6^2,$$

whence it is seen that in this case $|a_{ik}| = -1$, therefore generally

$$|a_{ik}| = -\Delta^3.$$

108. System of two special complexes*. If we have any second quadric, its complex equation may, as has been observed, be brought to the form $\phi = 0$, where

$$\Sigma\left(\frac{\partial \phi}{\partial x_i}\right)^2 \equiv 4\Delta'(x^2).$$

The discriminant $\chi(\rho)$ of $f - \rho\,.\,\phi$ *has only reciprocal roots*; for writing

$$B_{ik} = \Sigma_r b_{ir}\,.\,a_{kr}, \quad (r = 1, \ldots, 6),$$

we have

$-\Delta^3\,.\,\chi = |a_{ik}|\,.\,\chi = |\Delta - \rho\,.\,B_{ik}|$, Δ appearing only in the principal diagonal,

$-\Delta'^3\,.\,\chi = |b_{ik}|\,.\,\chi = |B_{ik} - \rho\,.\,\Delta'|$, Δ' " " " " ;

showing that if r is a root of $\chi = 0$, so also is r', where $r' = \frac{1}{r}\,.\,\frac{\Delta}{\Delta'}$; thus we have three pairs of roots r, r', each pair being connected by the equation $rr' = \frac{\Delta}{\Delta'}$.

If the quadrics have as equations in point-coordinates

$$U \equiv a_1\xi_1^2 + a_2\xi_2^2 + a_3\xi_3^2 + a_4\xi_4^2 = 0,$$
$$V \equiv b_1\xi_1^2 + b_2\xi_2^2 + b_3\xi_3^2 + b_4\xi_4^2 = 0,$$

* The following is an account of an investigation by Voss, "Die Liniengeometrie in ihrer Anwendung auf die Flächen zweiten Grades," *Math. Ann.* x.

then f and ϕ the complex equations of U and V expressed in Plücker coordinates are,

$$f \equiv a_1 a_2 p_{12}^2 + \ldots + a_2 a_3 p_{23}^2 = 0,$$
$$\phi \equiv b_1 b_2 p_{12}^2 + \ldots + b_2 b_3 p_{23}^2 = 0.$$

If now we make the substitutions*

$$\sqrt{a_1 a_2} \cdot p_{12} = \sigma \cdot p'_{12}, \quad \sqrt{a_1 a_3} \cdot p_{13} = \sigma \cdot p'_{13}, \quad \sqrt{a_1 a_4} \cdot p_{14} = \sigma \cdot p'_{14},$$
$$\sqrt{a_3 a_4} \cdot p_{34} = \sigma \cdot p'_{34}, \quad \sqrt{a_2 a_4} \cdot p_{42} = \sigma \cdot p'_{42}, \quad \sqrt{a_2 a_3} \cdot p_{23} = \sigma \cdot p'_{23};$$

we obtain
$$f \equiv p'^2_{12} + p'^2_{34} + \ldots,$$
$$\phi \equiv \frac{p'^2_{12}}{r_1} + \frac{p'^2_{34}}{r_1'} + \ldots,$$

where
$$r_1 = \frac{a_1 a_2}{b_1 b_2}, \quad r_1' = \frac{a_3 a_4}{b_3 b_4}, \quad r_1 r_1' = \frac{\Delta}{\Delta'}, \text{ \&c.};$$

and the discriminant equation $\chi(\rho) = 0$ becomes

$$\left(1 - \frac{\rho}{r_1}\right)\left(1 - \frac{\rho}{r_1'}\right) \ldots = 0;$$

i.e. r_1, r_1' &c. are the three pairs of reciprocal roots of $\chi(\rho) = 0$.

Returning to the general case for f and ϕ, expressing that $\chi(\rho) = 0$ is a reciprocal equation, we obtain

$$\chi(\rho) \equiv \Delta^3 - \rho A_1 \Delta^2 + \rho^2 A_2 \Delta - \rho^3 A_3 + \rho^4 A_2 \Delta' - \rho^5 A_1 \Delta'^2 + \rho^6 \Delta'^3 = 0 \quad \ldots\ldots(\text{I});$$

where A_1, A_2, A_3 are mutual Invariants of f and ϕ.

We shall now show the connexion of these Invariants with the Invariants θ, θ', Δ, Δ' of the two quadrics expressed in point-coordinates. For this purpose introduce in (I) the quantity z defined by the equation $z = \rho + \frac{1}{\rho}\frac{\Delta}{\Delta'}$; (I) then becomes

$$\Delta'^3 z^3 - A_1 \Delta'^2 z^2 + z(\Delta' A_2 - 3\Delta'^2 \Delta) - A_3 + 2A_1 \Delta \Delta' = 0 \quad \ldots\ldots\ldots\ldots(\text{II}).$$

The roots of the last equation are seen to be $\frac{a_1 a_2}{b_1 b_2} + \frac{a_3 a_4}{b_3 b_4}$ &c., and are therefore the quantities

$$\lambda_1 \lambda_2 + \lambda_3 \lambda_4, \quad \lambda_1 \lambda_3 + \lambda_2 \lambda_4, \quad \lambda_1 \lambda_4 + \lambda_2 \lambda_3,$$

where the λ_i are the roots of the discriminant of $U + \lambda V$, *i.e.* of the equation

$$\Delta' \lambda^4 + \theta' \lambda^3 + \Phi \lambda^2 + \theta \lambda + \Delta = 0 \quad \ldots\ldots\ldots\ldots\ldots\ldots\ldots\ldots(\text{III}).$$

But the equation whose roots are the above functions of the roots of the equation (III) is known to be

$$\Delta'^3 z^3 - \Delta'^2 \Phi z^2 + z(\Delta' \theta \theta' - 4\Delta \Delta'^2) - (\Delta' \theta^2 - 4\Delta \Delta' \Phi + \Delta \theta'^2) = 0 \quad \ldots\ldots\ldots(\text{IV}).$$

By comparison of (II) and (IV) we obtain

$$A_1 = \Phi,$$
$$A_2 = \theta\theta' - \Delta\Delta',$$
$$A_3 = \theta'^2 \Delta - 2\Phi \Delta \Delta' + \theta^2 \Delta'.$$

109. Covariant tetrahedral complex. The complex

$$\Sigma \frac{\partial f}{\partial x_i} \frac{\partial \phi}{\partial x_i} = 0$$

* This is merely a change to a new system p'_{ik} of line-coordinates, (Art. 81).

expresses the condition that the polars of a line x with regard to U and to V should intersect. It is thus a covariant complex of f and ϕ. If any line be taken in a face of the common self-conjugate tetrahedron of U and V, its polar lines pass through the opposite vertex, and therefore necessarily intersect. Hence any line in any of the four faces of this tetrahedron belongs to the complex, which is therefore tetrahedral. To it there belong the tangents to the curve of intersection of the two quadrics.

Two other covariant complexes of the system are

$$\Sigma b_{ik} \frac{\partial f}{\partial x_i} \frac{\partial f}{\partial x_k} = 0, \quad \Sigma a_{ik} \frac{\partial \phi}{\partial x_i} \frac{\partial \phi}{\partial x_k} = 0;$$

which respectively express that the polar of x with regard to U should touch V, and that the polar of x with regard to V should touch U.

110. The Complex of Battaglini or Harmonic Complex*. The locus of lines which meet two quadrics in points forming a harmonic range has been already investigated (Art. 89); it was seen that if the quadrics are $U = 0$, $V = 0$, where

$$U \equiv \Sigma a_{ik} \xi_i \xi_k, \quad V \equiv \Sigma b_{ik} \xi_i \xi_k,$$

the complex equation of U being $\Psi = 0$, then the required complex has as its equation $\Sigma b_{ik} \frac{\partial}{\partial a_{ik}} \Psi = 0$. This result may be obtained otherwise in the following manner.

If two conics have as equations in point-coordinates $S = 0$, $S' = 0$, and in line-coordinates $\Sigma = 0$, $\Sigma' = 0$, the tangential equation of any one of the pencil of conics $S + kS' = 0$ is $\Sigma + k\Phi + k^2\Sigma' = 0$. Each common tangent of $S + kS' = 0$, and $S - kS' = 0$, satisfies the equation $\Phi = 0$, which is also the condition that a line should be cut harmonically by S and S'†.

If U and V are two quadrics whose complex equations are respectively $\Psi = 0$, $\Psi' = 0$, the complex equation of $U + kV = 0$ is

$$\Psi + k\Psi_1 + k^2\Psi' = 0;$$

this equation shows that there are two quadrics of the *pencil* of quadrics (U, V) (*i.e.* those which pass through the curve of inter-

* For an interesting investigation of the characteristics of this complex when the quadrics have special relative positions, see the memoir, *Math. Ann.* Bd. XXIII. by Segre and Loria, "Sur les différentes espèces de complexes du 2e degré qui coupent harmoniquement deux surfaces du second ordre."

† Salmon's *Conic Sections*, 6th edition, pp. 307, 344.

section of U and V) which touch any given line. Any line of the complex $\Psi_1=0$ touches both $U+kV=0$ and $U-kV=0$; but a line which touches the last two quadrics *cuts U and V in four harmonic points*; for if any plane through such a line cuts the quadrics U, V in the conics whose equations, referred to a triangle of reference in their plane, are $S=0$, $S'=0$, this line will touch $S+kS'=0$, $S-kS'=0$, and hence cuts S and S' harmonically, *i.e.* it cuts U and V harmonically. It follows that $\Psi_1=0$ is the complex equation of lines which meet U and V in four harmonic points.

Taking as the equations of U and V

$$\Sigma a_{ik}\xi_i\xi_k=0, \quad \Sigma b_{ik}\xi_i\xi_k=0,$$

it is seen that
$$\Psi_1 \equiv \Sigma b_{ik}\frac{\partial}{\partial a_{ik}}\Psi=0.$$

If the quadrics are referred to their common self-conjugate tetrahedron, their equations are

$$\Sigma a_i\xi_i^2=0, \quad \Sigma b_i\xi_i^2=0;$$

hence, since
$$\Psi=\Sigma a_ia_k\,p_{ik}^2, \quad \Psi'=\Sigma b_ib_k\,p_{ik}^2,$$

therefore
$$\Psi_1\equiv\Sigma\,(a_ib_k+a_kb_i)\,p_{ik}^2.$$

By the process explained in Art. 81, the equation of this complex, which will be denoted by H^2, may be brought into either of the forms

$$a\,(p^2_{12}+p^2_{34})+b\,(p^2_{13}+p^2_{42})+c\,(p^2_{14}+p^2_{23})=0,$$

or
$$a\,(\pi^2_{12}+\pi^2_{34})+b\,(\pi^2_{13}+\pi^2_{42})+c\,(\pi^2_{14}+\pi^2_{23})=0.$$

This differs from the canonical form in Plücker coordinates of the general complex by the absence of product terms, the equation of the latter complex being

$$(\lambda_1-\lambda_2)\,(p^2_{12}+p^2_{34})+(\lambda_3-\lambda_4)\,(p^2_{13}+p^2_{42})+(\lambda_5-\lambda_6)\,(p^2_{14}+p^2_{23})$$
$$+\,2\,(\lambda_1+\lambda_2)\,p_{12}p_{34}+2\,(\lambda_3+\lambda_4)\,p_{13}p_{42}+2\,(\lambda_5+\lambda_6)\,p_{14}p_{23}=0.$$

The conditions therefore for a Harmonic complex are

$$\lambda_1+\lambda_2=\lambda_3+\lambda_4=\lambda_5+\lambda_6.$$

If the equation of the complex be brought to its normal form (Art. 87), $\Sigma\lambda_i=0$, and these conditions may be stated as follows: "the sums of two pairs of the quantities λ should be separately zero." It is easily seen that these conditions are satisfied for the complex $f(x)=0$ in its normal form when $A_3=0$, $A_5=0$, where A_3 and A_5

are the coefficients of μ^3 and μ respectively in the discriminant of $f(x) - \mu(x^2) = 0$.

111. The Tetrahedroid. The singular surface of the Harmonic complex is derived from the general case by the conditions $a_2 = b_2 = c_2$ which make $E = 0$ (Art. 82), and give as its equation

$$A\Sigma y^4 + 2B(y_1^2 y_2^2 + y_3^2 y_4^2) + 2C(y_1^2 y_3^2 + y_2^2 y_4^2) + 2D(y_1^2 y_4^2 + y_2^2 y_3^2) = 0;$$

with the condition

$$A(A^2 - B^2 - C^2 - D^2) + 2BCD = 0.$$

This surface is called the Tetrahedroid. On inserting the values of A, B, C and D given in Art. 82, it is seen that the equation of the surface may be written

$$y_1^2\left(y_1^2 + \frac{b^2 + c^2}{bc} y_2^2 + \frac{c^2 + a^2}{ca} y_3^2 + \frac{a^2 + b^2}{ab} y_4^2\right) + (ay_2^2 + by_3^2 + cy_4^2)\left(\frac{y_2^2}{a} + \frac{y_3^2}{b} + \frac{y_4^2}{c}\right) = 0.$$

The form of the equation shows that the section of the surface by the plane α_1 of the tetrahedron of reference consists of the sections by α_1 of C and C' respectively, where

$$C \equiv ay_2^2 + by_3^2 + cy_4^2 = 0$$

is the complex cone of A_1, and

$$C' \equiv \frac{y_2^2}{a} + \frac{y_3^2}{b} + \frac{y_4^2}{c} = 0$$

is the cone through A_1 and the complex conic of α_1; for the equation of this conic in plane-coordinates is seen to be

$$au_2^2 + bu_3^2 + cu_4^2 = 0,$$

and therefore in point-coordinates is

$$\frac{y_2^2}{a} + \frac{y_3^2}{b} + \frac{y_4^2}{c} = 0.$$

Writing

$$T \equiv y_1^2 + \left(\frac{b}{c} + \frac{c}{b}\right) y_2^2 + \left(\frac{c}{a} + \frac{a}{c}\right) y_3^2 + \left(\frac{a}{b} + \frac{b}{a}\right) y_4^2,$$

the equation of the singular surface Φ becomes

$$CC' + y_1^2 T = 0.$$

This form of the equation shows that Φ contains two families of ∞^1 twisted quartics, viz.,

$$C + \lambda y_1^2 = 0, \quad T - \lambda C' = 0;$$

and

$$C' + \lambda' y_1^2 = 0, \quad T - \lambda' C = 0.$$

Through any point of Φ there passes one curve of each family. Any curve τ of one species and any curve τ' of the other species lie on a quadric, since

$$C+\lambda y_1^2-\frac{1}{\lambda'}(T-\lambda C')\equiv\frac{\lambda}{\lambda'}(C'+\lambda' y_1^2)-\frac{1}{\lambda'}(T-\lambda' C).$$

There are two harmonic complexes which have a given tetrahedroid as singular surface; for when

$$\frac{b}{c}+\frac{c}{b},\ \frac{c}{a}+\frac{a}{c},\ \frac{a}{b}+\frac{b}{a}$$

are each given there are two sets of values for $a:b:c$; if one set is $\alpha:\beta:\gamma$, the other is

$$\frac{1}{\alpha}:\frac{1}{\beta}:\frac{1}{\gamma};$$

so that for the two complexes T is the same, while C and C' are interchanged.

By reference to the equations of Art. 82, in which E is to be taken as zero, it is seen that the singular points of the surface lie by fours in the coordinate planes; the singular tangent planes of the surface pass by fours through the vertices of the tetrahedron of reference; thus the coordinates of the singular points β in the plane α_1 are given by the equations

$$\beta_2^2:\beta_3^2:\beta_4^2=a(c^2-b^2):b(a^2-c^2):c(b^2-a^2),$$

and the coordinates u of the four singular tangent planes through A_1 are given by the same proportion. These four singular points are seen to be the intersections of the conics (α_1, C), (α_1, C').

112. There are ∞^1 pairs of quadrics such that for each pair the Harmonic complex is the same as the given complex. In the first place it is easy to see that the Harmonic complex for the quadrics

$$U+\sigma V,\ U-\sigma V \text{ is } \Psi-\sigma^2\Psi'=0,$$

where Ψ and Ψ' are the complex equations of U and V respectively; for if

$$U\equiv l_1y_1^2+l_2y_2^2+l_3y_3^2+l_4y_4^2,\ V\equiv m_1y_1^2+m_2y_2^2+m_3y_3^2+m_4y_4^2,$$

the complex Ψ_1 for $U+\sigma V$, $U-\sigma V$ is

$$\Sigma\,(\overline{l_i-\sigma m_i}\cdot\overline{l_k+\sigma m_k}+\overline{l_i+\sigma m_i}\cdot\overline{l_k-\sigma m_k})\,p_{ik}^2\equiv 2\Psi-2\sigma^2\Psi'.$$

Again denoting by F the quadric $C+\lambda y_1^2=0$, and by K the cone

$$C+\lambda y_1^2-\lambda(T-\lambda C')=0,$$

we find $$K \equiv \rho \left(\frac{y_2^2}{\lambda a - bc} + \frac{y_3^2}{\lambda b - ca} + \frac{y_4^2}{\lambda c - ab}\right)^*,$$

$$\rho = \frac{(\lambda a - bc)(\lambda b - ca)(\lambda c - ab)}{abc}.$$

From the preceding result, the Harmonic complex of

$$F + \sigma K,\ F - \sigma K,\ \text{is}\ \Phi - \sigma^2\rho^2 K_1 = 0;$$

where

$$\Phi \equiv \lambda a p^2_{12} + \lambda b p^2_{13} + \lambda c p^2_{14} + bc p^2_{34} + ca p^2_{42} + ab p^2_{23},$$

$$K_1 \equiv \frac{p^2_{23}}{(\lambda a - bc)(\lambda b - ca)} + \frac{p^2_{34}}{(\lambda b - ca)(\lambda c - ab)} + \frac{p^2_{42}}{(\lambda c - ab)(\lambda a - bc)},$$

hence

$$\Phi - \sigma^2\rho^2 K_1 \equiv \lambda a p^2_{12} + \lambda b p^2_{13} + \lambda c p^2_{14} + bc p^2_{34} + ca p^2_{42} + ab p^2_{23}$$

$$- \sigma^2 \frac{(\lambda a - bc)(\lambda b - ca)(\lambda c - ab)}{(abc)^2} (\overline{\lambda c - ab}\, p^2_{23} + \overline{\lambda b - ca}\, p^2_{42}$$

$$+ \overline{\lambda a - bc}\, p^2_{34}) \equiv \lambda . H^2,\ \text{if}\ \sigma^2 = \frac{-(abc)^2}{(\lambda a - bc)(\lambda b - ca)(\lambda c - ab)}.$$

Any pair $F - \sigma K$, $F + \sigma K$ intersect on a curve τ which lies on Φ.

If a line of the complex H^2 meets τ in P it must meet one of these two quadrics in a point consecutive to P, *i.e.* it must lie in the tangent plane at P to *one* of the pair; and the tangent planes to the quadrics at P form the two pencils of complex lines through P.

In like manner, by interchanging C and C', we obtain ∞^1 pairs of quadrics such that for each pair the harmonic complex is

$$H'^2 \equiv \frac{1}{a}(p^2_{12} + p^2_{34}) + \frac{1}{b}(p^2_{13} + p^2_{42}) + \frac{1}{c}(p^2_{14} + p^2_{23}) = 0.$$

The given Harmonic complex is also *the locus of lines through which four harmonic tangent planes can be drawn to any one pair of* ∞^1 *pairs of quadrics.*

For, starting with the complex equation

$$a(\pi^2_{12} + \pi^2_{34}) + b(\pi^2_{13} + \pi^2_{42}) + c(\pi^2_{14} + \pi^2_{23}) = 0,$$

and repeating the previous analysis, using *plane* instead of *point* coordinates, we arrive at the result just stated.

Let $F_1 + \mu L_1$, $F_1 - \mu L_1$ be a pair of these quadrics, then any complex line which lies on a common tangent plane π of the quadrics must lie in two consecutive tangent planes of one of

* From the values of the coordinates recently obtained for the four singular tangent planes through A_1, it is seen that K touches each of these planes.

them, *i.e.* must pass through one of the points of contact P of π with the quadrics; hence π is a singular plane of the complex, the centres of the pencils in it being the points of contact of π with the quadrics. Now through P two of the preceding quadrics $F+\sigma K$, $F-\sigma K$ will pass, and it was seen that π was the tangent plane to *one* of them at P, say to $F+\sigma K$, hence $F+\sigma K$, $F_1+\mu L_1$, being referred to a common self-conjugate tetrahedron and having a common tangent plane at a common point must *coincide**. Thus the second set of quadrics is the same as the first set but is *differently paired.*

113. Painvin's Complex. If in the quadrics, referred to plane coordinates, $\Sigma a_i u_i^2 = 0$, $\Sigma b_i u_i^2 = 0$,

we make $a_1 = a_2 = a_3 = 1$, $a_4 = 0$;

$$b_1 = a^2,\quad b_2 = b^2,\quad b_3 = c^2,\quad b_4 = -1,$$

and suppose that $u_4 = 1$, the harmonic complex degenerates into the locus of intersection of pairs of perpendicular tangent planes to an ellipsoid.

The equation of the harmonic complex

$$\Sigma \pi^2_{ik}(a_i b_k + a_k b_i) = 0$$

becomes in this case

$$Ap^2_{14} + Bp^2_{42} + Cp^2_{34} - p^2_{12} - p^2_{13} - p^2_{23} = 0,$$

if $A = b^2 + c^2$, $B = c^2 + a^2$, $C = a^2 + b^2$.

Hence, by Art. 82, or directly, by expressing that the section of the complex cone of the point (x, y, z) by a coordinate plane breaks up into two lines, we obtain as the equation of the singular surface

$$(x^2+y^2+z^2)(Ax^2+By^2+Cz^2) - \{A(B+C)x^2 + B(C+A)y^2 + C(A+B)z^2\} + ABC = 0,$$

which is seen to be the Wave Surface for the ellipsoid

$$\frac{x^2}{A} + \frac{y^2}{B} + \frac{z^2}{C} = 1.$$

The singular lines of a harmonic complex $\Sigma a_{ik} p^2_{ik} = 0$ belong to the tetrahedral complex

$$a_{12}a_{34}p_{12}p_{34} + a_{13}a_{42}p_{13}p_{42} + a_{14}a_{23}p_{14}p_{23} = 0.$$

In the case of Painvin's complex the last equation takes the form

$$(a^2+b^2)\,p_{12}p_{43} + (c^2+a^2)\,p_{13}p_{42} + (b^2+c^2)\,p_{14}p_{23} = 0,$$

which may be written in the form

$$c^2 p_{12}p_{43} + b^2 p_{13}p_{42} + a^2 p_{14}p_{23} = 0,$$

and is therefore the complex of normals for the given ellipsoid.

* See Sturm, *Lin. Geom.* Bd. III. S. 344.

CHAPTER VIII.

THE COSINGULAR COMPLEXES.

114. It has been seen that there is a singly infinite set of quadratic complexes which have a given Kummer surface as singular surface (Art. 84). This will now be shown independently, as follows:—using the coordinates of Klein, let $\Sigma\lambda_i x_i^2 = 0$ be any given quadratic complex C^2, *the complex*

$$\Sigma \frac{y_i^2}{\lambda_i + \mu} = 0,$$

where μ *has any definite value, has the same singular surface as* C^2.

For, denoting this complex by C_μ^2, if y_i is a singular line of C_μ^2, then $\dfrac{y_i}{\lambda_i + \mu}$ is a line, (Art. 76), and intersects y_i in a point of the singular surface of C_μ^2, the pencil of tangents thereat being

$$\left(y_i, \frac{y_i}{\lambda_i + \mu}\right).$$

Now if we write

$$x_i = \frac{y_i}{\lambda_i + \mu},$$

it follows that $\qquad \Sigma\lambda_i x_i^2 = 0, \quad \Sigma\lambda_i^2 x_i^2 = 0,$

hence x is a singular line of C^2 (Art. 76), and touches the singular surface of C^2 at a point for which the pencil of tangents is $(\overline{\lambda_i + \mu}\, x_i, x_i)$; but this is identical with the pencil $\left(y_i, \dfrac{y_i}{\lambda_i + \mu}\right)$; hence, the singular surfaces of C_μ^2 and C^2, having the same pencils of tangent lines, must be *identical*. The complexes C_μ^2 are said to be *cosingular*.

115. It appears that if y is the singular line of C_μ^2, at a point P of Φ, at which the singular line in C^2 is x; then

$$y_i = (\lambda_i + \mu)\, x_i;$$

by varying μ we obtain the pencil of tangents to Φ at P, hence each tangent to Φ at P is a singular line for *one* of the cosingular complexes C_μ^2.

If μ be eliminated between the equations

$$\Sigma \frac{y_i^2}{\lambda_i + \mu} = 0, \qquad \Sigma \frac{y_i^2}{(\lambda_i + \mu)^2} = 0,$$

we obtain the complex equation of Φ, *i.e.* the complex formed by its tangents.

Since

$$\Sigma \frac{y_i^2}{\lambda_i + \mu} = \frac{1}{\mu} \Sigma \frac{y_i^2}{1 + \frac{\lambda_i}{\mu}} = \frac{1}{\mu}\left\{\Sigma y_i^2 - \frac{1}{\mu} \Sigma \lambda_i y_i^2 + \ldots\right\}$$

$$= -\frac{1}{\mu^2} \Sigma \lambda_i y_i^2 + \ldots;$$

it is seen that C^2 is included in the series of complexes C_μ^2, and corresponds to the value $\mu = \infty$. The complexes C_μ^2 also include each fundamental complex *taken doubly*; as we see by taking successively $\mu + \lambda_1 = 0, \ldots \mu + \lambda_6 = 0$.

Through any line l there pass four complexes C_μ^2, (Art. 84), viz. those determined by the equation $\Sigma \frac{l_i^2}{\lambda_i + \mu} = 0$. If two of these values of μ, say μ_1 and μ_2, coincide, then since in that case

$$\Sigma \frac{l_i^2}{(\lambda_i + \mu_1)^2} = 0,$$

it follows that l is a singular line of the complex $C_{\mu_1}^2$, hence, *the singular lines of a complex C_μ^2 are its lines of intersection with a consecutive complex of the system.*

If a plane be drawn through l to touch Φ in O, the four singular lines corresponding to the above four complexes are the joins of O to the points of intersection of l and Φ (Art. 84).

It should be noticed that if two complexes

$$\Sigma k_i x_i^2 = 0, \quad \Sigma \lambda_i x_i^2 = 0,$$

are cosingular, we have

$$k_i = \frac{\alpha\lambda_i + \beta}{\gamma\lambda_i + \delta},$$

since
$$k_i = \frac{\frac{\alpha}{\gamma}(\gamma\lambda_i + \delta) + \beta - \frac{\alpha\delta}{\gamma}}{\gamma\lambda_i + \delta} = \frac{\alpha}{\gamma} + \frac{\beta\gamma - \alpha\delta}{\gamma^2} \cdot \frac{1}{\lambda_i + \frac{\delta}{\gamma}}.$$

116. Correspondence between lines of cosingular complexes*. If between the coordinates of two lines x and y, the following six equations exist,

$$x_i = \frac{y_i}{\sqrt{(\lambda_i + \mu)}},$$

it follows that x belongs to the complex $\sum_1^6 \lambda_i x_i^2 = 0$, or C^2, and y belongs to the cosingular complex $\sum_1^6 \frac{y_i^2}{\lambda_i + \mu} = 0$, or C_μ^2.

Thus the above equations establish between the lines of C^2 and C_μ^2 a (1, 1) correspondence, by aid of which many important properties of the quadratic complex can easily be demonstrated. A fundamental fact of the correspondence is the following:—if x and X are two lines of C^2 and y, Y their corresponding lines in C_μ^2, it is clear that $\Sigma x_i Y_i \equiv \Sigma y_i X_i$; hence, *if x intersects Y, then y intersects X.*

If $\Sigma x_i X_i = 0$, then $\Sigma \frac{y_i Y_i}{\lambda_i + \mu} = 0$, but y, Y will not, in general, intersect; if y and Y *do* intersect, *i.e.* if $\Sigma y_i Y_i = 0$, then since

$$\Sigma \frac{y_i^2}{\lambda_i + \mu} = 0, \qquad \Sigma \frac{y_i Y_i}{\lambda_i + \mu} = 0, \qquad \Sigma \frac{Y_i^2}{\lambda_i + \mu} = 0,$$

it follows that $\Sigma \frac{(y_i + \rho Y_i)^2}{\lambda_i + \mu} = 0$ for all values of ρ, or y and Y belong to the same pencil of lines of C_μ^2; and since

$$\Sigma y_i Y_i \equiv \Sigma (\lambda_i + \mu) x_i X_i = \Sigma \lambda_i x_i X_i,$$

it follows similarly that x and X belong to the same pencil of lines of C^2; *i.e. if a pair of lines of C^2 intersect, and also the corresponding pair of C_μ^2, each pair belongs to a pencil of lines of its own complex: to a pencil of either complex corresponds a pencil of the other.*

Any point P is the vertex of a complex cone of C^2 and of C_μ^2; the two loci of corresponding lines of C_μ^2 and C^2 respectively will not be cones (or conics), as just seen, and are such that any line of one locus meets all the lines of the other locus; the loci are therefore the *two sets of generators of the same quadric*, hence, *to any complex cone (or complex conic) of C^2 corresponds a regulus of C_μ^2 and conversely.* There thus arise ∞^3 reguli of C^2, the 'images' of the complex cones and of the complex conics of C_μ^2; they will be said to form a '*triplex*' of reguli of C^2 and will be denoted by

* See a paper by the author thus entitled, *Quarterly Journal* (1903).

Σ_μ, Σ_μ', where Σ_μ consists of the images of the complex cones of C_μ^2 and Σ_μ' of the images of the complex conics of C_μ^2.

Take any two non-intersecting lines x, X of C^2, and take the quantity μ determined from the equation

$$\mu = -\frac{\Sigma \lambda_i x_i X_i}{\Sigma x_i X_i}, \text{ or } \Sigma (\lambda_i + \mu) x_i X_i = 0,$$

as that which determines a cosingular complex C_μ^2; then to x and X will correspond two intersecting lines y and Y; these latter lines belong to one complex cone and one complex conic of C_μ^2; hence, *any two lines x, X of C^2 which do not intersect, determine one cosingular complex C_μ^2 in which the two corresponding lines y, Y intersect; and there are two reguli of C^2 through x, X, viz. the images of the complex cone and complex conic of C_μ^2 determined by y, Y.*

It has been seen that the complex cones and conics of any cosingular complex C_μ^2 have for images ∞^3 reguli of C^2 forming a triplex, and since there is a singly infinite number of complexes C_μ^2 *there is a quadruply infinite number of reguli of C^2* (Caporali, *Geometria*).

The directrices of the reguli of a given Σ_μ form ∞^3 reguli, which are the images of complex cones of C^2, and therefore reguli of C_μ^2. It is known that the polar line with regard to a fundamental complex, of any line of a quadratic complex C^2, belongs to C^2, (Art. 73); hence, the polars of the lines of a complex cone of C_μ^2 form a complex conic of C_μ^2, and from the equations of correspondence it is seen at once that *the polar line of y corresponds to the polar line of x*; hence, if a regulus of Σ_μ is polarized with regard to a fundamental complex, we obtain a regulus of Σ_μ'.

Through the vertex of a complex cone of C_μ^2 there pass ∞^2 planes, each of which contains a complex conic of C_μ^2, each conic having two lines in common with the cone; we have, in correspondence, ∞^2 reguli of C^2 belonging to a Σ_μ', which have two lines in common with one regulus ρ of Σ_μ; these ∞^2 reguli are said to form the '*field*' of ρ.

117. The complexes $\mathbf{R}_4^2$, $\mathbf{R}_4'^2$. If a complex $\Sigma a_i x_i = 0$ contains a regulus ρ of C^2 which belongs to Σ_μ, the complex $\Sigma \frac{a_i}{\sqrt{(\lambda_i + \mu)}} x_i = 0$ contains the cone of C_μ^2 which corresponds to ρ, hence the last complex must be *special, i.e.*

$$\Sigma \frac{a_i^2}{\lambda_i + \mu} = 0.$$

The complexes $\Sigma a_i x_i = 0$ which satisfy the last condition, μ having a given value, form a quadruply infinite system, which will be denoted by R_4^2; the system passes through all the reguli of a triplex, and two members of the system are contained in any 'pencil of linear complexes,' *i.e.* a set of complexes of the form

$$(\alpha x) + \rho(\beta x) = 0,$$

where ρ is variable.

In any one complex $(ax) = 0$, of a given R_4^2, are contained two singly infinite systems of reguli of Σ_μ, Σ_μ', viz. those which correspond to the complex cones and complex conics, whose vertices and planes respectively are united to the line

$$\frac{a_i}{\sqrt{(\lambda_i + \mu)}}.$$

Similarly, if a complex $\Sigma a_i x_i = 0$ passes through the directrices of a regulus ρ of Σ_μ forming a regulus of C_μ^2, the complex $\Sigma a_i \sqrt{(\lambda_i + \mu)}\, x_i = 0$ contains a cone of Σ and is therefore special, hence

$$\Sigma a_i^2 (\lambda_i + \mu) = 0.$$

This quadruply infinite system will be denoted by R'^2_4; if a linear complex belongs to an R_4^2 and to an R'^2_4 for the same value of μ, both $\frac{a_i}{\sqrt{(\lambda_i + \mu)}}$ and $a_i \sqrt{(\lambda_i + \mu)}$ are lines; denoting them by l and l', it follows that $l_i' = l_i(\lambda_i + \mu)$, hence l' is the polar line of l with regard to C^2, and the lines l, l' belong respectively to the quadratic complexes

$$\Sigma (\lambda_i + \mu)^2 l_i^2 = 0, \quad \Sigma \frac{l'^2_i}{(\lambda_i + \mu)^2} = 0.$$

118. The congruence [2, 2]. The lines common to C^2 and any given linear complex A, form a congruence which is of the second order and second class; for the lines of this congruence through any point P are the two intersections of the polar plane of P for A with the complex cone of P for C^2; in any plane π, the lines of the congruence are the tangents from the pole of π for A to the complex conic of π.

If the complex $\Sigma a_i x_i = 0$ is given, the equation $\Sigma \frac{a_i^2}{\lambda_i + \mu} = 0$

gives five values for μ, so that if, for instance, μ_1 be one of them, the complex cones of $C^2_{\mu_1}$ whose vertices are on the line $\frac{a_i}{\sqrt{(\lambda_i+\mu_1)}}$ correspond to ∞^1 reguli of C^2 which belong to $\Sigma a_i x_i = 0$, similarly for the complex conics of $C^2_{\mu_1}$ whose planes pass through this line; from the five lines we thus derive 10 *systems of* ∞^1 *reguli of the congruence* (C^2, A).

This method of correspondence thus enables us to investigate the congruence (C^2, A), by consideration of the *simpler* congruence consisting of the lines which belong to a quadratic complex and meet a given line.

The two systems which correspond to the complex cones and to the complex conics connected with the same line $\frac{a_i}{\sqrt{(\lambda_i+\mu)}}$, may be said to be *associated*; since each of these complex cones has two lines in common with each complex conic, it follows that any two reguli belonging respectively to two associated systems have always two generators in common. Conversely, if two reguli of different systems have two generators in common, those systems must be associated, for if the reguli ρ_1 and ρ_2 belong to two non-associated systems, say those connected with μ_1 and μ_2, the transformation $x_i = \frac{y_i}{\sqrt{(\lambda_i+\mu_1)}}$ turns ρ_1 into a *cone*, and ρ_2 into a *regulus* of $C^2_{\mu_1}$, which cannot have two generators in common with a cone. At the same time it is seen that, if ρ_1 and ρ_2 belong to two different and non-associated systems, they have *one* generator in common, for ρ_2 is transformed as above into a regulus of $C^2_{\mu_1}$ of which the line $\frac{a_i}{\sqrt{(\lambda_i+\mu_1)}}$ is a directrix, and this regulus will have one generator in common with each cone of $C^2_{\mu_1}$, whose vertex lies on this line.

The congruence (C^2, A) *contains* 16 *pencils of lines*; for, taking one of the roots μ_1 of the equation $\Sigma \frac{a_i^2}{\lambda_i+\mu} = 0$, to each line of $C^2_{\mu_1}$ which meets the line $\frac{a_i}{\sqrt{\lambda_i+\mu_1}}$ there corresponds one line of (C^2, A), and *vice versâ*; but the congruence $\left(C^2_{\mu_1}, \frac{a_i}{\sqrt{\lambda_i+\mu_1}}\right)$ contains 16 pencils, viz. the eight pencils of $C^2_{\mu_1}$ at the four points in which $\frac{a_i}{\sqrt{\lambda_i+\mu_1}}$ meets the singular surface, and the eight pencils in the four tangent planes β_i through $\frac{a_i}{\sqrt{\lambda_i+\mu_1}}$; hence, since a pencil of lines y corresponds to a pencil of lines x (Art. 116), there are 16 pencils in the congruence (C^2, A).

Though, in general, there pass through any point only two lines of a congruence (2, 2), it is now seen that there are 16 exceptional points, through each of which there pass ∞^1 lines of the congruence; such a point is called a *singular point* of the congruence. Hence there are 16 singular points in a congruence (2, 2); also there are 16 *singular planes*, each of which contains ∞^1 lines of the congruence which form a pencil.

In the congruence $\left(C_{\mu_1}{}^2, \frac{a_i}{\sqrt{\lambda_i+\mu_1}}\right)$, if the centres of the pencils in β_i are B_i, B_i', the pencil (B_i, β_i) has a line in common with five other pencils, viz. (B_i', β_i) and four pencils whose centres lie on $\frac{a_i}{\sqrt{\lambda_i+\mu_1}}$, (Art. 77); hence the corresponding pencil in (C^2, A) has a line in common with each of five other pencils, *i.e.*, *passes through the centres of these five pencils*, hence, *each singular plane contains six singular points*; and, similarly, *through each singular point pass six singular planes.*

Each complex cone of $C_{\mu_1}{}^2$ whose vertex is on $\frac{a_i}{\sqrt{\lambda_i+\mu_1}}$, contains one line of each of the eight pencils in the planes β_i; hence, *each regulus, of the corresponding system of reguli in* (C^2, A), *passes through eight singular points of the congruence*; and the sixteen singular points are divided into ten sets of eight points by the ten systems of reguli.

119. Focal surface of the congruence. The lines of a congruence (2, 2), or (C^2, A), which meet any line p form a ruled quartic of class I, having another directrix p', the polar line of p for A, and upon p and p' the lines of the congruence determine a (2, 2) correspondence of points; of this correspondence there are four branch points on both p and p' (Introd. xvii.), hence it occurs four times that two consecutive lines of the congruence intersect on p. The locus of the ∞^2 points of intersection of consecutive lines of the congruence is therefore a surface of the fourth degree; this surface is called the Focal Surface of the congruence.

If p belongs to the congruence (but does not pass through a singular point), the quartic surface formed by lines of the congruence which intersect p, is of a different character; for, any line y, which meets p, meets only one other generator of this surface, since in the plane (p, y) there is only *one* other line of the congruence, hence p is a *triple* line of the surface, which is of class XII, possessing a simple directrix and a double generator; so that p is met by two consecutive lines of the congruence in points P, P' respectively; and there is no other point of the focal surface upon p except P and P', for if Q were such a point, then through Q would pass three lines of the congruence, viz. p and the two (consecutive) lines through Q; hence p *touches* the focal surface at P and P', thus *the lines of the congruence are bitangents of the focal surface.*

Lastly, taking any line p through a singular point S of the congruence, the lines of the congruence which meet p consist of the pencil through S together with a ruled cubic of which p is the double directrix. The two generators through each point of p meet the single directrix of the ruled cubic in points Q, Q', so that there is determined upon the two directrices, a [1, 2] correspondence, which is therefore given by an equation of the form

$$xu + v = 0 \text{..............................(i)},$$

where u and v are quadratic expressions in y, the coordinate of a point Q, x being the coordinate of a point P upon p.

For points P in which p meets the focal surface, the points Q, Q' coincide, *i.e.* the quadratic equation (i) has equal roots, this gives two such points P upon p; from which we learn that, exclusive of S, p meets the focal surface in two points only, hence S is a double point of the focal surface.

The focal surface, being therefore of the fourth degree, and possessing 16 double points, *is a Kummer surface.*

120. Confocal congruences. We shall now investigate the complex represented by the equation

$$(\lambda x^2) + (\lambda a^2)(ax)^2 - 2(ax)(\lambda ax) = 0 \text{..............(I)}.$$

This complex meets the linear complex A, or $(ax) = 0$, in the congruence (C^2, A); it may be brought into its canonical form in the following manner.

Consider the system of complexes A, B_1, B_2, B_3, B_4, B_5, where

$$A = \Sigma a_i x_i, \quad B_1 = \Sigma b_{1i} x_i, \text{ ...... } B_5 = \Sigma b_{5i} x_i;$$

we take these six complexes as being in involution in pairs and choose their constants so that

$$\Sigma a_i^2 = 1, \quad \Sigma b_{1i}^2 = 1, \; \ldots\ldots, \; \Sigma b_{5i}^2 = 1.$$

Then, (Art. 28), we have

$$x_i \equiv a_i A + b_{1i} B_1 + b_{2i} B_2 + b_{3i} B_3 + b_{4i} B_4 + b_{5i} B_5.$$

We shall now suppose that $b_{ki} = \rho_k \dfrac{a_i}{\lambda_i + \mu_k}$, where μ_k is a root of the equation $\Sigma \dfrac{a_i^2}{\lambda_i + \mu} = 0$; this ensures the involution between each pair of complexes, since $\Sigma_i \dfrac{a_i^2}{(\lambda_i + \mu_k)(\lambda_i + \mu_{k'})} = 0$.

Taking the six complexes A, B_k as coordinate complexes and substituting for the x_i in (I), this equation assumes the form

$$\Sigma_i \lambda_i \{a_i A + \Sigma_k b_{ki} B_k\}^2 + (\lambda a^2) A^2 - 2A \Sigma_i \lambda_i a_i' \{a_i A + \Sigma_k b_{ki} B_k\} = 0.$$

Now since

$$\begin{aligned} \Sigma \lambda_i b_{1i} b_{2i} &= \Sigma (\lambda_i + \mu_1) b_{1i} b_{2i} - \mu_1 \Sigma b_{1i} b_{2i} \\ &= \rho_1 \rho_2 \Sigma \frac{a_i^2}{\lambda_i + \mu_2} - \mu_1 \Sigma \frac{a_i^2}{(\lambda_i + \mu_1)(\lambda_i + \mu_2)} \\ &= 0, \end{aligned}$$

the term $B_1 B_2$ disappears, similarly for all the product terms; hence the equation (I) becomes

$$B_1^2 \Sigma \lambda_i b_{1i}^2 + B_2^2 \Sigma \lambda_i b_{2i}^2 + B_3^2 \Sigma \lambda_i b_{3i}^2 + B_4^2 \Sigma \lambda_i b_{4i}^2 + B_5^2 \Sigma \lambda_i b_{5i}^2 = 0;$$

while since

$$\Sigma \lambda_i b_{1i}^2 = \Sigma (\lambda_i + \mu_1) b_{1i}^2 - \mu_1 \Sigma b_{1i}^2 = -\mu_1, \text{ \&c.},$$

the final form of the equation is

$$\sum_1^5 \mu_k B_k^2 = 0.$$

The complex represented by this equation has been seen to intersect A in the congruence (C^2, A).

Now, (Art. 83), the singular surface of the complex

$$\sum_2^6 (\lambda_i - \lambda_1) x_i^2 = 0,$$

is the focal surface for the six congruences

$$x_1 = 0, \quad \Sigma \frac{x_i^2}{\lambda_i - \lambda_1} = 0, \quad (i \neq 1),$$

$$\ldots\ldots\ldots\ldots\ldots\ldots\ldots\ldots\ldots\ldots$$

$$x_6 = 0, \quad \Sigma \frac{x_i^2}{\lambda_i - \lambda_6} = 0, \quad (i \neq 6).$$

We conclude therefore, in the first place, that *the singular surface of the complex*

$$\sum_1^5 \frac{B_k^2}{\mu_k} = 0 \qquad\text{(II)},$$

is the focal surface for the congruence

$$A = 0, \quad \sum_1^5 \mu_k B_k^2 = 0,$$

i.e. for the congruence (C^2, A).

Next, substituting in the equation of the complex (II) from the identical relation

$$-B_1^2 \equiv A^2 + B_2^2 + B_3^2 + B_4^2 + B_5^2,$$

we obtain as an equivalent form of (II)

$$-A^2 + B_2^2 . \frac{\mu_1 - \mu_2}{\mu_2} + B_3^2 \frac{\mu_1 - \mu_3}{\mu_3} + B_4^2 \frac{\mu_1 - \mu_4}{\mu_4} + B_5^2 \frac{\mu_1 - \mu_5}{\mu_5} = 0;$$

therefore the singular surface of $\Sigma \frac{B_k^2}{\mu_k} = 0$ is the focal surface of the congruence

$$B_1 = 0, \quad -A^2 + \Sigma \frac{\mu_k}{\mu_1 - \mu_k} . B_k^2 = 0, \quad (k \neq 1), \text{ \&c.};$$

the five congruences confocal with (C^2, A) *are therefore seen to be*

$$B_1 = 0, \quad -A^2 + \Sigma \frac{\mu_k}{\mu_1 - \mu_k} . B_k^2 = 0, \quad (k \neq 1),$$

....

$$B_5 = 0, \quad -A^2 + \Sigma \frac{\mu_k}{\mu_5 - \mu_k} . B_k^2 = 0, \quad (k \neq 5);$$

their focal surface being the singular surface of $\Sigma \frac{B_k^2}{\mu_k} = 0$, *where*

$$A = \Sigma a_i x_i, \quad B_k = \Sigma b_{ki} x_i, \quad b_{ki} = \rho_k . \frac{a_i}{\lambda_i + \mu_k}, \quad \frac{1}{\rho_k^2} = \sum_i . \frac{a_i^2}{(\lambda_i + \mu_k)^2};$$

provided that the μ_k *are the solutions of the equation*

$$\Sigma \frac{a_i^2}{\lambda_i + \mu} = 0.$$

Again consider the complex

$$\Sigma \frac{x_i^2}{\lambda_i + \mu_1} + B_1^2 \Sigma \frac{b_{1i}^2}{\lambda_i + \mu_1} - 2B_1 \Sigma \frac{b_{1i} x_i}{\lambda_i + \mu_1} = 0 \qquad\text{(III)}.$$

This complex clearly contains the congruence $(B_1, C_{\mu_1}^2)$; on substitution for the x_i in terms of $A, B_1 \dots B_5$, the equation assumes the form

$$\Sigma \frac{1}{\lambda_i + \mu_1} (a_i A + b_{1i} B_1 + \ldots + b_{5i} B_5)^2 + B_1^2 \Sigma \frac{b_{1i}^2}{\lambda_i + \mu_1}$$
$$- 2B_1 \Sigma \frac{b_{1i}}{\lambda_i + \mu_1} (a_i A + \ldots + b_{5i} B_5) = 0 \ldots\ldots \text{(IV)}.$$

Now since we have

$$\Sigma \frac{a_i^2}{(\lambda_i + \mu_1)(\lambda_i + \mu_k)} = 0, \quad \Sigma \frac{a_i^2}{(\lambda_i + \mu_1)(\lambda_i + \mu_j)} = 0,$$

therefore $$\Sigma \frac{a_i^2}{(\lambda_i + \mu_1)(\lambda_i + \mu_k)(\lambda_i + \mu_j)} = 0,$$

provided that $$\mu_k \neq \mu_j \neq \mu_1.$$

Hence it is easily seen that all the terms of the equation (IV) disappear except those involving the squares of B_2, B_3, B_4, B_5, and the equation takes the form

$$B_2^2 \rho_2^2 \Sigma \frac{a_i^2}{(\lambda_i + \mu_1)(\lambda_i + \mu_2)^2} + \ldots + B_5^2 \rho_5^2 \Sigma \frac{a_i^2}{(\lambda_i + \mu_1)(\lambda_i + \mu_5)^2} = 0.$$

The complex represented by the latter equation, therefore, contains the congruence $(B_1, C_{\mu_1}^2)$; but since

$$\Sigma \frac{a_i^2}{(\lambda_i + \mu_1)(\lambda_i + \mu_2)} - \Sigma \frac{a_i^2}{(\lambda_i + \mu_2)^2} = (\mu_2 - \mu_1) \Sigma \frac{a_i^2}{(\lambda_i + \mu_1)(\lambda_i + \mu_2)^2},$$

it is seen that

$$\frac{1}{\mu_1 - \mu_2} = \rho_2^2 \Sigma \frac{a_i^2}{(\lambda_i + \mu_1)(\lambda_i + \mu_2)^2}; \ \&\text{c.}$$

The equation of the last complex therefore becomes

$$\sum_2^5 \frac{B_k^2}{\mu_1 - \mu_k} = 0;$$

and the congruence

$$B_1 = 0, \quad B_1^2 + \sum_2^5 \frac{\mu_1}{\mu_1 - \mu_k} B_k^2 = 0,$$

is identical with the congruence $(B_1, C_{\mu_1}^2)$.

But $$B_1^2 + \sum_2^5 \frac{\mu_1}{\mu_1 - \mu_k} B_k^2 \equiv -A^2 + \sum_2^5 \frac{\mu_k}{\mu_1 - \mu_k} B_k^2,$$

therefore by the preceding it follows that *the five congruences confocal with* (A, C^2) *are the congruences* $(B_1, C_{\mu_1}^2) \ldots (B_5, C_{\mu_5}^2)$.

It should be observed that since the congruence (A, C^2) is identical with $A = 0$, $C^2 + AA' = 0$, where A' is any linear complex, it follows that any congruence (2, 2) is contained in ∞^6 quadratic complexes.

The lines of a congruence (2, 2) which meet any given line l form a ruled quartic of class I; sixteen of these lines meet any given quartic curve c^4; hence the lines of the congruence which meet c^4 form a ruled surface whose degree is sixteen. If c^4 is a section of Φ, the focal surface of the congruence, by any plane π, the two generators of this ruled surface through each point of c^4 *coincide*, and the surface degenerates into two coincident surfaces of degree *eight*. Denoting by S this surface of the eighth degree, it is clear that S touches Φ along c^4, the other curve of intersection of S and Φ being formed by the second focal points on the generators of S, while S and Φ touch also along this latter curve. Since these two curves of contact form the sole intersection of S and Φ it follows that the order of the latter curve must be *twelve*. The points of intersection of this curve and π consist partly of the four points of contact with Φ of the lines of the congruence in π, and partly of points of contact of such lines as meet Φ in four consecutive points.

Hence, *there is a curve of order* 8 *on* Φ *at each of whose points there is a tangent of* Φ *which has four-point contact with* Φ*.

Since a Kummer surface is the focal surface for six congruences (2, 2) it follows that on this surface there are six curves of four-point contact.

121. The quartic surface (C^2, A, A′). The lines common to a quadratic complex C^2 and two linear complexes A and A', where A and A' are $(ax)=0$, $(a'x)=0$, respectively, will in general form a ruled quartic of class I, whose double directrices are the common polar lines of A and A'. If this quartic surface splits up into two reguli, since they have two common directrices, they must have also two common generators, (Art. 58); *i.e., each belongs to the field of the other;* hence, there is some cosingular complex C_μ^2 in which these reguli correspond to a complex cone and complex conic having two lines in common. It follows that the lines $\dfrac{a_i}{\sqrt{\lambda_i+\mu}}$, $\dfrac{a_i'}{\sqrt{\lambda_i+\mu}}$ *must intersect.*

The conditions required, therefore, in order that the surface (C^2, A, A') should consist of two reguli, are

$$\Sigma\frac{a_i^2}{\lambda_i+\mu}=0,\quad \Sigma\frac{a_i a_i'}{\lambda_i+\mu}=0,\quad \Sigma\frac{a_i'^2}{\lambda_i+\mu}=0;$$

* See Sturm, *Liniengeom.* Bd. II. S. 42.

hence, for all values of ρ, $\Sigma \frac{(a_i + \rho a_i')^2}{\lambda_i + \mu} = 0$, *i.e.*, the system of two terms, (Art. 51), determined by A and A', belongs entirely to one R_4^2.

The foregoing condition may also be expressed as follows:—the equation $\Sigma \frac{(a_i + \rho a_i')^2}{\lambda_i + \mu} = 0$ is satisfied for the two values of ρ which make $A + \rho A'$ a special complex, *i.e.* by the common polar lines of A and A'; hence, *the common polar lines of A and A' must belong to the same cosingular complex C_μ^2 and be directrices of some regulus of Σ_μ.*

122. Projective formation of C^2. If three linear complexes A, A', A'' give a regulus ρ of C^2, we have the six equations

$$\left.\begin{array}{lll} \Sigma \frac{a_i^2}{\lambda_i + \mu} = 0, & \Sigma \frac{a_i'^2}{\lambda_i + \mu} = 0, & \Sigma \frac{a_i''^2}{\lambda_i + \mu} = 0 \\ \Sigma \frac{a_i a_i'}{\lambda_i + \mu} = 0, & \Sigma \frac{a_i a_i''}{\lambda_i + \mu} = 0, & \Sigma \frac{a_i' a_i''}{\lambda_i + \mu} = 0 \end{array}\right\} \text{.........(1)};$$

they state that the three lines

$$\frac{a_i}{\sqrt{(\lambda_i + \mu)}}, \quad \frac{a_i'}{\sqrt{(\lambda_i + \mu)}}, \quad \frac{a_i''}{\sqrt{(\lambda_i + \mu)}}$$

meet in a point or lie in a plane, thus the complex cone of this point or complex conic of this plane, for C_μ^2, corresponds to ρ.

If the first two lines pass through a given point P, then in any given plane π which does not pass through P, there is one line satisfying the conditions, viz. the intersection with π of the plane through the two lines. Let ρ be the regulus of C^2 which corresponds to the complex cone of P for C_μ^2, and ρ' that which corresponds to the complex conic of π for C_μ^2, our result states that any two complexes A, A' through ρ, and any third complex A'' through ρ', intersect in a new regulus of C^2, provided that the preceding equations are satisfied.

Let now the equations of A, A', A'' be

$$\left.\begin{array}{l} \Sigma a_i x_i \equiv \alpha L + \beta M + \gamma N = 0 \\ \Sigma a_i' x_i \equiv \alpha' L + \beta' M + \gamma' N = 0 \\ \Sigma a_i'' x_i \equiv \alpha'' L' + \beta'' M' + \gamma'' N' = 0 \end{array}\right\} \text{...........(2)},$$

where

$L = 0$, $M = 0$, $N = 0$ are three given linear complexes through ρ,
$L' = 0$, $M' = 0$, $N' = 0$ ....ρ'.

Then since ρ and ρ' are reguli of Σ_μ, Σ_μ' respectively, the first four above conditions (1) are satisfied independently of the values of α, β, &c.; and if A and A' are given, the ratios $\alpha'' : \beta'' : \gamma''$ are determined by the equations

$$\Sigma \frac{a_i a_i''}{\lambda_i + \mu} = 0, \quad \Sigma \frac{a_i' a_i''}{\lambda_i + \mu} = 0;$$

hence, ρ being a given regulus of Σ_μ and ρ' of Σ_μ', *any* two complexes A, A' through ρ, determine a third complex A'' through ρ', such that the complexes A, A', A'' intersect in a regulus of C^2, (which belongs to the field of ρ); this is called *the projective formation* of C^2.

Each of three complexes L, M, N through the given regulus ρ contains two undetermined constants; similarly for the complexes L', M', N'; taking L as $\Sigma l_i x_i$, &c., we are therefore at liberty to suppose the following equations to exist between the constants

$$\Sigma \frac{l_i m_i'}{\lambda_i + \mu} = \Sigma \frac{l_i n_i'}{\lambda_i + \mu} = \Sigma \frac{m_i l_i'}{\lambda_i + \mu} = \Sigma \frac{m_i n_i'}{\lambda_i + \mu}$$
$$= \Sigma \frac{n_i l_i'}{\lambda_i + \mu} = \Sigma \frac{n_i m_i'}{\lambda_i + \mu} = 0 \quad \text{......(3).}$$

The two equations

$$\Sigma \frac{a_i a_i''}{\lambda_i + \mu} = \Sigma \frac{a_i' a_i''}{\lambda_i + \mu} = 0$$

now assume the form

$$\left.\begin{aligned} \alpha\alpha'' \Sigma \frac{l_i l_i'}{\lambda_i + \mu} + \beta\beta'' \Sigma \frac{m_i m_i'}{\lambda_i + \mu} + \gamma\gamma'' \Sigma \frac{n_i n_i'}{\lambda_i + \mu} = 0 \\ \alpha'\alpha'' \Sigma \frac{l_i l_i'}{\lambda_i + \mu} + \beta'\beta'' \Sigma \frac{m_i m_i'}{\lambda_i + \mu} + \gamma'\gamma'' \Sigma \frac{n_i n_i'}{\lambda_i + \mu} = 0 \end{aligned}\right\} \text{...(4).}$$

Eliminating the variable quantities α, β, γ, &c. between (2) and (4), we obtain the equation of C^2 in the form

$$\frac{LL'}{\Sigma \frac{l_i l_i'}{\lambda_i + \mu}} + \frac{MM'}{\Sigma \frac{m_i m_i'}{\lambda_i + \mu}} + \frac{NN'}{\Sigma \frac{n_i n_i'}{\lambda_i + \mu}} = 0.$$

From the foregoing process we observe that the equation of any quadratic complex C^2 may be brought to the form just given, if L, M, N are three complexes through a regulus ρ of C^2, and L', M', N' are three complexes through a regulus ρ' of C^2, where ρ and ρ' belong respectively to a Σ_μ and the corresponding Σ_μ', *provided that the equations* (3) *are satisfied.* The geometrical significance of these latter equations is that the six lines

$$\frac{l_i}{\sqrt{(\lambda_i + \mu)}}, \quad \frac{m_i}{\sqrt{(\lambda_i + \mu)}}, \text{ \&c.}$$

form a tetrahedron. Hence, to get the equation of C^2 in the form now obtained for it, *we take any tetrahedron and multiply the coordinates of its edges by the quantities* $\sqrt{(\lambda_i+\mu)}$; thus each edge gives rise to one of the six linear complexes L, M, N, L', M', N'. Hence the equation of C^2 may be brought to this form in ∞^{13} ways.

The ∞^3 reguli of Σ_μ and Σ_μ' respectively, are now seen to be given by the equations

$$\left.\begin{aligned} L &= \rho M' - \frac{c}{a}\sigma N', \\ M &= \tau N' - \frac{a}{b}\rho L', \\ N &= \sigma L' - \frac{b}{c}\tau M', \end{aligned}\right\} \text{ and } \left\{\begin{aligned} L' &= \rho_1 M - \frac{c}{a}\sigma_1 N, \\ M' &= \tau_1 N - \frac{a}{b}\rho_1 L, \\ N' &= \sigma_1 L - \frac{b}{c}\tau_1 M; \end{aligned}\right.$$

where $$\frac{1}{a}:\frac{1}{b}:\frac{1}{c} = \Sigma\frac{l_i l_i'}{\lambda_i+\mu} : \Sigma\frac{m_i m_i'}{\lambda_i+\mu} : \Sigma\frac{n_i n_i'}{\lambda_i+\mu};$$

by giving all values to $\rho, \sigma, \tau, \rho_1, \sigma_1, \tau_1$.

123. Caporali's Theorem. It has been seen that the equation of any general quadratic complex can be formed by aid of a tetrahedron in the manner described. An application of this method will now be given to prove the theorem of Caporali, that *any congruence* (2, 2) *is contained in* 40 *tetrahedral complexes**.

Taking L, or $\Sigma l_i x_i$, as any given linear complex, the equation $\Sigma\frac{l_i^2}{\lambda_i+\mu}=0$ gives five values of μ, which correspond to the five pairs of associated reguli of the congruence (C^2, L). Take one of these values of μ, say μ_1, this determines a line $\frac{l_i}{\sqrt{(\lambda_i+\mu_1)}}$ with which are connected the vertices of the complex cones, (and the planes of the complex conics) connected with this pair of systems of reguli of (C^2, L).

Through the line $\frac{l_i}{\sqrt{(\lambda_i+\mu)}}$ there pass four tangent planes $\beta_1, \beta_2, \beta_3, \beta_4$ to the singular surface; let B_1 and B_1' be the centres of pencils of C^2 in β_1, B_2, B_2' in β_2, &c. (Art. 77), and let the line joining B_1 and B_2 be $\frac{l_i'}{\sqrt{(\lambda_i+\mu_1)}}$, this determines L'; again of the eight pencils of C^2 through the points A_1, A_2, A_3, A_4 in

* This theorem is due to Caporali, *Sui complessi e sulle congruenze di 2° grado*, Atti dei Lyncei, (1877-1878).

which $\frac{l_i}{\sqrt{(\lambda_i+\mu_1)}}$ meets the singular surface, each such pencil passes through *one* of the points B_1, B_1'; B_2, B_2', and so on, *i.e. two* of these pencils whose centres may be denoted by A_1 and A_2 pass through both B_1 and B_2 (see Table, Art. 77). Hence the lines A_1B_1, B_1A_2, A_2B_2, B_2A_1 form a twisted quadrilateral formed of lines of C^2, in which any two which intersect belong to the same pencil of C^2.

Taking these lines respectively as

$$\frac{m_i}{\sqrt{(\lambda_i+\mu_1)}}, \quad \frac{n_i}{\sqrt{(\lambda_i+\mu_1)}}, \quad \frac{m_i'}{\sqrt{(\lambda_i+\mu_1)}}, \quad \frac{n_i'}{\sqrt{(\lambda_i+\mu_1)}},$$

the quantities m_i, n_i, m_i', n_i', must be *coordinates of their corresponding lines of* $C^2_{\mu_1}$, *therefore these latter lines must form a twisted quadrilateral, in which any two lines which meet, form part of a pencil of* $C^2_{\mu_1}$ (Art. 116).

Hence the form of equation of C^2 derived from the quadrilateral $A_1B_1A_2B_2$ being

$$\frac{LL'}{\Sigma\frac{l_il_i'}{\lambda_i+\mu_1}} + \frac{MM'}{\Sigma\frac{m_im_i'}{\lambda_i+\mu_1}} + \frac{NN'}{\Sigma\frac{n_in_i'}{\lambda_i+\mu_1}} = 0,$$

the intersection of C^2 with $L=0$ gives the *tetrahedral complex*

$$\frac{MM'}{\Sigma\frac{m_im_i'}{\lambda_i+\mu_1}} + \frac{NN'}{\Sigma\frac{n_in_i'}{\lambda_i+\mu_1}} = 0.$$

Now the line $\frac{l_i}{\sqrt{(\lambda_i+\mu)}}$ may be determined in five ways, and for each such line there are 24 lines $\frac{l_i'}{\sqrt{(\lambda_i+\mu)}}$, since the centres of the eight pencils (B_i, β_i), (B_j, β_j) may be joined in 24 ways.

Hence the equation of C^2 may be written in 5×24, or 120 ways, in each of which the result of putting $L=0$ gives a tetrahedral complex.

Moreover *any* form of equation of C^2,

$$aLL' + bMM' + cNN' = 0,$$

which complies with this condition, must be derived from *one* of these 120 tetrahedra; for if m_i, n_i, m_i', n_i' are two pairs of opposite edges of the tetrahedron of such a tetrahedral complex,

$$\left.\begin{aligned} \sigma M + N = 0 \quad &\text{(i)} \\ M' - \sigma \frac{c}{b} N' = 0 \quad &\text{(ii)} \\ L = 0 \quad & \end{aligned}\right\} \text{ and } \left\{\begin{aligned} \tau M + N' = 0 \quad &\text{(iii)} \\ M' - \tau \frac{c}{b} N = 0 \quad &\text{(iv)} \\ L = 0 \quad & \end{aligned}\right.$$

give two systems of reguli of (C^2, L), and since (i), (ii), (iii), and (iv) taken simultaneously are equivalent to only *three* equations, giving a regulus ρ, it follows that any regulus of one system has two lines in common with any regulus of the other, viz. the intersection of ρ and L. Hence these two systems are associated, (Art. 118), and correspond to the same value of μ, say μ_1.

The complex $\sigma M + N = 0$ is *special*, having for directrix a line of the pencil (m_i, n_i); the lines of this pencil, therefore, being directrices of reguli of (C^2, L) belong to $C_{\mu_1}{}^2$ and meet $\dfrac{l_i}{\sqrt{\lambda_i + \mu_1}}$; similarly for the pencils

$$(m_i', n_i'),\ (m_i, n_i'),\ (m_i', n_i).$$

Hence, as before, the lines

$$\frac{m_i}{\sqrt{(\lambda_i + \mu_1)}},\ \frac{n_i}{\sqrt{(\lambda_i + \mu_1)}},\ \frac{m_i'}{\sqrt{(\lambda_i + \mu_1)}},\ \frac{n_i'}{\sqrt{(\lambda_i + \mu_1)}}$$

form a twisted quadrilateral in which any two lines which intersect belong to a pencil of C^2, while these four lines all meet the line whose coordinates are

$$\frac{l_i}{\sqrt{\lambda_i + \mu_1}}.$$

Again, if $\qquad aLL' + bMM' + cNN' = 0$

is an equation of the required form, and K, K' are the special complexes whose directrices are the remaining pair of opposite edges of the tetrahedron formed by the lines m_i, n_i, m_i', n_i', there exists an identical relation of the form

$$\alpha KK' + \beta MM' + \gamma NN' = 0;$$

so that by eliminating in turn MM' and NN' between the last two equations, we derive two other forms of the equation of C^2 of the required type, thus each tetrahedral complex which contains (C^2, L) gives rise to three of the 120 stated forms of the equation of C^2, *i.e. there are* 40 *tetrahedral complexes which contain* (C^2, L).

124. Condition for (1, 1) correspondence in any coordinate system. We will now consider the analytical conditions to be satisfied, in order that the equations $y_i = \sum_k a_{ik} x_k$ may give a (1, 1) correspondence between cosingular complexes, for *any* coordinate system.

The equations $$y_i = \Sigma a_{ik} x_k \quad \text{.........................(1)},$$
where x_i and y_i are both coordinates of lines, give rise to two quadratic complexes X^2, Y^2 when we substitute from (1) in $\omega(y)=0$ for the y_i in terms of the x_i, and similarly for the x_i in $\omega(x)=0$.

If x, y; x', y' be any two pairs of corresponding lines, then if x and x' intersect, and also y and y', $y_i + \lambda y_i'$ is a line for all values of λ, *i.e.* $x_i + \lambda x_i'$ is a line of X^2 for all values of λ; hence x, x' belong to the same pencil of X^2, and y, y' to the same pencil of Y^2; thus, for two lines of X^2 which intersect and do not belong to the same pencil, the corresponding lines of Y^2 do not intersect; while to a pencil of X^2 there corresponds a pencil of Y^2.

It will now be shown that if
$$\Sigma x_i \frac{\partial \omega(y')}{\partial y_i'} \equiv \Sigma x_i' \frac{\partial \omega(y)}{\partial y_i} \quad \text{................(2)},$$
i.e. if, whenever x meets y', y meets x', *the complexes* X^2, Y^2 *are cosingular*. For if P is any point on the singular surface of X^2, to the two pencils of X^2 through P will correspond two pencils of Y^2 having a common line, while to the complex cone of Y^2 through P will correspond a regulus of X^2 such that each line of it meets the above two pencils of Y^2; hence this regulus must break up into a pair of pencils, and therefore the complex cone of Y^2 through P will consist of two pencils, that is, P is a point in the singular surface of both X^2 and Y^2.

If $\omega(x) \equiv \Sigma a_{ik} x_i x_k$, by equating the coefficients of $x_i x_k'$ on each side of (2), we obtain as the necessary conditions
$$\sum_r a_{ir} a_{rk} = \sum_r a_{kr} a_{ri},$$
for all values of i and k.

Denoting these expressions by A_{ik}, A_{ki}, we have, $A_{ik} = A_{ki}$, and on multiplying the equations (1) by $a_{i1}, \ldots, a_{i6}$, we obtain
$$\tfrac{1}{2} \frac{\partial \omega}{\partial y_i} = \sum_r A_{ir} x_r,$$
and the equations (1) are equivalent to the following
$$\frac{\partial \omega}{\partial y_i} = \frac{\partial \Phi}{\partial x_i}, \quad \Phi \equiv \Sigma A_{ik} x_i x_k,$$

Hence any quadratic expression Φ in the six variables x_i, gives rise to a (1, 1) correspondence between the lines of two cosingular quadratic complexes*.

125. Equation of the complex referred to a special tetrahedron. In any singular tangent plane π of the singular surface of C^2 let the poles of the fundamental complexes C_α, C_β, C_γ be A, B, C respectively. Then through BC, CA, AB there pass singular tangent planes π_A, π_B, π_C whose intersection D is also a point of the closed system determined by the three complexes C_α, C_β, C_γ, (Arts. 26, 61). This point D is therefore a double point of the surface (Art. 82), and the tetrahedron $ABCD$ is such that each vertex is a double point and each face a singular tangent plane of the singular surface.

Taking this tetrahedron as the one of reference, the equation of the complex assumes the form

$$A^2 + 2Lp_{12}p_{34} + 2Mp_{13}p_{42} + 2Np_{14}p_{23} = 0.$$

For, since all the lines of the complex in a singular plane of the Kummer surface form one pencil, (Art. 80), and similarly all those through a double point of the surface, hence, if the equation of the complex be

$$f(p_{12}, p_{13}, p_{14}, p_{23}, p_{34}, p_{42}) = 0,$$

these eight conditions reduce f to a perfect square, save as to terms

$$Lp_{12}p_{34} + Mp_{13}p_{42} + Np_{14}p_{23};$$

for, let the centres of the pencils in the coordinate planes be P_1, P_2, P_3 and P_4, respectively, and the planes of the pencils through the vertices π_1, π_2, π_3 and π_4, then, since the points A_2, P_1, P_3, P_4 lie in π_2, the line A_2P_1 meets P_3P_4; hence the four lines A_1A_2, A_3A_4, P_1P_2, P_3P_4 are intersected by A_2P_1, similarly they are intersected by A_1P_2, A_4P_3, A_3P_4; hence they belong to the same regulus. There is therefore one linear complex A in which A_1A_2, P_3P_4 and A_3A_4, P_1P_2 are two pairs of polar lines; each of the foregoing eight pencils of f belongs to A.

By identifying the polar plane for A of each vertex A_i with the two (coincident) planes which form the complex cone of A_i for f, and proceeding similarly for the coordinate planes, it easily follows that

$$f \equiv A^2 + 2Lp_{12}p_{34} + 2Mp_{13}p_{42} + 2Np_{14}p_{23}.$$

* It is easily seen that no (1, 1) correspondence which is not thus formed can lead to two cosingular complexes.

126. The complex A is of the form

$$a_2(p_{12}-\alpha p_{34})+a_4(p_{13}-\beta p_{42})+a_6(p_{14}-\gamma p_{23})=0,$$

but we may take such multiples of the point-coordinates as will make $\alpha=\beta=\gamma=1$, (Art. 81), this will not affect the form of the invariable relation $p_{12}p_{34}+p_{13}p_{42}+p_{14}p_{23}\equiv 0$.

The equation of the complex now assumes the form

$$f(x)\equiv -(a_2x_2+a_4x_4+a_6x_6)^2 \\ +L(x_1^2+x_2^2)+M(x_3^2+x_4^2)+N(x_5^2+x_6^2)=0\ldots\text{(I.)},$$

where, as usual, $x_1=p_{12}+p_{34}$, $ix_2=p_{12}-p_{34}$, &c., with reference to the coordinate system now adopted.

The equations which connect a line x of this complex with a line y of a cosingular complex are the following:—

$$\left.\begin{aligned} y_2&=ax_2+fx_4+ex_6, & y_1&=x_1\sqrt{(L+\rho)}\\ y_4&=fx_2+bx_4+dx_6, & y_3&=x_3\sqrt{(M+\rho)}\\ y_6&=ex_2+dx_4+cx_6, & y_5&=x_5\sqrt{(N+\rho)}\end{aligned}\right\}\ldots\ldots\ldots(1),$$

where

$$\left.\begin{aligned} L+\rho-a_2^2&=a^2+f^2+e^2, & -a_2a_4&=af+fb+ed,\\ M+\rho-a_4^2&=f^2+b^2+d^2, & -a_4a_6&=ef+bd+cd,\\ N+\rho-a_6^2&=e^2+d^2+c^2, & -a_6a_2&=ae+fd+ce.\end{aligned}\right\}\ldots(2).$$

For this transformation (1) gives, as has been seen, (Art. 124), two cosingular complexes; also $\Sigma(y_i^2)\equiv f(x)+\rho\Sigma(x_i^2)$, while

$$\Sigma(x_i^2)\equiv(K_2y_2+K_4y_4+K_6y_6)^2+\frac{y_1^2+y_2^2}{L+\rho}+\frac{y_3^2+y_4^2}{M+\rho}+\frac{y_5^2+y_6^2}{N+\rho}\ldots\text{(II.)};$$

provided that values can be found for K_2, K_4, K_6 which make coexistent the equations

$$\left.\begin{aligned} \Delta^2\left(K_2^2+\frac{1}{L+\rho}\right)&=A^2+F^2+E^2, & \Delta^2K_2K_4&=AF+FB+ED,\\ \Delta^2\left(K_4^2+\frac{1}{M+\rho}\right)&=F^2+B^2+D^2, & \Delta^2K_4K_6&=EF+BD+CD,\\ \Delta^2\left(K_6^2+\frac{1}{N+\rho}\right)&=E^2+D^2+C^2, & \Delta^2K_6K_2&=AE+FD+CE,\end{aligned}\right\}(3);$$

where Δ is $\begin{vmatrix} a & f & e\\ f & b & d\\ e & d & c\end{vmatrix}$ and A, B, &c., its first minors.

Now it is easily found from (2) that

$$L+\rho=-\frac{EF+BD+CD}{ef+bd+cd},$$

and from the last equations, that

$$\frac{\Delta^2}{L+\rho} = -\frac{\Delta^2(ef+bd+cd)}{EF+BD+CD},$$

hence the equations (3) are coexistent.

To determine the manner in which ρ enters into the coefficients K_2, K_4, K_6, we notice that (II) becomes, for six values of ρ, the square of a linear function of the y_i, since the six fundamental complexes taken doubly form part of a cosingular system; this can only happen if K_2, K_4, K_6 become infinite together, or if K_2 is infinite for $\rho+L=0$, K_4 for $\rho+M=0$, K_6 for $\rho+N=0$; also $\rho=\infty$ gives the given complex (I).

All these conditions are satisfied by the form

$$\left(\frac{a_2}{\rho+L}y_2+\frac{a_4}{\rho+M}y_4+\frac{a_6}{\rho+N}y_6\right)^2$$
$$+\frac{(\rho-\rho_1)(\rho-\rho_2)(\rho-\rho_3)}{(\rho+L)(\rho+M)(\rho+N)}\left\{\frac{y_1^2+y_2^2}{\rho+L}+\frac{y_3^2+y_4^2}{\rho+M}+\frac{y_5^2+y_6^2}{\rho+N}\right\}=0,$$

where

$$a_2^2=\frac{(L+\rho_1)(L+\rho_2)(L+\rho_3)}{(M-L)(N-L)},\quad a_4^2=\frac{(M+\rho_1)(M+\rho_2)(M+\rho_3)}{(L-M)(N-M)},$$
$$a_6^2=\frac{(N+\rho_1)(N+\rho_2)(N+\rho_3)}{(L-N)(M-N)}.$$

This is the form of the equation of the cosingular complexes referred to the stated tetrahedron.

127. Involution of tangent linear complexes. *The tangent linear complexes of any line l, with regard to the four cosingular complexes through it, are mutually in involution; e.g.* for the complexes corresponding to μ_1 and μ_2 the tangent linear complexes are

$$\Sigma\left(\frac{l_i}{\lambda_i+\mu_1}+\nu l_i\right)x_i=0,\quad \Sigma\left(\frac{l_i}{\lambda_i+\mu_2}+\nu' l_i\right)x_i=0,$$

and since $$\Sigma\frac{l_i^2}{\lambda_i+\mu_1}=0,\quad \Sigma\frac{l_i^2}{\lambda_i+\mu_2}=0,$$

we have by subtraction

$$\Sigma\frac{l_i^2}{(\lambda_i+\mu_1)(\lambda_i+\mu_2)}=0,$$

hence the two complexes are in involution.

Moreover the involution determined on l by one pair of these complexes, *e.g.* $C_{\mu_1}^2$ and $C_{\mu_2}^2$, is the same as that determined by the other pair, $C_{\mu_3}^2$ and $C_{\mu_4}^2$; for the double points of the first involution are the points in which the *two lines* $\frac{l_i}{\lambda_i+\mu_1}+k\frac{l_i}{\lambda_i+\mu_2}$ meet l, while the double points of the second involution are the intersections of the *two lines* $\frac{l_i}{\lambda_i+\mu_3}+k'\frac{l_i}{\lambda_i+\mu_4}$ with l; and, since each of the first two lines meets each of the second, it follows that the double points of each involution are the vertices of this twisted quadrilateral which lie on l.

A tangent plane through l to Φ will contain a pencil, to which l belongs, of each of these four cosingular complexes; denoting them by C_1^2, C_2^2, C_3^2, and C_4^2 and the tangent linear complexes of l for them by T_1, T_2, T_3, T_4 respectively, the pencil which belongs to C_1^2 also belongs to T_1, and so on; if the four points in which l meets Φ are A_1, A_2, A_3, A_4, and the four tangent planes through it to Φ are β_1, β_2, β_3 and β_4, the notation of the points A may be so arranged that

(A_1, β_1) belongs to T_1 and C_1^2,
(A_2, β_1) .... T_2 and C_2^2,
(A_3, β_1) .... T_3 and C_3^2,
(A_4, β_1) .... T_4 and C_4^2.

Now let β_2 be that plane whose pole for T_1 is A_2, then will A_1 be its pole in T_2 (since T_1 and T_2 are in involution), also A_4 is its pole for T_3 and A_3 for T_4; similarly for β_3 and β_4, and we have the following scheme of pencils which belong respectively to the four complexes C_i^2,

C_1^2	(A_1, β_1)	(A_2, β_2)	(A_3, β_3)	(A_4, β_4),
C_2^2	(A_2, β_1)	(A_1, β_2)	(A_4, β_3)	(A_3, β_4),
C_3^2	(A_3, β_1)	(A_4, β_2)	(A_1, β_3)	(A_2, β_4),
C_4^2	(A_4, β_1)	(A_3, β_2)	(A_2, β_3)	(A_1, β_4).

128. The lines common to three of the cosingular complexes which pass through l, form a ruled surface R_{123} of the 16th degree which passes through l; for the lines of this surface which meet any line b are given by the equations

$$\Sigma\frac{x_i^2}{\lambda_i+\mu_1}=0,\quad \Sigma\frac{x_i^2}{\lambda_i+\mu_2}=0,\quad \Sigma\frac{x_i^2}{\lambda_i+\mu_3}=0,\quad (x^2)=0,\quad (bx)=0,$$

and are therefore 16 in number.

Corresponding tangent linear complexes are

$$\Sigma \frac{l_i x_i}{\lambda_i + \mu_1} = 0, \quad \Sigma \frac{l_i x_i}{\lambda_i + \mu_2} = 0, \quad \Sigma \frac{l_i x_i}{\lambda_i + \mu_3} = 0 \ldots\ldots \text{(i)};$$

if these equations are taken simultaneously, they determine a regulus which contains l and the line consecutive to l in R_{123}, *i.e. the regulus touches R_{123} along l*; now the tangent planes of this regulus along l are determined by l itself and the directrices of the special complexes of the 'system of three terms' determined by (i); but the complex $\Sigma \frac{l_i x_i}{\lambda_i + \mu_4} = 0$ is in involution with this system, hence these directrices belong to the latter complex, (Art. 58), therefore, *the surface of intersection of three cosingular complexes C_1^2, C_2^2, C_3^2 which contain a given line l is touched along l by the complex cones of the points of l for C_4^2.*

129. Conics determined in a plane by cosingular complexes. The cosingular complexes determine in any given plane a system of conics; four of these conics can be drawn to touch any line in the plane, since four complexes pass through the line.

If in the equation $\Sigma \frac{x_i^2}{\lambda_i + \mu} = 0$, we write $x_i = a_i + kb_i$, we obtain

$$\Sigma \frac{a_i^2}{\lambda_i + \mu} + 2k\Sigma \frac{a_i b_i}{\lambda_i + \mu} + k^2 \Sigma \frac{b_i^2}{\lambda_i + \mu} = 0.$$

If, now, (a, b) is a point on the complex conic corresponding to μ, the roots of this equation in k must be equal, *i.e.*

$$\Sigma \frac{a_i^2}{\lambda_i + \mu} \Sigma \frac{b_i^2}{\lambda_i + \mu} - \left(\Sigma \frac{a_i b_i}{\lambda_i + \mu}\right)^2 = 0.$$

Let $P_{ik} = (a_i b_k - a_k b_i)^2$, then we easily see that $\sum_k P_{ik} = 0$, since $\sum_k a_k b_k = 0$, and the equation to determine μ may be written

$$\Sigma \frac{P_{ik}}{(\lambda_i + \mu)(\lambda_k + \mu)} = 0.$$

In this equation the coefficient of μ^4 is $\sum_i \sum_k P_{ik}$ which is zero, and the coefficient of μ^3 is $\sum_i \sum_k P_{ik} \{\Sigma\lambda - (\lambda_i + \lambda_k)\}$, which is also zero; thus the equation for μ reduces to a quadratic; hence, through any point of the given plane there pass two conics of the system.

130. Elliptic coordinates of a line. The parameters μ of the four cosingular complexes of C^2 which pass through a line y may be taken as the coordinates of this line*, and if

$$f(\mu)=(\lambda_1+\mu)\ldots(\lambda_6+\mu),$$

we have

$$\rho \,.\, y_i^2=\frac{(\lambda_i+\mu_1)(\lambda_i+\mu_2)(\lambda_i+\mu_3)(\lambda_i+\mu_4)}{f'(-\lambda_i)} \quad \ldots\ldots \text{(i)};$$

for since

$$\Sigma\frac{1}{f'(-\lambda_i)}=\Sigma\frac{\lambda_i}{f'(-\lambda_i)}=\Sigma\frac{\lambda_i^2}{f'(-\lambda_i)}=\Sigma\frac{\lambda_i^3}{f'(-\lambda_i)}$$
$$=\Sigma\frac{\lambda_i^4}{f'(-\lambda_i)}=0 \quad \ldots\ldots\ldots\ldots\ldots\ldots \text{(ii)},$$

it is easily verified that y belongs to the complexes corresponding to μ_1, μ_2, μ_3, and μ_4, where the latter quantities are the roots of

$$\Sigma\frac{y_i^2}{\lambda_i+\mu}=0;$$

we shall denote these complexes by C_1^2, C_2^2, C_3^2, C_4^2.

The quantities $\dfrac{1}{\sqrt{f'(-\lambda_i)}}$ are coordinates of the singular lines of C^2 which satisfy the equations $(\lambda^3x^2)=0$; $(\lambda^4x^2)=0$, (Art. 80), *i.e.* the singular lines of the third order.

Taking one of the four complexes as C^2, which corresponds to $\mu=\infty$, (Art. 115), the lines x of C^2 are given by the equations

$$\rho \,.\, x_i^2=\frac{(\lambda_i+\mu_1)(\lambda_i+\mu_2)(\lambda_i+\mu_3)}{f'(-\lambda_i)},$$

and the *singular* lines of C^2 by the equations

$$\rho \,.\, x_i^2=\frac{(\lambda_i+\mu_1)(\lambda_i+\mu_2)}{f'(-\lambda_i)} \quad \ldots\ldots\ldots\ldots\ldots \text{(iii)};$$

for the lines determined thereby satisfy the equations $(\lambda x^2)=0$, $(\lambda^2x^2)=0$, for all values of μ_1 and μ_2.

The lines $\rho \,.\, x_i^2=\dfrac{\lambda_i+\mu}{f'(-\lambda_i)}$ satisfy *also* the equation $(\lambda^3x^2)=0$, and hence are the tangents to the principal tangent curve determined by C^2 on Φ, (Art. 80), *i.e.* the singular lines of the second order.

If in equations (i), μ_1 and μ_2 have *given* values, and $\mu_3=\mu_4=\mu$, we obtain the 32 pencils of lines

$$\sigma \,.\, y_i=\pm(\lambda_i+\mu)\sqrt{\frac{(\lambda_i+\mu_1)(\lambda_i+\mu_2)}{f'(-\lambda_i)}} \quad \ldots\ldots \text{(iv)};$$

* This method is due to Klein, see *Math. Ann.* II.

these pencils are the tangent lines of Φ at the points of contact of the 32 singular lines

$$\pm\sqrt{\frac{(\lambda_i+\mu_1)(\lambda_i+\mu_2)}{f'(-\lambda_i)}} \text{ of } C^2,$$

i.e. a tangent of Φ is $(\lambda_i+\mu)x_i$, where x_i is a singular line of C^2. The quantities μ_1 and μ_2 may be regarded as determining a point on Φ.

If in (iv) μ_1 be taken as constant and μ_2 vary, the pencils of lines y belong to C_1^2, and touch Φ, but are not singular lines of C_1^2, hence they are the tangents to Φ at the points of the principal tangent curve whose tangents are singular lines of C_1^2 of the second order; these latter tangents are obtained by putting $\mu=\mu_1$; hence, *the tangents to the principal tangent curve of* Φ *related to the complex* C_1^2 *are obtained by putting* $\mu_3=\mu_4=\mu_1$ *and varying* μ_2: *taking* $\mu_1=$ *constant, and* $\mu_2=\mu_3=\mu_4$, *gives the other principal tangents of* Φ *at the points of this curve.*

If x and x' are the singular lines of C^2 at the points of contact P and P' of a bitangent line of Φ, $x_k=x'_k$ except for one value of k, say i, for which $x_i=-x_i'$, (Art. 83); hence $x_k^2=x_k'^2$, and the values of μ_1 and μ_2 in (iii) are the same for each singular line; but the tangents y_i at P being given by the equations

$$\rho\,.\,y_i^2=(\lambda_i+\mu)^2(\lambda_i+\mu_1)(\lambda_i+\mu_2)/f'(-\lambda_i),$$

we see that P lies on the principal tangent curve of C_1^2 and also on that of C_2^2, and the same holds for P', so that these points P and P' lie on the same two principal tangent curves.

When two pairs of values of μ in (i) are equal, *e.g.* $\mu_3=\mu_1$, $\mu_4=\mu_2$, x is a singular line of both the complexes C_1^2 and C_2^2, and would therefore seem to be a bitangent line of Φ, but in this case

$$\rho\,.\,x_i=\frac{(\lambda_i+\mu_1)(\lambda_i+\mu_2)}{\sqrt{f'(-\lambda_i)}};$$

whence x meets the three lines

$$\frac{1}{\sqrt{f'(-\lambda_i)}},\quad \frac{\lambda_i}{\sqrt{f'(-\lambda_i)}},\quad \frac{\lambda_i^2}{\sqrt{f'(-\lambda_i)}},$$

which themselves belong either to a sheaf or to a plane pencil; hence x is any line in a singular tangent plane or through a double point of Φ.

The equation $(dy^2)=0$, when expressed in terms of the coordinates μ_i of the line y, becomes

$$d\mu_1^2\frac{(\mu_1-\mu_2)(\mu_1-\mu_3)(\mu_1-\mu_4)}{f(\mu_1)}+d\mu_2^2\frac{(\mu_2-\mu_1)(\mu_2-\mu_3)(\mu_2-\mu_4)}{f(\mu_2)}$$
$$+d\mu_3^2\frac{(\mu_3-\mu_1)(\mu_3-\mu_2)(\mu_3-\mu_4)}{f(\mu_3)}+d\mu_4^2\frac{(\mu_4-\mu_1)(\mu_4-\mu_2)(\mu_4-\mu_3)}{f(\mu_4)}=0.$$

If μ_3 and μ_4 are constant, we have as the differential equation of the curves of the congruence (C_3^2, C_4^2)

$$d\mu_1\sqrt{\frac{(\mu_1-\mu_3)(\mu_1-\mu_4)}{f(\mu_1)}}=d\mu_2\sqrt{\frac{(\mu_2-\mu_3)(\mu_2-\mu_4)}{f(\mu_2)}};$$

if $\mu_3=\mu_4$, the differential equation of the curves whose tangents are singular lines of C_3^2, is seen to be

$$\frac{d\mu_1(\mu_1-\mu_3)}{\sqrt{f(\mu_1)}}=\frac{d\mu_2(\mu_2-\mu_3)}{\sqrt{f(\mu_2)}}.$$

131. Bitangent linear complexes. Of the ∞^1 tangent linear complexes $\Sigma(\lambda_i+\mu)x_iy_i=0$, of C^2, which have in common with C^2 not only x but all lines of C^2 consecutive to x, there are six which are *bitangent*, viz. those obtained by writing successively $\lambda_1+\mu=0, \ldots\ldots, \lambda_6+\mu=0$. For the complex

$$\overline{\lambda_2-\lambda_1}\,x_2y_2+\overline{\lambda_3-\lambda_1}\,x_3y_3+\overline{\lambda_4-\lambda_1}\,x_4y_4+\overline{\lambda_5-\lambda_1}\,x_5y_5+\overline{\lambda_6-\lambda_1}\,x_6y_6=0$$

"touches" C^2 both in x and in the line all of whose coordinates are the same as x except x_1, *i.e.* the polar of x for the complex $x_1=0$: denote these six complexes by $T_1, T_2, T_3, T_4, T_5, T_6$.

If A be any linear complex, $(ax)=0$, it will have a pair of polar lines z in common with T_1, of which the coordinates are given by the equations

$$\left.\begin{aligned}\sigma.z_1&=\mu a_1,\\ \sigma.z_2&=(\lambda_2-\lambda_1)x_2+\mu a_2,\\ \sigma.z_3&=(\lambda_3-\lambda_1)x_3+\mu a_3,\\ \sigma.z_4&=(\lambda_4-\lambda_1)x_4+\mu a_4,\\ \sigma.z_5&=(\lambda_5-\lambda_1)x_5+\mu a_5,\\ \sigma.z_6&=(\lambda_6-\lambda_1)x_6+\mu a_6\end{aligned}\right\}\ldots\ldots\ldots\ldots\ldots\ldots\text{(i)};$$

where μ has either of the values obtained by expressing that $(z^2)=0$. The locus of the lines z is a quadratic complex S_1^2; for, eliminating the x_i from these equations we find

$$\frac{(z_1a_2-z_2a_1)^2}{\lambda_2-\lambda_1}+\frac{(z_1a_3-z_3a_1)^2}{\lambda_3-\lambda_1}+\frac{(z_1a_4-z_4a_1)^2}{\lambda_4-\lambda_1}$$
$$+\frac{(z_1a_5-z_5a_1)^2}{\lambda_5-\lambda_1}+\frac{(z_1a_6-z_6a_1)^2}{\lambda_6-\lambda_1}=0.$$

We obtain in this way six quadratic complexes S_i^2, which we shall prove to be *cosingular**, having the focal surface of (C^2, A) as singular surface.

For, from the equations (i) we deduce

$$\tfrac{1}{2}\sigma\frac{\partial S_1^2}{\partial z_i} = \sigma a_1 . \frac{z_i a_1 - z_1 a_i}{\lambda_i - \lambda_1} = a_1^2 x_i, \quad (i = 2, 3, \ldots 6),$$

hence
$$\tfrac{1}{2}\sigma\frac{\partial S_1^2}{\partial z_1} = -\sigma\sum_2^6 a_i . \frac{z_i a_1 - z_1 a_i}{\lambda_i - \lambda_1} = -a_1\sum_2^6 a_i x_i$$
$$= a_1^2 x_1, \text{ if } \sum_1^6 a_i x_i = 0.$$

So that if x belongs to A, $\dfrac{\partial S_1^2}{\partial z}$ is a line, viz. x, and z is a singular line of S_1^2.

Hence x is a tangent to the singular surface of S_1^2 at the point of contact of the singular line z. The same thing applies to the other line, z', associated with x by the equations (i). Therefore, any line x of the congruence (C^2, A) is a *bitangent* of the singular surface of S_1^2.

This result holds, similarly, for the other five complexes $S_2^2, \ldots S_6^2$.

Hence these complexes have as their common singular surface the focal surface of the congruence (C^2, A).

132. Principal Surfaces†. The tangent linear complexes of a line x which belongs to a complex $F = 0$ of degree n, being

$$(yf) + \mu(yx) = 0, \quad \text{(Art. 74)} \quad \ldots\ldots\ldots\ldots\text{(i)},$$

where $f_i \equiv \dfrac{\partial F}{\partial x_i}$, the tangent complex corresponding to a consecutive line $x + dx$ is

$$\{y(df + \mu dx + x d\mu)\} + \{y(f + \mu x)\} = 0 \quad \ldots\ldots\ldots\text{(ii)},$$

where the dx_i are connected by the equations

$$(xdx) = 0, \quad (fdx) = 0, \quad (kdx) = 0,$$

in which $(kx) = 1$, (the quantities k_i being constant), is an equation arbitrarily assumed between the coordinates x_i of any line‡.

If the complexes (i) and (ii) are the *same*, we must have that

$$df_i + \mu dx_i + x_i d\mu = dt\,(f_i + \mu x_i) \quad \ldots\ldots\ldots\ldots\text{(iii)}.$$

* This theorem was communicated to the author by Mr J. H. Grace.

† These surfaces are of interest from their analogy with lines of curvature in four dimensions; see Art. 228.

‡ This method is due to Voss, see *Math. Ann.* IX. "Ueber Complexe und Congruenzen."

This gives nine equations between $dx_1, \ldots\ldots, dx_6, d\mu, dt$; but since $(xdf)=0$, they are equivalent to eight equations. Eliminating the differentials we obtain

$$\begin{vmatrix} f_{11}+\mu & f_{12} & f_{13} & f_{14} & f_{15} & f_{16} & x_1 & f_1+\mu x_1 \\ \cdots & \cdots & \cdots & \cdots & \cdots & \cdots & \cdots & \cdots \\ \cdots & \cdots & \cdots & \cdots & \cdots & \cdots & \cdots & \cdots \\ f_{61} & f_{62} & f_{63} & f_{64} & f_{65} & f_{66}+\mu & x_6 & f_6+\mu x_6 \\ k_1 & k_2 & k_3 & k_4 & k_5 & k_6 & 0 & 0 \\ x_1 & x_2 & x_3 & x_4 & x_5 & x_6 & 0 & 0 \end{vmatrix} = 0.$$

Now multiplying the last column by $n-1$ and subtracting from it the first six columns multiplied respectively by $x_1, \ldots x_6$ and the seventh by $\mu(n-2)$, we find

$$\begin{vmatrix} f_{11}+\mu & f_{12} & f_{13} & f_{14} & f_{15} & f_{16} & x_1 & 0 \\ \cdots & \cdots & \cdots & \cdots & \cdots & \cdots & \cdots & \cdots \\ f_{61} & f_{62} & f_{63} & f_{64} & f_{65} & f_{66}+\mu & x_6 & 0 \\ k_1 & k_2 & k_3 & k_4 & k_5 & k_6 & 0 & -(kx) \\ x_1 & x_2 & x_3 & x_4 & x_5 & x_6 & 0 & 0 \end{vmatrix} = 0;$$

or, finally,

$$\begin{vmatrix} f_{11}+\mu & f_{12} & f_{13} & f_{14} & f_{15} & f_{16} & x_1 \\ \cdots & \cdots & \cdots & \cdots & \cdots & \cdots & \cdots \\ \cdots & \cdots & \cdots & \cdots & \cdots & \cdots & \cdots \\ f_{61} & \cdots & \cdots & \cdots & \cdots & f_{66}+\mu & x_6 \\ x_1 & \cdots & \cdots & \cdots & \cdots & x_6 & 0 \end{vmatrix} = 0.$$

This is an equation to determine μ which is of the third degree, since the coefficient of μ^4 is equal to $n(n-1)F$, which is zero; hence, there are in general three finite and distinct values of μ, say μ_1, μ_2, μ_3; each of them gives one set of values of dx_i, hence *there are three lines of F consecutive to x for each of which one tangent linear complex is the same as one of x.* Thus starting from x we may proceed to the one of the three consecutive lines which corresponds to μ_1 and then from that line to the one which corresponds to $\mu_1+d\mu_1$, and so on; thus we have a singly infinite set of lines forming a ruled surface; such a surface is called a Principal Surface of the Complex: in each line x of F three Principal Surfaces intersect.

In the case of the quadratic complex C^2, or $(\lambda x^2)=0$, the determinant for μ is

$$\begin{vmatrix} \lambda_1+\mu & 0 & 0 & 0 & 0 & 0 & x_1 \\ 0 & \lambda_2+\mu & \cdots & \cdots & \cdots & \cdots & x_2 \\ \cdots & \cdots & \cdots & \cdots & \cdots & \cdots & \cdots \\ 0 & \cdots & \cdots & \cdots & \cdots & \lambda_6+\mu & x_6 \\ x_1 & \cdots & \cdots & \cdots & \cdots & x_6 & 0 \end{vmatrix} = 0, \text{ i.e. } \Sigma \frac{x_i^2}{\lambda_i+\mu} = 0.$$

Hence *the values of* μ *are those which give the three complexes through* x *cosingular to* C^2; so that if dx_i' are the increments of the x_i corresponding to μ_1, dx_i'' corresponding to μ_2, dx_i''' corresponding to μ_3, the equations (iii) become in this case

$$dx_i' + \frac{x_i d\mu_1}{\lambda_i+\mu_1} = dt \,.\, x_i \quad \text{.................(iv)},$$

hence $$\Sigma \frac{x_i dx_i'}{\lambda_i+\mu_2} + d\mu\, \Sigma \frac{x_i^2}{(\lambda_i+\mu_1)(\lambda_i+\mu_2)} = dt\, \Sigma \frac{x_i^2}{\lambda_i+\mu_2},$$

therefore $$\Sigma \frac{x_i dx_i'}{\lambda_i+\mu_2} = 0, \text{ similarly } \Sigma \frac{x_i dx_i'}{\lambda_i+\mu_3} = 0,$$

i.e. the line $x_i + dx_i'$ belongs to $C_{\mu_2}^2$ and $C_{\mu_3}^2$; hence *the Principal Surfaces for* C^2 *are the ruled surfaces* R_{12}, R_{23}, R_{31}; *i.e. the intersections of* C^2 *with two of the three quadratic complexes through* x *cosingular with* C^2.

133. Involutory position of two lines. It is to be noticed that from equations (iv) it follows that

$$(dx'dx'')=0, \quad (dx''dx''')=0, \quad (dx'''dx')=0:$$

we shall show that this holds for any complex.

In equations (iii) let the line x be taken as the edge A_1A_4 of the tetrahedron of reference, and let the equation $(kdx)=0$ be $dx_6=0$, then since A_1A_4 has the coordinates $(0, 0, 0, 0, 1, -i)$, the equation $(xdx)=0$ becomes $dx_5 - i dx_6 = 0$, hence $dx_5=0$.

In the present case, therefore, the equations (iii) take the form

$$df_1+\mu dx_1 = dt\,.\,f_1, \quad df_2+\mu dx_2 = dt\,.\,f_2, \quad df_3+\mu dx_3 = dt\,.\,f_3, \quad df_4+\mu dx_4 = dt\,.\,f_4,$$

or,

$$\left.\begin{array}{l} (f_{11}+\mu)\,dx_1 + f_{12}\,dx_2 + f_{13}\,dx_3 + f_{14}\,dx_4 = dt\,.\,f_1, \\ \cdots\cdots\cdots\cdots\cdots\cdots\cdots\cdots \\ \cdots\cdots\cdots\cdots\cdots\cdots\cdots\cdots \\ f_{14}\,dx_1 + f_{24}\,dx_2 + f_{34}\,dx_3 + (f_{44}+\mu)\,dx_4 = dt\,.\,f_4, \\ f_1\,dx_1 + f_2\,dx_2 + f_3\,dx_3 + f_4\,dx_4 = 0. \end{array}\right\}$$

together with the last equation.

These equations show that $\frac{dx_1}{dt}$, $\frac{dx_2}{dt}$, $\frac{dx_3}{dt}$, $\frac{dx_4}{dt}$ are proportional to the coordinates of a point in which a quadric of the pencil

$$f_{11}\,\xi_1^2 + 2f_{12}\,\xi_1\xi_2 + \ldots\ldots + \mu(\xi_1^2+\xi_2^2+\xi_3^2+\xi_4^2) = 0,$$

touches the plane

$$f_1\xi_1 + f_2\xi_2 + f_3\xi_3 + f_4\xi_4 = 0.$$

It is easily seen that there are three such quadrics* and that any two points of contact are conjugate with regard to any quadric of the pencil, and hence with regard to $\xi_1^2+\xi_2^2+\xi_3^2+\xi_4^2=0$; therefore

$$(dx'\,dx'')=0, \quad (dx''\,dx''')=0, \quad (dx'''\,dx')=0.$$

To determine the property expressed by these equations we notice that any plane π through A_1A_4, having coordinates $(0, \pi_2, \pi_3, 0)$, meets the line joining the points a_i and β_i in the point $a_i+\lambda\beta_i$, where

$$\lambda=-\frac{\pi_2 a_2+\pi_3 a_3}{\pi_2\beta_2+\pi_3\beta_3};$$

and meets the line joining a_i' to β_i' in the point $a_i'+\mu\beta_i'$, where

$$\mu=-\frac{\pi_2 a_2'+\pi_3 a_3'}{\pi_2\beta_2'+\pi_3\beta_3'};$$

then, eliminating π_2 and π_3 we obtain a relation between λ and μ which is *symmetrical* if

$$\beta_2 a_3'-\beta_3 a_2'+a_3\beta_2'-a_2\beta_3'=0.$$

The last equation may be written

$$(\beta_2+a_3)(\beta_2'+a_3')-(\beta_2-a_3)(\beta_2'-a_3')+(\beta_3-a_2)(\beta_3'-a_2')-(\beta_3+a_2)(\beta_3'+a_2')=0.$$

Now if the points a_i and a_i' are consecutive to A_1 and the points β_i and β_i' to A_4, then taking $p_{ik}=a_i\beta_k-a_k\beta_i$, $x_1=p_{12}+p_{34}$, &c.; it is clear that

$$dp_{12}=\beta_2, \quad dp_{34}=a_3, \quad dp_{13}=\beta_3, \quad dp_{42}=-a_2, \text{ \&c.,}$$

and the last equation becomes

$$dx_1dx_1'+dx_2dx_2'+dx_3dx_3'+dx_4dx_4'=0;$$

this equation is therefore the condition that the planes through the line x should meet the two lines $x+dx$, $x+dx'$ in pairs of points *ultimately forming an involution.* The two lines consecutive to x are then said to have an Involutory position with regard to each other.

* Since the sections of the quadrics $U+\lambda V=0$ by any plane form a *pencil* of conics through the four points in which the curve (U, V) meets the plane; and *three* of these conics consist of a pair of lines. Moreover the diagonals of the complete quadrilateral formed by the four points form a self-conjugate triangle with reference to any conic through the four points.

CHAPTER IX.

POLAR LINES, POINTS, AND PLANES.

134. Polar lines. The polar line l' of a line l with reference to a complex C_μ^2, cosingular to C^2, has coordinates $l_i - k\dfrac{l_i}{\lambda_i + \mu}$, (Art. 79), where

$$k = \frac{2\Sigma \dfrac{l_i^2}{\lambda_i + \mu}}{\Sigma \dfrac{l_i^2}{(\lambda_i + \mu)^2}};$$

l' coincides with l for values of μ which make $k = 0$, *i.e.* for the four values given by

$$\Sigma \frac{l_i^2}{\lambda_i + \mu} = 0.$$

Regarding μ as variable, the lines l' generate a ruled surface, on which l is therefore a fourfold line; to find the degree of this surface, we determine the number of lines l' which meet a given line a, *i.e.* for which $(al') = 0$, or, for which

$$(al) - k\Sigma \frac{a_i l_i}{\lambda_i + \mu} = 0;$$

if l and a intersect this equation reduces to

$$k\Sigma \frac{a_i l_i}{\lambda_i + \mu} = 0.$$

Hence there are eight values of μ, viz. the four obtained from the equation $k = 0$, and the four from $\Sigma \dfrac{a_i l_i}{\lambda_i + \mu} = 0$; the degree of the surface is therefore eight.

An exceptional case is that of a line which is common to four fundamental complexes, *e.g.* for either of the lines

$$x_1 = x_2 = x_3 = x_4 = 0,$$

the polar of *one* of these lines, with regard to any complex included in the series $(kx^2)=0$, is *the other*; these are the only lines, in the case of the *general* complex, for which the polar relationship is reciprocal.

There are nine lines l for which l' is the polar with regard to C^2; for since $\rho l_i' = l_i(1-k\lambda_i)$, where $k=\frac{2(\lambda l^2)}{(\lambda^2 l^2)}$, the values of k are those given by the equation

$$\Sigma \frac{l_i'^2}{(1-k\lambda_i)^2}=0,$$

and are nine in number.

These lines form a closed system; for any one of them being given, l' is determined, and hence the other eight lines l.

135. Between the ten lines consisting of l' and the nine lines l for which l' is the polar for a quadratic complex the following remarkable relation exists*.

In any quadratic complex $C_{\mu_1}^2$, or $\Sigma \frac{x_i^2}{\lambda_i+\mu_1}=0$, the lines x for which a given line l^{I} is the polar for $C_{\mu_1}^2$, are afforded by the equations

$$\rho . l_i^{\text{I}} = x_i\left(1-\frac{k}{\lambda_i+\mu_1}\right),$$

or, if $\mu=\mu_1-k$,

$$\rho . l_i^{\text{I}}(\lambda_i+\mu_1) = x_i(\lambda_i+\mu).$$

The equation to determine μ is $\Sigma \frac{\{l_i^{\text{I}}(\lambda_i+\mu_1)\}^2}{(\lambda_i+\mu)^2}=0$, which is of the tenth degree, and of which one root is μ_1; let the others be denoted by $\mu_2, \mu_3, \dots \mu_{10}$, then between l^{I} and the nine lines $l^{\text{II}} \dots l^{\text{X}}$, of which l^{I} is the polar for $C_{\mu_1}^2$, we have the equations

$$\rho . l_i^{\text{I}}(\lambda_i+\mu_1) = l_i^{\text{II}}(\lambda_i+\mu_2) = l_i^{\text{III}}(\lambda_i+\mu_3) = \dots = l_i^{\text{X}}(\lambda_i+\mu_{10}).$$

Again starting with l^{II}, we find as the nine lines of which l^{II} is the polar for $C_{\mu_2}^2$, those given by the equations

$$\sigma . l_i^{\text{II}}(\lambda_i+\mu_2) = x_i(\lambda_i+\mu),$$

where μ is one of the roots of the equation

$$\Sigma \frac{\{l_i^{\text{II}}(\lambda_i+\mu_2)\}^2}{(\lambda_i+\mu)^2}=0;$$

an equation of the tenth degree of which one root is μ_2.

Inserting in this equation for $l_i^{\text{II}}(\lambda_i+\mu_2)$, its value

$$\rho . l_i^{\text{I}}(\lambda_i+\mu_1),$$

we obtain, to determine μ, the *same equation* as before.

* This theorem is due to W. Stahl; *Crelle's Journal* (1883).

Hence we conclude that there are 10 complexes $C_{\mu_1}{}^2, \dots C_{\mu_{10}}{}^2$ cosingular with C^2 such that

for $C_{\mu_1}{}^2$ the line $l^{\rm I}$ is the polar of $l^{\rm II}, l^{\rm III}, \dots l^{\rm X}$;

... $C_{\mu_2}{}^2$ .... $l^{\rm II}$ .... $l^{\rm I}, l^{\rm III}, \dots l^{\rm X}$;

and so on; and having given any line l and any quadratic complex, nine other lines, and nine other cosingular complexes are determined having these relations.

136. Corresponding loci of polar lines. From the connexion between the coordinates of a line x and its polar x' for C^2, it is clear that when x' describes a complex of degree n, x describes a complex of degree $3n$, since x_i' is proportional to a cubic expression in the coordinates x_i.

If x' belongs to the linear complex $(ay) = 0$, the locus of x is

$$(ax) - k(\lambda ax) = 0,$$

i.e.

$$(ax)(\lambda^2 x^2) - 2(\lambda x^2)(\lambda ax) = 0,$$

a cubic complex, to which the singular lines of C^2 belong.

137. *If x describes a plane pencil, its polar x' describes a ruled cubic.*

For let $x_i = a_i + \mu b_i$, where a and b are the lines of C^2 in the given pencil, then

$$\rho x_i' = a_i + \mu b_i - k\lambda_i(a_i + \mu b_i),$$

hence writing

$$(\lambda^2 a^2) = A, \quad (\lambda^2 b^2) = B, \quad (\lambda^2 ab) = C, \quad (\lambda ab) = D,$$

we have

$$4\mu D - k(A + 2\mu C + \mu^2 B) = 0,$$

therefore

$$\sigma \,.\, x_i' = (a_i + \mu b_i)(A + 2\mu C + \mu^2 B) - 4\mu D\lambda_i a_i - 4\mu^2 D\lambda_i b_i;$$

from which it follows that any line meets three of the lines x', hence the locus of x' is a ruled cubic whose directrices are the lines common to the complexes

$$(ax) = 0, \quad (bx) = 0, \quad (\lambda ax) = 0, \quad (\lambda bx) = 0.$$

These lines* are the polar of the point (a, b) for the complex

* The relation between two intersecting lines α, β which are connected by the equation $(\lambda\alpha\beta) = 0$ should be noticed; it states that the polar line of α for C^2 meets β, and *vice versâ*; hence α and β are conjugate lines both with regard to the complex conic of the plane (α, β), and with regard to the complex cone of the point (α, β).

There are two lines x which satisfy the equations

$$(x\alpha) = 0, \quad (x\beta) = 0, \quad (\lambda\alpha x) = 0, \quad (\lambda\beta x) = 0, \quad (\alpha\beta) = 0;$$

conic of the plane (a, b), and the polar line for the plane (a, b) of the complex cone of the point (a, b); since the latter line meets the two lines a and b (which are special positions of x'), it is therefore the double directrix of the surface.

If the given pencil contains a singular line of C^2 the locus of x' is a regulus. For if b is a singular line, then $B = 0$.

Two polar lines l and l' determine a congruence $(lx) = 0$, $(l'x) = 0$; if l describes a pencil $a + \mu b$, these congruences give rise to the complex

$$(ax)(\lambda bx) - (bx)(\lambda ax) = 0.$$

If l' is the polar of l and P any point on l, then if the lines x describe the pencil (P, l'), l is the double directrix of the locus of x'; for here, since $(xl') = 0$, and also

$$(xl') = \Sigma x_i l_i (1 - k\lambda_i) = -k(\lambda lx),$$

therefore $$(\lambda lx) = 0;$$

moreover $$(lx') = \Sigma l_i x_i (1 - \rho\lambda_i) = -\rho(\lambda lx) = 0;$$

therefore all the lines x' meet l. But there are two lines x' in the pencil (P, l'), viz. the two lines of C^2, hence l is the double directrix of the locus of x'. Through each point P' of l there pass two lines x', to which there correspond two lines x of (P, l'); we deduce that *if P and P' are any two points on any line l, there are two planes through l for whose complex conics P and P' are conjugate points.*

138. *When x' describes a sheaf of centre P, x will describe a congruence of order 2 and class 3.* For the locus of x is clearly that which is formed by the polar lines of P with regard to the complex conics of the planes through P; and if P' be any point, there are two planes through PP' for which P and P' are conjugate with regard to their complex conic; hence, through any point P' we can draw two lines of the locus of x. To find the *class* of the locus of x (or the number of lines x in any plane), we proceed as follows:—

the one which lies in the plane (α, β), being conjugate both to α and β, is the polar of the point (α, β) for the complex conic of the plane (α, β); similarly the line x which passes through the point (α, β) is seen to be the polar line of the complex cone of the point (α, β) for the plane (α, β).

If in addition to $(\alpha\beta) = 0$, $(\lambda\alpha\beta) = 0$, we have $(\lambda^2\alpha\beta) = 0$, the polar lines for C^2 of α and β will intersect.

Consider the locus of x when x' meets two given lines a and b; it is obtained from the equations

$$(ax) = k\,(a\lambda x),\quad (bx) = k\,(b\lambda x),\quad k = \frac{2\,(\lambda x^2)}{(\lambda^2 x^2)};$$

they give a congruence whose lines are included in the congruence

$$\left.\begin{aligned}(ax)(b\lambda x) - (bx)(a\lambda x) = 0\\ (ax)(\lambda^2 x^2) - 2\,(\lambda x^2)(a\lambda x) = 0\end{aligned}\right\} \quad \ldots\ldots\ldots\ldots\ldots\text{(i)}.$$

The congruence given by these equations is of degree and class 6, and excluding the lines of the congruence $(ax) = 0$, $(a\lambda x) = 0$, we obtain as the *required* congruence, one of degree and class 5. Now suppose a and b to intersect, then we may divide the locus of x' into two portions, viz. the sheaf (a, b) and the plane system (a, b). To the former corresponds a congruence K of lines x, which is of order 2, from above, and to the latter corresponds another congruence K' of lines x, which must therefore be of order 3.

By duality we see that the order of K is equal to the class of K', and conversely, hence K is a congruence (2, 3) and K' is a congruence (3, 2); so that, *when x' describes a sheaf, x describes a congruence* (2, 3); *when x' describes a plane system, x describes a congruence* (3, 2).

Since the polar line of a line of C^2 coincides with it, the point (a, b) is "singular" for the complex K, *i.e.* all the lines of the complex cone of this point belong to K; similarly the tangents of the complex conic of the plane (a, b) belong to K'.

Thus through any point Q we have two lines x, and corresponding to them their polars x' through P; taking another position of P, we have again two lines x' corresponding to two other lines of the sheaf Q.

Hence, *the locus of lines x' corresponding to a sheaf of lines x is a congruence K_1 of order* 2, and similarly it is seen to be of *class* 3. Reciprocally, *if x describe a plane system, x' describes a congruence K_1' of order* 3 *and class* 2.

139. *If x' describes a plane pencil, x will describe a ruled surface whose degree is* 7. For it has been seen that the lines x, whose polars x' meet any line q, form a cubic complex Q^3, to which the singular lines of C^2 belong. The intersection of this complex Q^3 with the congruence given by the previous equations (i), (Art. 138), is a ruled surface of degree $2\,.\,3\,.\,2\,.\,3 = 36$; while the intersection of Q^3 and the congruence $(ax)=0$, $(a\lambda x)=0$ is a ruled sextic; hence the lines x of Q^3 whose polars x' meet the lines a and b form a ruled surface of degree 30.

But the singular lines of C^2 which belong to the congruence (i), are given by the equations

$$(\lambda x^2)=0, \quad (\lambda^2x^2)=0, \quad (ax)(b\lambda x)-(bx)(a\lambda x)=0,$$

and hence form a ruled surface of degree 16. We conclude therefore, that *the lines x whose polars x' meet the three lines q, a and b, (thus forming a regulus), are the generators of a ruled surface of degree* 14.

If a and b intersect each other the regulus formed by x' becomes two pencils, and the locus of x becomes two surfaces of degree 7.

140. Polar planes and points of the complex. Denoting by x, y, z, w the four singular lines through any point P, we have the series of identical equations (Art. 13),

$$Ax_i+By_i+Cz_i+Dw_i=0 \quad \ldots\ldots\ldots\ldots\ldots\ldots\text{(i)},$$

the line α common to the planes of xy and zw, has therefore the coordinates Ax_i+By_i, (or Cz_i+Dw_i), (Art. 13); similarly β which is common to the planes xz and yw, is Ax_i+Cz_i; and γ which is common to the planes xw and yz, is Ax_i+Dw_i.

From (i), the equations

$$\left.\begin{array}{lll}(\lambda\alpha\beta)=0, & (\lambda\beta\gamma)=0, & (\lambda\gamma\alpha)=0,\\ (\lambda^2\alpha\beta)=0, & (\lambda^2\beta\gamma)=0, & (\lambda^2\gamma\alpha)=0\end{array}\right\} \quad \ldots\ldots\ldots\text{(ii)},$$

at once follow, *e.g.*

$$\begin{aligned}(\lambda\alpha\beta)&=\Sigma\lambda_i(Ax_i+By_i)(Ax_i+Cz_i)\\ &=BC(\lambda yz)+CA(\lambda zx)+AB(\lambda xy)^*=0.\end{aligned}$$

Now it was seen, (Art. 137), that (P being the point (α, β)) the two solutions of the equations,

$$(u\alpha)=0, \quad (\lambda\alpha u)=0, \quad (u^2)=0,$$
$$(u\beta)=0, \quad (\lambda\beta u)=0, \quad (\alpha\beta)=0,$$

are the polar of P for the complex conic of the plane (α, β), and the polar line of the plane (α, β) for the complex cone of P; in the present case the solutions are obviously γ_i and $\lambda_i\gamma_i+\kappa''\gamma_i$, *i.e.* γ, and its polar line for C^2; hence it is seen that the polar line

* This arises from squaring each side of the equations $-Dw_i=Ax_i+By_i+Cz_i$, multiplying by λ_i, and adding.

for C^2 of any one of the lines α, β, γ is the polar line of P for the complex conic of the plane of the other two; denoting these polar lines by α', β', γ' it is clear that they intersect each other; *e.g.*

$$(\alpha'\beta') = \Sigma\,(\lambda\alpha_i + \kappa\alpha_i)(\lambda\beta_i + \kappa'\beta_i) = 0;$$

the plane of the polar lines of α, β, γ is called *the polar plane of P for the complex C^2*.

Again from (i) we derive the system of equations

$$A\,(\lambda_i x_i + \mu x_i) + B\,(\lambda_i y_i + \mu y_i) + C\,(\lambda_i z_i + \mu z_i) + D\,(\lambda_i w_i + \mu w_i) = 0;$$

so that the four lines $\lambda x + \mu x, \ldots\ldots\ldots, \lambda w + \mu w$ are seen to lie, for any given value of μ, on the same regulus ρ. Now the line α' is $\lambda_i\,(Ax_i + By_i) + \kappa\,(Ax_i + By_i)$ or $A\,(\lambda x + \kappa x) + B\,(\lambda y + \kappa y)$ where $(\lambda^2 xy) + 2\kappa\,(\lambda xy) = 0$, and, in consequence, $\lambda x + \kappa x, \lambda y + \kappa y$ intersect in a point O_1; thus for $\mu = \kappa$ the regulus ρ becomes two plane pencils which have one common line, viz. α', which therefore passes through O_1 and O_1', the centres of these two pencils.

The line PO_1 is seen to be the intersection of the planes $(x, \lambda x)\,(y, \lambda y)$, *i.e.* the intersection of the tangent planes of Φ, at the points of contact of x and y respectively. The lines $\lambda x + \mu x$, $\lambda y + \mu y$ only meet for $\mu = \infty$, or $\mu = \kappa$; they determine on PO_1 an involution, of which P and O_1 are the double points.

The lines $\lambda x + \mu x, \lambda y + \mu' y$ will meet on PO_1 provided that

$$(\mu + \mu')(\lambda xy) + (\lambda^2 xy) = 0, \text{ or } \mu + \mu' = 2\kappa.$$

If such a point of intersection lies on Φ, and v is the singular line associated with the point, we have

$$\tau v_i = \rho\,(\lambda_i x_i + \mu x_i) + \sigma\,(\lambda_i y_i + \mu' y_i) + u_i,$$

where u is the line PO_1; together with the condition that u_i, $\lambda_i x_i + \mu x_i$ and $\lambda_i y_i + \mu' y_i$ belong to the linear complex λv; (which involves that this linear complex should be *special* and therefore $(\lambda^2 v^2) = 0$); therefore we have for the determination of the lines v, *i.e.* the singular lines associated with the respective points in which PO_1 meets Φ, the equations

$$\mu + \mu' = 2\kappa,$$

$$\rho\,(\lambda^2 xu) + \sigma\,(\lambda^2 yu) + (\lambda u^2) = 0,$$

$$\rho\,(\lambda^3 x^2) + \sigma\,\{(\lambda^3 xy) + (\mu + \mu')(\lambda^2 xy) + \mu\mu'(\lambda xy)\} + (\lambda^2 xu) = 0,$$

$$\rho\,\{(\lambda^3 xy) + (\mu + \mu')(\lambda^2 xy) + \mu\mu'(\lambda xy)\} + \sigma\,(\lambda^3 y^2) + (\lambda^2 yu) = 0.$$

Since these equations are symmetrical with regard to μ and μ', it follows that if $\lambda x + \mu x, \lambda y + \mu' y$ meet in a point A_1 of Φ, then will $\lambda x+\mu' x$, $\lambda y+\mu y$ meet in a point A_2 of Φ; the lines v for these points will have the same pair of values for ρ and σ, and differ only through the interchange of μ and μ'.

Now the lines $\lambda x+\mu x$, $\lambda x+\kappa x$, $\lambda x+\mu' x$, x form a harmonic pencil, since $\mu+\mu'=2\kappa$, therefore

$$\frac{2}{PO_1}=\frac{1}{PA_1}+\frac{1}{PA_2};$$

in a precisely similar manner we obtain

$$\frac{2}{PO_1}=\frac{1}{PA_3}+\frac{1}{PA_4};$$

therefore $$\frac{4}{PO_1}=\frac{1}{PA_1}+\frac{1}{PA_2}+\frac{1}{PA_3}+\frac{1}{PA_4},$$

i.e., the point O_1 lies on the third polar (polar plane) of P with regard to Φ; and this being true for each point O, we see that *the polar plane of P with regard to C^2 is the third polar of P with regard to Φ.*

141. The four singular lines which lie in any plane π are connected by an equation similar in form to (i), and in that case the lines α, β, γ are the diagonals of the complete quadrilateral formed by them. The equations (ii), (Art. 140), again hold; the first three of them assert that α, β, γ form a self-conjugate triangle with regard to the complex conic of their plane, the latter three show that the polar lines for C^2 of α, β, γ are concurrent. The point in which these polar lines meet is called the *pole of π with regard to C^2*; by duality, this point is seen to be the third polar, (polar point), of the plane π with regard to Φ, considered as a surface of the fourth class.

The point P for which π is the polar plane is *not the pole* of π; corresponding to any plane π there are 11 points for which π is the polar plane, viz. the 27 intersections of the first polars with regard to Φ of any three points of π diminished by the number of double points of Φ, *i.e.* $27-16=11$.

In the present case from consideration of the equations

$$(u\alpha)=0, \quad (\lambda\alpha u)=0, \quad (u^2)=0,$$
$$(u\beta)=0, \quad (\lambda\beta u)=0, \quad (\alpha\beta)=0;$$

if the vertices of the triangle formed by α, β, γ be denoted by A, B and C, the polar line of π with regard to the complex cone of A is the polar line of α for C^2, and similarly for the complex cones of B and C.

142. The diameters of the complex. The polars with regard to C^2 of lines in the plane at infinity are called *diameters*, (Art. 79). If P is any point on a line l, the polar plane π of l with regard to the complex cone of P passes through l', the polar of l for C^2; hence if l lie in any plane ϵ, the polar line a for ϵ with regard to the complex cone of P lies in π, *i.e.* meets l'. If ϵ is the plane at infinity a is the *axis* of a complex cylinder, and we have, since l' becomes a diameter d, *every diameter d is met by the axes of all complex cylinders which meet the line δ to which d is polar, (i.e. which are parallel to the direction determined by δ).*

143. The Centre of the complex. It has been seen, (Art. 141), that in any plane π there is one triangle, self-conjugate for the complex conic of C^2 in π, and such that the polar lines for C^2 of its sides intersect in one point, the pole of π for C^2.

Taking the tetrahedron thus formed as that of reference, the pole of π being A_4, it is easy to see that the conditions just stated cause the equation of C^2 to assume the form

$$a_{12}p^2_{12} + a_{13}p^2_{13} + a_{14}p^2_{14} + a_{23}p^2_{23} + a_{34}p^2_{34} + a_{42}p^2_{42} + 2Lp_{12}p_{34}$$
$$+ 2Mp_{13}p_{42} + 2Np_{14}p_{23} + 2Rp_{14}p_{42} + 2Sp_{42}p_{34} + 2Tp_{34}p_{14} = 0;$$

for the method of Art. 79 shows that if p'_{ik} is the polar line of p_{ik} for any quadratic complex $f(p_{ik}) = 0$, p'_{ik} is given by the equations

$$\rho \cdot p'_{12} = \kappa p_{12} + \tfrac{1}{2}\frac{\partial f}{\partial p_{34}}, \text{ \&c.},$$

and expressing that A_3A_4 is the polar of A_1A_2, &c., we obtain the form stated.

Hence, if l_{ik} is any line in the plane α_4, since $l_{14} = l_{42} = l_{34} = 0$, its polar line l'_{ik} is given by the equations

$$\left.\begin{array}{lll} \rho \cdot l'_{12} = \kappa l_{12} + Ll_{12}, & \rho \cdot l'_{13} = \kappa l_{13} + Ml_{13}, & \rho \cdot l'_{14} = a_{23}l_{23} \\ \rho \cdot l'_{34} = a_{12}l_{12}, & \rho \cdot l'_{42} = a_{13}l_{13}, & \rho \cdot l'_{23} = \kappa l_{23} + Nl_{23} \end{array}\right\} \ldots\text{(i)}.$$

These equations do not involve the coefficients R, S, T; hence the polar of any line of the plane system α_4 is the same for ∞^3 quadratic complexes.

The equation $\quad a_{23}p^2_{23} + a_{13}p^2_{13} + a_{12}p^2_{12} = 0,$

is satisfied by the tangents of the complex conic c^2 of α_4; if the lines l, m are *conjugate* for c^2, then, since $l - \mu m$, $l + \mu m$ are tangents of c^2, we have

$$a_{23}l_{23}m_{23} + a_{13}l_{13}m_{13} + a_{12}l_{12}m_{12} = 0 \ldots\ldots\ldots\ldots\text{(ii)};$$

if, in addition, their polar lines l', m' intersect each other, we have from (i) that

$$Na_{23}l_{23}m_{23} + Ma_{13}l_{13}m_{13} + La_{12}l_{12}m_{12} = 0 \quad \ldots\ldots\text{(iii)},$$

hence, *there is one line m, conjugate to l for c^2, such that the polars of l and m intersect.*

From equations (ii) and (iii) we derive

$$(\kappa+N)\,a_{23}l_{23}m_{23}+(\kappa+M)\,a_{13}l_{13}m_{13}+(\kappa+L)\,a_{12}l_{12}m_{12}=0,$$

i.e.

$$l'_{12}m'_{34}+l'_{13}m'_{42}+l'_{23}m'_{14}=0,$$

but this is the condition, (Art. 4), that the plane through A_4 and l' should contain the point in which m' meets α_4, hence, *the plane* (l', m') *passes through* A_4.

The polars for C^2 of the lines of the pencil (l, m) form a ruled cubic ρ^3, (Art. 137), whose simple directrix is n the polar line of the point (l, m) for c^2; its double directrix d being the polar line of the plane (l, m) for the complex cone of the point (l, m). Let A and B be the points in which n meets c^2; then upon n two involutions are determined, one consisting of points P, Q which divide AB harmonically, the other of the pairs of points P, P'; Q, Q', &c., in which the two generators through the points of d meet n. Now A and B are the double points of the first involution and they form a pair of conjugate points in the second; hence the involutions are *harmonic*, (Introd. v.). It follows that P', Q' divide AB harmonically; hence, if the plane of the generators through P and P' meets d in L, while the plane of the generators through Q and Q' meets d in L', the points L, L' are conjugate points of an involution upon d.

If p, q are the generators of ρ^3 through P and Q, the planes (p, n), (q, n) therefore form a pair in an involution of planes; *the double planes of this involution are* α_4 *and the plane* (A_4, n); hence, *the planes* (p, n), (q, n) *are harmonically divided by the planes* α_4 *and* (A_4, n). Two such polar lines p, q, will be called *conjugate.*

If α_4 is taken as the plane at infinity, its pole for C^2 is termed by Plücker the Centre of the complex; and it follows from what has just been seen, that *if p, q are two conjugate diameters, the planes through p parallel to q and through q parallel to p are equidistant from the centre of the complex; and if p, q, r are three mutually conjugate diameters, (i.e., meeting the plane at infinity in points which form a self-conjugate triangle for its complex conic), the centre of the parallelepiped, which has p, q, r for non-intersecting edges, is the centre of the complex.*

CHAPTER X.

REPRESENTATION OF A COMPLEX BY THE POINTS OF THREE-DIMENSIONAL SPACE.

144. A (1, 1) correspondence can be established between the lines of a quadratic complex C^2 and the points of space by aid of formulae due to Klein and Nöther*. For if the edges A_1A_2, A_1A_3 of the tetrahedron of reference be taken to be two lines of a pencil of lines of C^2, the equation of C^2 assumes the form

$$p_{12}b + p_{13}a + \phi = 0,$$

where a and b are linear functions, and ϕ a quadratic function, of the four other line coordinates. We may therefore write

$$\begin{aligned}
\nu \,.\, p_{14} &= (ax_3 - bx_2)\, x_1,\\
\nu \,.\, p_{42} &= (ax_3 - bx_2)\, x_2,\\
\nu \,.\, p_{34} &= (ax_3 - bx_2)\, x_3,\\
\nu \,.\, p_{23} &= (ax_3 - bx_2)\, x_4,\\
\nu \,.\, p_{12} &= \phi \,.\, x_2 - ax_1x_4,\\
\nu \,.\, p_{13} &= -\,\phi x_3 + bx_1x_4\,;
\end{aligned}$$

where in a, b and ϕ we suppose p_{14} replaced by x_1, p_{42} by x_2, &c.

The coordinates p_{ik} will then satisfy the identity

$$p_{12}p_{34} + p_{13}p_{42} + p_{14}p_{23} \equiv 0,$$

and the equation of the complex. It is seen that each line of the complex defines in general one point x_i, and *vice versâ*.

To the lines of a congruence (L, C^2) will correspond the points of the cubic surface

$$(ax_3 - bx_2)\sum_{1}^{4} l_i x_i + l_5(\phi \,.\, x_2 - ax_1x_4) + l_6(-\,\phi \,.\, x_3 + bx_1x_4) = 0.$$

This cubic surface being denoted by σ^3 and the quadric

* *Gött. Nach.* 1869. The developments of this chapter are due to Caporali, see *Geometria.*

$ax_3 - bx_2 = 0$ by ρ, it is clear that σ^3 and ρ have in common the generator of ρ given by $l_5 x_2 - l_6 x_3 = l_5 a - l_6 b = 0$, together with a quintic curve q^5; this curve q^5 will be called the *fundamental* curve of the representation. All the cubic surfaces σ^3 pass through q^5.

If in the linear complex L we have $l_5 = l_6 = 0$, it will contain the lines $p_{14} = p_{42} = p_{34} = p_{23} = 0$, *i.e.* those which form the pencil (A_1, α_4); this pencil being called the *fundamental pencil*, it is seen that to the points of a plane there will correspond the lines of a congruence (C^2, L), where L contains the fundamental pencil.

To any point of ρ, or $ax_3 - bx_2 = 0$, corresponds a line belonging to the fundamental pencil, determined by the equation

$$x_3 p_{12} + x_2 p_{13} = 0;$$

so that to all the points of the generator $a = \mu b$, $x_2 = \mu x_3$ there corresponds the *same line* of the fundamental pencil. Denoting for convenience the region occupied by the lines p_{ik} by Λ, and that occupied by the points x_i' by S, it is seen that while to a point of S there corresponds in general one line of Λ, to a point x_i of q^5 there will correspond the singly infinite set of lines

$$\frac{p_{14}}{x_1} = \frac{p_{42}}{x_2} = \frac{p_{34}}{x_3} = \frac{p_{23}}{x_4};$$

these lines form the pencil (A_1, α_4) together with another pencil which has one line in common with it; the latter pencil must therefore have its centre on the curve of intersection of α_4 with the singular surface, and its plane passing through A_1.

Any line of (A_1, α_4) meets the singular surface in three points distinct from A_1, and therefore belongs to three of the preceding pencils (which correspond to points of q^5), *hence, each of the generators of ρ of the system $a = \mu b$, $x_2 = \mu x_3$, meets q^5 three times.* Since any plane through a generator of this system meets q^5 five times, it follows that each generator of the *other system* meets q^5 *twice*, such a pair of points will be called *conjugate*.

To a line of S corresponds the ruled surface common to C^2 and two linear complexes which contain the fundamental pencil, and therefore a ruled cubic; since the line meets ρ twice, the ruled cubic contains two lines of (A_1, α_4), *i.e.* its double directrix passes through A_1 and its single directrix lies in α_4.

If the line of S meets q^5 in the point P, the pencil which corresponds to P breaks off from the ruled cubic, *i.e. to a line of S which meets q^5 once there corresponds a regulus of Λ.*

If the line meets q^5 twice there will correspond to it in Λ a pencil, hence, *to a chord of q^5 there corresponds a pencil of C^2. To two pencils of C^2 which have a common line there correspond intersecting chords of q^5.* Conversely to a pencil of C^2 corresponds a line of S which meets q^5 twice. Since any line of C^2 belongs to four pencils of C^2, four chords of q^5 can be drawn through any point.

To any curve c in S there will correspond a ruled surface τ in Λ, while to each intersection of τ and any special linear complex L corresponds an intersection of c and the cubic surface σ^3 connected with L, *i.e.*, *the degree of τ is equal to the number of intersections of c and σ^3 diminished by the number of intersections of c and q^5, hence, to any conic of S which meets q^5 four times there corresponds a regulus of C^2.*

There are ∞^1 such conics in any plane, or ∞^4 conics in all, hence there are ∞^4 reguli of C^2*, (Art. 116).

The reguli which contain one line of (A_1, a_4) correspond in S to lines which intersect q^5.

Conversely, any regulus ρ of C^2 being given by the equations

$$\Sigma a_{ik} p_{ik} = \Sigma b_{ik} p_{ik} = \Sigma c_{ik} p_{ik} = 0,$$

we may eliminate p_{12}, p_{13} and obtain a complex which contains ρ and the fundamental pencil; thus the curve c corresponding to ρ is plane; also it is a conic, for since any other complex $\Sigma d_{ik} p_{ik} = 0$, for which $d_{12} = d_{13} = 0$, has two lines in common with ρ, the corresponding plane has two points in common with c; thus c is a conic, which by the foregoing is seen to intersect q^5 four times.

If y_i is a point of Λ on p_{ik}, we have

$$y_2 p_{34} + y_3 p_{42} + y_4 p_{23} = 0, \quad \text{(Art. 3)},$$

therefore

$$y_2 x_3 + y_3 x_2 + y_4 x_4 = 0;$$

so that the curve of S which corresponds to the complex cone of Λ whose vertex is y_i, is a conic which lies in the plane whose equation is that just given, *i.e. to the complex cones of Λ correspond conics in planes through A_1.*

Similarly if u_i is a plane through p_{ik} we have

$$u_1 p_{14} + u_2 p_{24} + u_3 p_{34} = 0,$$

hence *to the complex conics of Λ correspond conics in planes which pass through A_4.*

* This result was discovered by Caporali from the above considerations.

145. The reguli of a congruence [2, 2]. To the lines of a congruence (L, C^2) correspond the points of a cubic surface σ^3; of the 27 lines of this cubic surface, one, r, is the common generator of ρ and σ^3, this line r is met by 10 others $p_1, \dots p_{10}$*, each of these lines p_i meets ρ a second time and therefore meets q^5 *once*; every plane section of σ^3 through one of these lines consists of the line and a conic which meets q^5 four times; these 10 systems of ∞^1 conics are therefore the 'images' of the 10 systems of ∞^1 reguli of (L, C^2). Each of the remaining 16 lines meets ρ twice, in points which do not lie on r, *i.e.* in points of q^5; it therefore corresponds to one of the 16 pencils of the congruence.

The lines p_i form five pairs of intersecting lines; if p, p' are such a pair, since the point (p, p') lies on σ^3, the conics in any

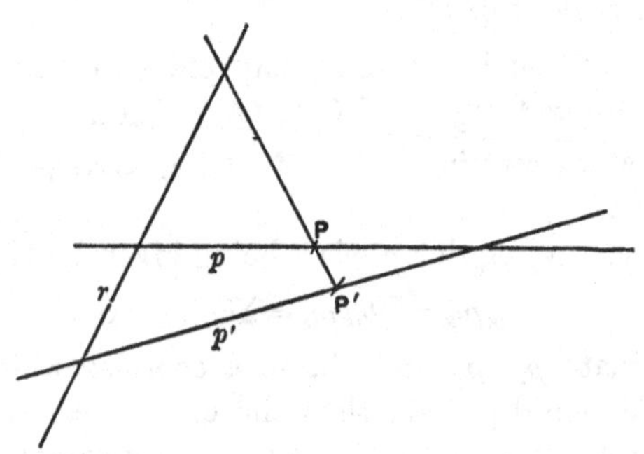

Fig. 8.

two planes through p and p' respectively have two points in common, and therefore are the images of reguli of two associated systems of (L, C^2), (Art. 118), one system belonging to a Σ_μ and the other to the connected Σ_μ'.

If P, P' are the points outside r in which p and p' meet ρ, the line PP' is a generator of ρ (since it contains three points of ρ), hence P and P' are conjugate points of q^5.

Two reguli of the same triplex Σ_μ belong to the same linear complex A, viz. that which corresponds to the line $\dfrac{a_i}{\sqrt{\lambda_i + \mu}}$ joining the vertices of the cones (or planes of the conics) which correspond to the two reguli in C_μ^2, (Art. 116); hence the two reguli belong to the same system of reguli of (A, C^2); it follows that the images of the two reguli are conics in planes through the same line p, *i.e.*

* Salmon, *Geom. of Three Dimensions*, 3rd Ed., p. 465.

the planes of these conics pass through the same point P of q^5; also the conics in planes through p', *i.e.* whose planes pass through P', are the images of the associated system of reguli of (A, C^2).

Thus, in each of the ∞^2 planes through any point P of q^5, the ∞^1 conics through the other four points of q^5 form the images of the ∞^3 reguli of a triplex Σ_μ, while the conics arising similarly from the conjugate point P' are the images of the reguli of Σ_μ'.

Since the singular lines are the complete intersection of C^2 with a quadratic complex, it follows that any pencil of C^2 contains *two*, any regulus of C^2 *four*, and any ruled cubic surface of C^2 *six* singular lines. Hence, of the locus in S corresponding to the singular lines, any line in S contains six points, therefore this surface is of the sixth degree. The surface has q^5 as double curve, for to each point P of q^5 there correspond two singular lines, viz. those in the pencils which correspond to P.

That the fundamental curve in S is of the fifth degree may be seen directly as follows :—a linear complex A, or $\Sigma a_{ik} p_{ik}=0$, through the fundamental pencil, gives rise, by aid of the (1, 1) correspondence of cosingular complexes, to a special linear complex A' with directrix a' through the pencil (P, π) corresponding to the fundamental pencil in a cosingular complex C_μ^2; and there are five pencils of the congruence (C_μ^2, a') which have one line in common with (P, π), *e.g.* if a' passes through P the pencils are the other pencil of (C_μ^2, a') through P together with four pencils in tangent planes to the singular surface through a'; hence there are five pencils of (C^2, A) each of which has one line in common with the fundamental pencil; *i.e.* the plane $\Sigma a_i x_i=0$ meets the fundamental curve in five points.

When there is a double line the curve q^5 has a double point.* For, to a double line l of C^2 corresponds a double line l' in C_μ^2 (for since C^2 contains a line which belongs to ∞^1 pencils so also must C_μ^2) and if A contains l then will A' contain l', hence two of the four points of intersection of a' and the singular surface coincide, similarly for two tangent planes through a'; therefore any linear complex through the fundamental pencil and the double line contains only four distinct pencils of C^2 which possess one line of the fundamental pencil, thus any plane through the point L which corresponds to l meets q^5 in only four distinct points, hence L is a double point of q^5.

146. Representation of the congruence [2, 2] by the points of a plane. The analytical basis of the representation of the lines of a congruence (2, 2) by the points of a singular plane of the congruence, depends upon the theorem of Caporali that any congruence (2, 2) is contained in a tetrahedral complex. That

* See Chapter XI.

there are 40 such tetrahedral complexes has already been shown, (Art. 123); Caporali's proof in a modified form will now be given.

If C^2 and A, or $\Sigma a_{ik} p_{ik}$, be the given quadratic and linear complexes which contain the congruence, it is clear that

$$C^2 + AA' = 0$$

is a quadratic complex which contains the congruence, whatever the complex A' may be. Let us take as the vertex A_4 and plane α_4 of reference, a singular point and singular plane of the congruence, and choose the coefficients of A' so that the squared terms disappear from the equation $C^2 + AA' = 0$.

The last equation now becomes

$$\begin{aligned} p_{23}(a_1 p_{14} + a_2 p_{24} + a_3 p_{34}) + p_{13}(b_1 p_{14} + b_2 p_{24} + b_3 p_{34}) \\ + p_{12}(c_1 p_{14} + c_2 p_{24} + c_3 p_{34}) + d_1 p_{13} p_{12} + d_2 p_{12} p_{23} + d_3 p_{23} p_{13} \\ + e_1 p_{24} p_{34} + e_2 p_{34} p_{42} + e_3 p_{42} p_{14} = 0. \end{aligned}$$

Moreover since α_4 is a singular plane of the congruence, the conic of this complex in α_4, which is obtained by writing zero for every p_{ik} in the equation either of whose suffixes is 4, must break up into two pencils one of which is that of A for α_4. But $d_1 p_{13} p_{12} + d_2 p_{12} p_{23} + d_3 p_{23} p_{13}$ cannot contain $a_{12} p_{12} + a_{13} p_{13} + a_{14} p_{14}$ as a factor, unless $d_1 = d_2 = d_3 = 0$.

Similarly since A_4 is a singular point, $e_1 = e_2 = e_3 = 0$. Thus the complex $C^2 + AA' = 0$ becomes

$$\begin{aligned} p_{23}(a_1 p_{14} + a_2 p_{24} + a_3 p_{34}) + p_{13}(b_1 p_{14} + b_2 p_{24} + b_3 p_{34}) \\ + p_{12}(c_1 p_{14} + c_2 p_{24} + c_3 p_{34}) = 0. \end{aligned}$$

Also the coefficients of p_{23}, p_{13}, p_{12}, equated to zero, give special linear complexes whose directrices lie in α_4; since the edges A_2A_3, A_3A_1, A_1A_2 of the tetrahedron of reference have not yet been determined, they may be taken as these respective directrices, which involves that

$$a_2 = a_3 = b_1 = b_3 = c_1 = c_2 = 0;$$

and $$C^2 + AA' \equiv a_1 p_{23} p_{14} + b_2 p_{13} p_{24} + c_3 p_{12} p_{34} = 0$$

is a tetrahedral complex.

Since ten singular points lie outside each singular plane, this gives $16 \times 10 = 160$ tetrahedra, but each tetrahedron being thus taken four times we arrive at 40 as the number of tetrahedral complexes which contain any congruence (2, 2).

The coordinates p_{ik} of a line of the congruence thus satisfy the three equations

$$p_{12}p_{34} + p_{13}p_{42} + p_{14}p_{23} = 0 \quad \text{.................(i)},$$
$$ap_{12}p_{34} + bp_{13}p_{42} + cp_{14}p_{23} = 0 \quad \text{.................(ii)},$$
$$\Sigma \alpha_{ik} p_{ik} = 0 \quad \text{.................(iii)}.$$

We may therefore write

$$\left.\begin{array}{ll} p_{14} = \rho x_1, & p_{23} = \sigma (b-c) x_2 x_3, \\ p_{24} = \rho x_2, & p_{31} = \sigma (c-a) x_3 x_1, \\ p_{34} = \rho x_3, & p_{12} = \sigma (a-b) x_1 x_2; \end{array}\right\} \quad \text{...........(iv)},$$

where ρ/σ is determined from substitution in (iii), giving

$$\rho \{\alpha_{14}x_1 + \alpha_{24}x_2 + \alpha_{34}x_3\} + \sigma \{\alpha_{23}(b-c)x_2x_3 + \alpha_{31}(c-a)x_3x_1 + \alpha_{12}(a-b)x_1x_2\} = 0.$$

Hence denoting the coefficient of ρ by R and that of σ by S, we obtain

$$p_{14} : p_{24} : p_{34} : p_{23} : p_{31} : p_{12}$$
$$= Sx_1 : Sx_2 : Sx_3 : R(c-b)x_2x_3 : R(a-c)x_1x_3 : R(b-a)x_1x_2 \text{...(v)}.$$

Through each point P of a singular plane of a congruence (C^2, A) there pass two lines of the congruence *one* of which passes through the pole of the plane for A, thus the point P determines one line of the congruence not in the plane, hence to the ∞^2 points of the plane correspond uniquely the ∞^2 lines of the congruence, thus (iv) establishes a (1, 1) correspondence between any line of the congruence and the point where it meets α_4. The only exceptions arise from the six singular points in α_4, which are the vertices A_1, A_2, A_3, the pole of A for α_4, and the intersections of $R=0$ and $S=0$.

The lines common to (C^2, A) and any other linear complex $A' \equiv \Sigma a'_{ik}p_{ik}$, form a ruled quartic of class 1, which will be denoted by ρ^4. This quartic is represented, from (v), by the cubic curve

$$S(a'_{14}x_1 + a'_{24}x_2 + a'_{34}x_3) + R\{(c-b)a'_{23}x_2x_3 + (a-c)a'_{31}x_1x_3 + (b-a)a'_{12}x_1x_2\} = 0.$$

The directrices of ρ^4 are the common polar lines of A and A'. By aid of A the equation of A' can be deprived of one of its terms and hence by varying A' we obtain ∞^4 surfaces ρ^4. Two such surfaces intersect in four lines, since this is the number of lines common to one quadratic and three linear complexes. Four lines of (C^2, A) determine *one* surface ρ^4. Through the lines common to two surfaces ρ^4, viz. (C^2, A, A'), (C^2, A, A''), there pass ∞^1 surfaces ρ^4, viz. those given by $C^2=0$, $A=0$, $A'+\mu A''=0$.

Since every linear complex contains one line of each pencil of (C^2, A) it is clear that each surface ρ^4 passes through the sixteen singular points and touches the sixteen singular planes.

If one of the common polar lines of A and A' meets a pencil of (C^2, A) so must the other, hence the plane of this pencil breaks off from ρ^4 and we have a ruled cubic ρ^3 of which the double directrix passes through a singular point of (C^2, A) and the other lies in a singular plane: there are thus ∞^2 such ruled cubics.

This method of representation may be employed in connexion with any singular plane σ of the congruence. Let S be the centre of the pencil of (C^2, A) in σ and $S_1 \ldots S_5$ the other singular points in σ, (Art. 118), having $\sigma_1 \ldots \sigma_5$ as their respective planes. Since any line of σ meets four generators of ρ^4 of which one belongs to (S, σ), the curve corresponding to ρ^4 in σ must be a cubic c^3, as has already been seen from the analytical formulae; since ρ^4 passes through $S_1 \ldots S_5$, so also must c^3.

If the double directrix of a ruled cubic ρ^3 of (C^2, A) passes through S its simple directrix lies in σ. The trace of ρ^3 on σ is therefore a line: to these ∞^2 ruled cubics there correspond the lines of σ. If the simple directrix passes through S_1 the double directrix must lie in σ_1, and the surface ρ^3 breaks up into a regulus together with the pencil (S_1, σ_1): hence, five systems of reguli are represented by the five pencils $(S_1, \sigma) \ldots (S_5, \sigma)$.

If a surface ρ^4 breaks up into two reguli, one of them contains a line through S, hence its trace must be a line which must pass through one of the other singular points in σ, say S_1; the trace of the other regulus is therefore a conic through S_2, S_3, S_4, S_5; so that five systems of reguli of (C^2, A) are represented by conics through four of the points $S_1 \ldots S_5$.

Finally to the five pencils $(S_1, \sigma_1) \ldots (S_5, \sigma_5)$ correspond their centres; while, since a surface ρ^3 and the pencil (S, σ) constitute a surface ρ^4 whose representative locus of the third degree passes through $S_1 \ldots S_5$, it follows that the pencil (S, σ) is represented by the conic through the points $S_1 \ldots S_5$.

The traces of the 10 pencils whose centres do not lie in σ being the lines joining in pairs the five points $S_1 \ldots S_5$, the latter 10 lines will represent these pencils.

147. Representation of the lines of a linear complex by points of space. If (A, α), (A', α') be respectively any pencil of lines and their polar lines for a given linear complex, these pencils will have AA' in common. Let (O), (O') be any two sheaves, then we may establish a collineation between the lines of (O) and the points of α, and between the lines of (O') and the points of α', in such a way, that to each pair of corresponding lines p, p' of (A, α), (A', α') there corresponds the *same plane* π through OO'. Then to two points P, P' on p, p' respectively, will correspond two lines OQ, $O'Q$ in π. This point Q will then represent the line PP' of the linear complex. The correspondence is in general of the (1, 1) character, but AA', being the join of *any* two of its points, will be represented by the intersection of *any* two lines of the pencils (O, ϵ), (O', ϵ), in the plane ϵ which corresponds to AA'. Thus AA' is represented by any point of ϵ.

Moreover, the *pencil* of complex lines whose centre is any point S of AA' is represented by the *point of intersection* of the lines OR, $O'R$ which correspond to S in the two collineations; the locus of R is therefore a conic c^2 in ϵ. The ∞^2 complex lines which intersect any line p of α, are represented by the points of the plane through O which corresponds to p; similarly for any line in α'.

Let the complex be referred to a tetrahedron of which the vertices A_3, A_2 are A and A' respectively, the planes α_1, α_4 are α and α' respectively, and the edges A_2A_1, A_3A_4 are any pair of polar lines; the equation of the complex will then be of the form

$$p_{12} = p_{34}.$$

In the region occupied by the representative points, let the plane which corresponds to A_2A_1 and A_3A_4 be taken to be $x_1 = 0$, the plane ϵ as $x_3 = 0$, the plane, through O corresponding to A_2A_4, as $x_2 = 0$, the plane, through O' corresponding to A_1A_3, as $x_4 = 0$.

We may then write

$$\rho \cdot p_{12} = \rho \cdot p_{34} = x_1, \quad \rho \cdot p_{13} = x_2, \quad \rho \cdot p_{14} = x_3, \quad \rho \cdot p_{42} = x_4,$$

from which we derive

$$\rho \cdot p_{23} = -\frac{x_1^2 + x_2x_4}{x_3},$$

hence

$$p_{12} : p_{13} : p_{14} : p_{23} : p_{34} : p_{42} = x_1x_3 : x_2x_3 : x_3^2 : -(x_1^2 + x_2x_4) : x_1x_3 : x_3x_4.$$

To the lines of a pencil of the complex there correspond the points of a line, and since each such pencil contains one line which meets AA', the former line meets the conic c^2 in ϵ: the equation of c^2 is seen to be

$$x_3 = 0, \quad x_1^2 + x_2 x_4 = 0.$$

To a linear congruence corresponds a quadric which passes through c^2; to the two reguli of the quadric correspond the two systems of pencils of the congruence.

To any congruence (2, 2) corresponds a quartic surface having c^2 as double curve; to the sixteen pencils of the congruence correspond sixteen lines of the quartic which meet c^2. Taking for the edge A_1A_4 of the tetrahedron of reference a line of the quadratic complex, the term p^2_{23} disappears from its equation and the quartic surface becomes a cubic surface through c^2. The lines which meet c^2 and touch this surface correspond to pencils whose centres and planes are Focal points* and Focal planes of the congruence. The sixteen lines of the cubic which meet c^2 correspond to the sixteen pencils of the congruence.

If c^2 be taken as the sphere-circle, we obtain the method of representation due to Lie which is discussed in Chapter XII.

* See Chapter XIV.

CHAPTER XI.

THE GENERAL EQUATION OF THE SECOND DEGREE.

148. It was stated in Chapter VI. that the equations

$$f(x) \equiv \Sigma a_{ik} x_i x_k = 0, \quad \omega(x) \equiv \Sigma \alpha_{ik} x_i x_k = 0,$$

of a quadratic complex, can in general be brought, by linear transformation, to the form

$$(\lambda x^2) = 0, \quad (x^2) = 0.$$

This will now be shown, while it will be seen that other forms of the equation of the complex arise when the equation

$$| a_{ik} + \lambda \alpha_{ik} | = 0,$$

i.e. the discriminant equation of $f + \lambda\omega$, has equal roots.

The method which has been followed is due to Darboux*, and was given by him in elucidation of the results of Weierstrass†.

Let $f \equiv \Sigma a_{ik} x_i x_k$ and $\phi \equiv \Sigma \alpha_{ik} x_i x_k$, be any two quadratic expressions in n variables; we shall consider the quantity

$$\frac{1}{\Delta(\lambda)} \begin{vmatrix} a_{11} + \lambda\alpha_{11} & \ldots & a_{1n} + \lambda\alpha_{1n} & X_1 \\ a_{21} + \lambda\alpha_{21} & \ldots & a_{2n} + \lambda\alpha_{2n} & X_2 \\ \ldots & \ldots & \ldots & \ldots \\ \ldots & \ldots & \ldots & \ldots \\ a_{n1} + \lambda\alpha_{n1} & \ldots & a_{nn} + \lambda\alpha_{nn} & X_n \\ X_1 & \ldots & X_n & 0 \end{vmatrix},$$

where the X_i are variables not yet defined, and $\Delta(\lambda)$ is the determinant $| a_{ik} + \lambda\alpha_{ik} |$.

* See "Mémoire sur la théorie algébrique des formes quadratiques," G. Darboux, *Liouville* (1874).

† "Zur Theorie der bilinearen und quadratischen Formen," *Berliner Monatsberichte* (1868). For a sketch of some methods of reduction of quadratic forms due to Kronecker and Jordan see "The reduction of quadratic forms and of linear substitution" by Prof. T. J. I'A. Bromwich, *Quarterly Journal* (1901). A discussion of the concomitants of linear and quadratic complexes will be found in a memoir by Prof. Forsyth, "Systems of quaternariants that are algebraically complete," *Camb. Phil. Trans.* vol. XIV.

We have, by use of partial fractions

$$\frac{1}{\Delta(\lambda)}\begin{vmatrix} a_{11}+\lambda\alpha_{11} \ldots X_1 \\ \cdots\cdots\cdots\cdots \\ \cdots\cdots\cdots\cdots \\ a_{n1}+\lambda\alpha_{n1} \ldots X_n \\ X_1 \ldots X_n \quad 0 \end{vmatrix} = \Sigma \frac{1}{\Delta'(\lambda_i)(\lambda-\lambda_i)} \begin{vmatrix} a_{11}+\lambda_i\alpha_{11} \ldots X_1 \\ \cdots\cdots\cdots\cdots \\ \cdots\cdots\cdots\cdots \\ a_{n1}+\lambda_i\alpha_{n1} \ldots X_n \\ X_1 \ldots X_n \quad 0 \end{vmatrix},$$

where the λ_i are the roots of $\Delta(\lambda)=0$, supposed to be all different.

Now let $\qquad X_i = \frac{1}{2}\frac{\partial F}{\partial x_i}$, where $F = f + \lambda\phi$,

then
$$X_i = \sum_k (a_{ik} + \lambda\alpha_{ik}) x_k;$$
hence, solving for the x_i
$$\Delta(\lambda) . x_i = \sum_k A_{ik} X_k,$$
where A_{ik} is the coefficient of $a_{ik}+\lambda\alpha_{ik}$ in $\Delta(\lambda)$; therefore
$$\Delta(\lambda) . \Sigma x_i X_i = \Sigma A_{ik} X_i X_k;$$
thus
$$\Delta(\lambda) . F = \Sigma A_{ik} X_i X_k,$$
hence
$$F = -\frac{1}{\Delta(\lambda)}\begin{vmatrix} a_{11}+\lambda\alpha_{11} \;\ldots\ldots\; X_1 \\ \cdots\cdots\cdots\cdots \\ a_{n1}+\lambda\alpha_{n1} \;\ldots\ldots\; X_n \\ X_1 \;\ldots\; X_n \quad 0 \end{vmatrix}$$
$$= -\Sigma \frac{1}{\Delta'(\lambda_i)(\lambda-\lambda_i)} \begin{vmatrix} a_{11}+\lambda_i\alpha_{11} \;\ldots\ldots\; X_1 \\ \cdots\cdots\cdots\cdots \\ a_{n1}+\lambda_i\alpha_{n1} \;\ldots\ldots\; X_n \\ X_1 \;\ldots\; X_n \quad 0 \end{vmatrix}.$$

Again the determinant

$$\begin{vmatrix} a_{11}+\lambda_i\alpha_{11}, & \ldots\ldots, & \frac{1}{2}\frac{\partial f}{\partial x_1}+\frac{1}{2}\lambda\frac{\partial\phi}{\partial x_1} \\ \cdots\cdots & \cdots\cdots & \cdots\cdots \\ a_{n1}+\lambda_i\alpha_{n1}, & \ldots\ldots, & \frac{1}{2}\frac{\partial f}{\partial x_n}+\frac{1}{2}\lambda\frac{\partial\phi}{\partial x_n} \\ \frac{1}{2}\frac{\partial f}{\partial x_1}+\frac{1}{2}\lambda\frac{\partial\phi}{\partial x_1}, & \ldots\ldots, & 0 \end{vmatrix},$$

since it arises from bordering a zero determinant, is a perfect square; by subtracting the first n rows multiplied respectively by

$x_1, \ldots x_n$ from the last row, and similarly the first n columns from the last column, the terms of the last row and column become

$$\tfrac{1}{2}(\lambda-\lambda_i)\frac{\partial\phi}{\partial x_1}, \ldots\ldots \tfrac{1}{2}(\lambda-\lambda_i)\frac{\partial\phi}{\partial x_n}$$

together with a term whose coefficient is zero; hence the determinant has the value $(\lambda-\lambda_i)^2 U_i^2$, where U_i, which is a linear function of the variables $x_1 \ldots x_n$, does not involve λ. It follows that

$$F = \Sigma \frac{(\lambda-\lambda_i)\, U_i^2}{\Delta'(\lambda_i)},$$

or

$$f + \lambda\phi = -\Sigma \frac{\lambda_i U_i^2}{\Delta'(\lambda_i)} + \lambda\Sigma \frac{U_i^2}{\Delta'(\lambda_i)},$$

whence

$$f = -\Sigma \frac{\lambda_i U_i^2}{\Delta'(\lambda_i)}, \quad \phi = \Sigma \frac{U_i^2}{\Delta'(\lambda_i)}.$$

149. To deal with the case in which $\Delta(\lambda)=0$ has equal roots, we investigate in the first place the properties of certain determinants which are of importance in the solution of the problem of the expression of a general quadratic in n variables as the sum of n squares.

The determinants Φ_p in question have the form

$$\Phi_p = \begin{vmatrix} a_{11} \ldots\ldots a_{1n} & y_{11} \ldots\ldots y_{1p} \\ \ldots\ldots & y_{21} \ldots\ldots y_{2p} \\ \ldots\ldots\ldots\ldots & \ldots\ldots\ldots\ldots \\ a_{n1} \ldots\ldots a_{nn} & y_{n1} \ldots\ldots y_{np} \\ y_{11} \ldots\ldots y_{n1} & 0 \ldots\ldots 0 \\ \ldots\ldots\ldots\ldots & \ldots\ldots\ldots\ldots \\ y_{1p} \ldots\ldots y_{np} & 0 \ldots\ldots 0 \end{vmatrix}$$

where the a_{ik} are the coefficients of a given quadratic expression f, Φ_0 is the discriminant of f, and the y's form p sets of n variables: it is clear that Φ_p is a linear function of the minors of Φ_0 of order p. Now if any quadratic form $\Psi\,(y_1 \ldots\ldots y_m)$ is identically zero, so are all its partial derivatives, hence so also is

$$x_1\frac{\partial\Psi}{\partial y_1} + \ldots\ldots + x_m\frac{\partial\Psi}{\partial y_m};$$

applying this result to the p sets of variables y in Φ_p, it is seen that if Φ_p is identically zero for *all* values of the quantities y, so also is

$$\begin{vmatrix} a_{11} \ldots\ldots a_{1n} & y_{11} \ldots\ldots y_{1p} \\ \ldots\ldots\ldots\ldots & \ldots\ldots\ldots\ldots \\ a_{n1} \ldots\ldots a_{nn} & y_{n1} \ldots\ldots y_{np} \\ x_{11} \ldots\ldots x_{n1} & 0 \ldots\ldots 0 \\ \ldots\ldots\ldots\ldots & \ldots\ldots\ldots\ldots \\ x_{1p} \ldots\ldots x_{np} & 0 \ldots\ldots 0 \end{vmatrix};$$

now the values 1, -1, and zero may be assigned to the variables x, y so as to make this determinant equal to *any* minor of Φ_0 of order p; hence if Φ_p is identically zero, so are all these minors of order p, and conversely. Since the minors of order $p-1$ are linear functions of the minors of order p, it follows that if Φ_p is identically zero so also are $\Phi_{p-1}, \ldots\ldots \Phi_0$.

In Φ_p let us suppose the a_{ik} replaced by $a_{ik}+\lambda a'_{ik}$, then $\Phi_0=\Delta(\lambda)$. Moreover let λ_i be a multiple root of $\Delta(\lambda)=0$, so that Φ_0 contains $(\lambda-\lambda_i)^{\nu_0}$ as a factor; let Φ_1 contain $(\lambda-\lambda_i)^{\nu_1}$ as a factor, then *every minor of the first order* of Φ_0 will contain $(\lambda-\lambda_i)^{\nu_1}$ as a factor; similarly if Φ_2 contains $(\lambda-\lambda_i)^{\nu_2}$, and so on. This may be indicated by the equation

$$(\lambda-\lambda_i)^{\nu_0}=(\lambda-\lambda_i)^{\nu_0-\nu_1}(\lambda-\lambda_i)^{\nu_1-\nu_2}\ldots\ldots;$$

so that in passing from the determinant $\Delta(\lambda)$, or Φ_0, to its first minors, the factor $(\lambda-\lambda_i)^{\nu_0-\nu_1}$ is lost, in passing from the first minors to the second minors of Φ_0 the factor $(\lambda-\lambda_i)^{\nu_1-\nu_2}$ is lost, and so on. These factors $(\lambda-\lambda_i)^{\nu_0-\nu_1}$, $(\lambda-\lambda_i)^{\nu_1-\nu_2}$ &c. were called by Weierstrass the Elementary Divisors of $\Delta(\lambda)$.

150. The following properties of the Elementary Divisors* will now be proved:

$$\text{(i)}\quad \nu_0>\nu_1>\nu_2\ldots\ldots;$$

$$\text{(ii)}\quad \nu_0-\nu_1>\nu_1-\nu_2\ldots\ldots.$$

To see that (i) holds, we notice that since by hypothesis Φ_1 contains $(\lambda-\lambda_i)^{\nu_1}$ as a factor, and hence each first minor of Φ_0, therefore $\dfrac{d\Phi_0}{d\lambda}$ must contain $(\lambda-\lambda_i)^{\nu_1}$ as a factor, *i.e.* $\nu_0>\nu_1$. In a similar way it is seen that $\nu_1>\nu_2$, and so on. A theorem in determinants which is also of use in the sequel will enable us to prove the second property. The theorem referred to is the following: if A is any determinant $|a_{ik}|$

$$\frac{dA}{da_{ik}}\cdot\frac{dA}{da_{rs}}-\frac{dA}{da_{is}}\cdot\frac{dA}{da_{rk}}=A\frac{d^2A}{da_{ik}da_{rs}};$$

for it is known from the theory of determinants† that

$$\begin{vmatrix} A_{ik} & A_{is} \\ A_{rk} & A_{rs} \end{vmatrix}=A\times\text{coefficient of}\begin{vmatrix} a_{ik} & a_{is} \\ a_{rk} & a_{rs} \end{vmatrix}\text{in } A;$$

from which the required result at once follows.

* For a full discussion of the Elementary Divisors see Muth, *Theorie der Elementartheiler*.

† Scott, *Theory of Determinants*, Chapter V.

In particular we have

$$A\frac{d^2A}{da_{n-1,\,n-1}\,da_{n,\,n}}=\frac{dA}{da_{n-1,\,n-1}}\frac{dA}{da_{n,\,n}}-\frac{dA}{da_{n-1,\,n}}\frac{dA}{da_{n,\,n-1}}.$$

Applying the last result to the determinant

$$\begin{vmatrix} a_{11}+\lambda\alpha_{11}\ldots\ldots a_{1n}+\lambda\alpha_{1n} & y_{11} & y_{12} \\ \ldots\ldots\ldots\ldots\ldots\ldots\ldots\ldots\ldots\ldots & & \\ a_{n1}+\lambda\alpha_{n1}\ldots\ldots a_{nn}+\lambda\alpha_{nn} & y_{n1} & y_{n2} \\ y_{11}\ldots\ldots\ldots\ldots\ldots y_{n1} & 0 & 0 \\ y_{12}\ldots\ldots\ldots\ldots\ldots y_{n2} & 0 & 0 \end{vmatrix}$$

we obtain $\Phi_0\Phi_2=\Psi_1\Psi_2-\Psi_3\Psi_4$, where Ψ_1 is the result of omitting the $n+1$th row and column in Φ_2 and therefore is divisible by $(\lambda-\lambda_i)^{\nu_1}$, similarly for Ψ_2, Ψ_3, Ψ_4, hence

$$\nu_0+\nu_2\geqslant 2\nu_1,\ \ i.e.\ \ \nu_0-\nu_1\geqslant\nu_1-\nu_2.$$

151. The theorem in determinants just considered affords a means of expressing any quadratic form $f\equiv\Sigma a_{ik}x_ix_k$ as the sum of n squares. For this purpose two new determinants R_p and A_p are introduced, where

$$R_p=\begin{vmatrix} a_{11} & \ldots\ldots & a_{1n} & y_{11} & \ldots\ldots & y_{1,\,n-p} & X_1 \\ \ldots & & & & & & \ldots \\ a_{n1} & \ldots\ldots & a_{nn} & y_{n1} & \ldots\ldots & y_{n,\,n-p} & X_n \\ y_{11} & \ldots\ldots & y_{n1} & 0 & \ldots\ldots & & 0 \\ \ldots & & & & & & \ldots \\ y_{1,\,n-p} & \ldots\ldots & y_{n,\,n-p} & 0 & \ldots\ldots & & 0 \\ X_1 & \ldots\ldots & X_n & 0 & \ldots\ldots & & 0 \end{vmatrix},$$

$$A_p=\begin{vmatrix} a_{11} & \ldots\ldots & a_{1n} & y_{11} & \ldots\ldots & y_{1,\,n-p} & X_1 \\ \ldots & & & & & & \ldots \\ a_{n1} & \ldots\ldots & a_{nn} & y_{n1} & \ldots\ldots & y_{n,\,n-p} & X_n \\ y_{11} & \ldots\ldots & y_{n1} & 0 & \ldots\ldots & & 0 \\ \ldots & & & & & & \ldots \\ y_{1,\,n-p} & \ldots\ldots & y_{n,\,n-p} & 0 & \ldots\ldots & & 0 \\ y_{1,\,n-p+1} & \ldots\ldots & y_{n,\,n-p+1} & 0 & \ldots\ldots & & 0 \end{vmatrix},$$

in which the X_i are variable quantities not yet defined.

Taking the determinant R_{p-1} as the A of last article, we deduce that

$$\Phi_{n-p}R_{p-1}=\Phi_{n-p+1}R_p-A_p^2,$$

whence

$$\frac{R_{p-1}}{\Phi_{n-p+1}}=\frac{R_p}{\Phi_{n-p}}-\frac{A_p^2}{\Phi_{n-p}\Phi_{n-p+1}};$$

by addition of all the equations so formed we have

$$\frac{R_0}{\Phi_n} = \frac{R_n}{\Phi_0} - \sum_1^n \frac{A_i^2}{\Phi_{n-i}\Phi_{n-i+1}};$$

but $R_0 \equiv 0$*, therefore

$$\frac{R_n}{\Phi_0} = \sum_1^n \frac{A_i^2}{\Phi_{n-i}\Phi_{n-i+1}}.$$

Now writing for X_i the value $\frac{1}{2}\frac{\partial f}{\partial x_i}$, R_n becomes $-\Phi_0 . f$†; therefore

$$f = -\sum_1^n \frac{A_i^2}{\Phi_{n-i}\Phi_{n-i+1}}.$$

152. Consider again the expression

$$F(X) = \frac{-1}{\Delta(\lambda)} \begin{vmatrix} a_{11} + \lambda\alpha_{11}, & \ldots & a_{n1} + \lambda\alpha_{n1}, & X_1 \\ \ldots & \ldots & \ldots & \ldots \\ a_{n1} + \lambda\alpha_{n1}, & \ldots & a_{nn} + \lambda\alpha_{nn}, & X_n \\ X_1 & \ldots & X_n & 0 \end{vmatrix}$$

in which the X_i may be *any* variable quantities; $F(X)$ may be expressed in the form

$$\sum_i \sum_k \left(\frac{Z_1}{\lambda - \lambda_i} + \frac{Z_2}{(\lambda - \lambda_i)^2} \quad \ldots + \frac{Z_k}{(\lambda - \lambda_i)^k}\right)$$

if $(\lambda - \lambda_i)^k$ is a factor of $\Delta(\lambda)$. By a theorem due to Lagrange and easily proved, the aggregate of all the fractions corresponding to the root λ_i is equal to the coefficient of $\frac{1}{h}$ in the expansion of

$$\frac{F(X_1 \ldots X_n, \lambda_i + h)}{\lambda - \lambda_i - h}$$

in powers of h.

Now $$F(X_1 \ldots X_n, \lambda_i + h) = -\frac{R_n}{\Phi_0}$$

if in R_n and Φ_0 we suppose the a_{ik} replaced by $a_{ik} + (\lambda_i + h)\alpha_{ik}$; hence, by the last article,

$$F(X_1 \ldots X_n, \lambda_i + h) = -\sum_1^n \frac{A^2_{n-p+1}}{\Phi_{p-1}\Phi_p},$$

* Since it contains a square of $(n+1)^2$ zeros.

† See Scott, *Determinants*, Chap. XI.

where

$$A_{n-p+1} = \begin{vmatrix} a_{11}+(\lambda_i+h)\alpha_{11}, \dots & a_{n1}+(\lambda_i+h)\alpha_{n1}, & y_{11} \dots y_{1,p-1} & X_1 \\ \dots\dots & & & \\ a_{n1}+(\lambda_i+h)\alpha_{n1}, \dots & a_{nn}+(\lambda_i+h)\alpha_{nn}, & y_{n,1} \dots y_{n,p-1} & X_n \\ y_{11} \dots\dots & y_{n,1} & 0 \dots\dots & 0 \\ \dots\dots & & & \\ y_{1,p-1} \dots\dots & y_{n,p-1} & 0 \dots\dots & 0 \\ y_{1,p} \dots\dots & y_{n,p} & 0 \dots\dots & 0 \end{vmatrix}.$$

On making the substitution

$$X_i = \frac{1}{2}\frac{\partial(f+\lambda\phi)}{\partial x_i},$$

it follows that $$F(X) = f + \lambda\phi,$$

so that

$$f + \lambda\phi = \sum_i \left(\text{coefficient of } \frac{1}{h} \text{ in } \frac{-1}{\lambda-\lambda_i-h}\sum_{p=1}^{p=n}\frac{A^2_{n-p+1}}{\Phi_p\Phi_{p-1}}\right).$$

Again, if in A_{n-p+1} we subtract from the last column the first n columns multiplied respectively by $x_1, \dots, x_n$, the terms of the last column become

$$\tfrac{1}{2}(\lambda-\lambda_i-h)\frac{\partial\phi}{\partial x_1}, \dots \tfrac{1}{2}(\lambda-\lambda_i-h)\frac{\partial\phi}{\partial x_n}, U_1, \dots, U_p,$$

hence $$A_{n-p+1} = (\lambda-\lambda_i-h)B_{n,p} + C_{n,p},$$

where $C_{n,p}$ is the aggregate of the terms in $U_1, \dots, U_p$.

Now, since the coefficients of the U_i are linear functions of the minors of order $p-1$ of Φ_0, while, by hypothesis, each minor of order $p-1$ of $|a_{ik}+\lambda\alpha_{ik}|$ contains the factor $(\lambda-\lambda_i)^{\nu_{p-1}}$, it follows that $C_{n,p}$ contains $h^{\nu_{p-1}}$ as a factor. Similarly $B_{n,p}$ is seen to contain h^{ν_p} as a factor. Moreover Φ_{p-1} and Φ_p contain $h^{\nu_{p-1}}$ and h^{ν_p} respectively as factors; hence

$$\frac{-A^2_{n-p+1}}{(\lambda-\lambda_i-h)\Phi_p\Phi_{p-1}} = -\frac{(h^{\nu_p}B'_{n,p}(\lambda-\lambda_i-h)+h^{\nu_{p-1}}C'_{n,p})^2}{(\lambda-\lambda_i-h)h^{\nu_p+\nu_{p-1}}\Delta_p.\Delta_{p-1}},$$

where

$$\Phi_p = h^{\nu_p}\Delta_p,\ \Phi_{p-1} = h^{\nu_{p-1}}\Delta_{p-1},\ B_{n,p} = h^{\nu_p}B'_{n,p},\ C_{n,p} = h^{\nu_{p-1}}C'_{n,p}.$$

In the development of this expression in powers of h, the only term which can give negative powers of h is

$$-\frac{(\lambda-\lambda_i-h)h^{2\nu_p}(B'_{n,p})^2}{\Delta_p\Delta_{p-1}h^{\nu_p+\nu_{p-1}}},$$

writing $\nu_{p-1}-\nu_p = e_p$, we have therefore to find the coefficient of $\frac{1}{h}$ in

$$-\left\{\frac{B'_{n,p}}{\sqrt{\Delta_p\Delta_{p-1}}}\right\}^2 \cdot \frac{\lambda-\lambda_i-h}{h^{e_p}}.$$

Expand $\dfrac{B'_{n,p}}{\sqrt{\Delta_p \Delta_{p-1}}}$ in ascending powers of h, then since $B'_{n,p}$ contains $x_1, \ldots, x_n$ linearly

$$\frac{B'_{n,p}}{\sqrt{\Delta_p \Delta_{p-1}}} = \xi_1 + \xi_2 h + \xi_3 h^2 + \ldots,$$

where the ξ_i are linear functions of the x_i; hence the coefficient of $\dfrac{1}{h}$ in

$$-\frac{A^2_{n-p+1}}{(\lambda - \lambda_i - h)\,\Phi_p \Phi_{p-1}}$$

is

$$-(\lambda - \lambda_i)(\xi_1 \xi_{e_p} + \xi_2 \xi_{e_p - 1} + \ldots + \xi_{e_p-1}\xi_2 + \xi_{e_p}\xi_1) + \xi_1 \xi_{e_p-1} + \ldots + \xi_{e_p-1}\xi_1.$$

Therefore the term

$$\frac{-A_n^2}{(\lambda - \lambda_i - h)\,\Phi_0 \Phi_1}$$

gives a coefficient of $\dfrac{1}{h}$ of this form, introducing $\nu_0 - \nu_1$ such variables ξ; the term

$$\frac{-A^2_{n-1}}{(\lambda - \lambda_i - h)\,\Phi_1 \Phi_2}$$

brings in $\nu_1 - \nu_2$ more such variables, &c.: the total number of variables thus introduced in connexion with the root λ_i of $\Delta(\lambda) = 0$ is therefore

$$\nu_0 - \nu_1 + \nu_1 - \nu_2 + \ldots = \nu_0.$$

Proceeding successively to each root of $\Delta(\lambda) = 0$, we see that, since $\Sigma \nu_0 = n$, *the total number of variables* ξ *is* n; also F, *i.e.* $f + \lambda\phi$ is equal to the aggregate of sets of terms of the type

$$-(\lambda - \lambda_i)(\xi_1 \xi_{e_p} + \xi_2 \xi_{e_p-2} + \ldots + \xi_{e_p}\xi_1) + \xi_1 \xi_{e_p-1} + \ldots + \xi_{e_p-1}\xi_1,$$

one such set being contributed by each elementary divisor $(\lambda - \lambda_i)^{e_p}$.

Hence

$$\left.\begin{aligned} f &= \Sigma\,\{\lambda_i(\xi_1\xi_{e_p} + \ldots + \xi_{e_p}\xi_1) + \xi_1\xi_{e_p-1} + \ldots + \xi_{e_p-1}\xi_1\} \\ \phi &= -\Sigma\,(\xi_1\xi_{e_p} + \ldots + \xi_{e_p}\xi_1) \end{aligned}\right\} \ldots\text{(A)}.$$

It is to be observed that $\Sigma e_p = n$. If in the case of a multiple root λ_i, it is *not the case* that $\lambda - \lambda_i$ is a factor of all the first minors of $\Delta(\lambda)$, then $\nu_1 = 0$, and there is only one elementary divisor connected with λ_i, viz. $(\lambda - \lambda_i)^{\nu_0}$.

153. Applying the several results thus obtained to the case of the quadratic complex, $n=6$*, $f=0$ is the equation of the complex and $\phi\equiv\omega=0$ is the identical relation; while $\Sigma e_p=6$.

The first case which occurs is that in which each e_p is unity; both f and ω then consist merely of squares; this complex, the general case in which $\Delta(\lambda)=0$ has six different roots, is denoted by the symbol [111111].

The equation $\Sigma e_p=6$ can be satisfied by sets of positive integers e_p in eleven ways, and the corresponding complexes are denoted by [111111], [11112], [1113], [1122], [114], [123], [222], [15], [24], [33], [6]: their equations will shortly be given. Each type contains a certain number of sub-cases, since two (or more) numbers e_p may refer to the same root λ_i of $\Delta(\lambda)=0$, (*i.e.* when $\nu_1\neq 0$); thus for instance if λ_1 is a triple root of $\Delta(\lambda)=0$, or $\Phi_0=0$, while $\lambda-\lambda_1$ is a factor of Φ_1, the elementary divisor $(\lambda-\lambda_1)^2$ is lost in passing from Φ_0 to Φ_1; two of the numbers e_p are 2 and 1, all the other roots of $\Delta(\lambda)=0$ being supposed distinct from each other and from λ_1; this case is denoted by [111(12)], the numbers e_p which refer to the same root λ_1 being enclosed in a single bracket. Thus again in the complex [11112], the determinant $\Delta(\lambda)$ has a factor $(\lambda-\lambda_i)^2$, and $\lambda-\lambda_i$ is not a factor of all its first minors; while in [1111(11)] $(\lambda-\lambda_i)^2$ is a factor of $\Delta(\lambda)$, and $\lambda-\lambda_i$ is a factor of all its first minors.

154. Arbitrary constants of a canonical form. When a root λ_i of the discriminant of $f+\lambda\omega$ is connected with only *one* elementary divisor (*i.e.* $\nu_1=0$, Art. 152), the arbitrary variables y disappear from the variables ξ connected with this elementary divisor. For in this case $(\lambda-\lambda_i)^{e_p}$ is a factor of Φ_0 but $\lambda-\lambda_i$ is not a factor of all the first minors of Φ_0, *i.e.* Φ_1 does not contain $\lambda-\lambda_i$ as a factor; thus the part of $f+\lambda\omega$ contributed by this elementary divisor is equal to the coefficient of $\frac{1}{h}$ in the development of $-\left(\frac{B_{n,1}}{\sqrt{\Delta_0\Phi_1}}\right)^2\frac{(\lambda-\lambda_i-h)}{h^{e_p}}$, where $\Phi_0=h^{e_p}\Delta_0$,

$$B_{n,1}=\begin{vmatrix} a_{11}+(\lambda_i+h)\alpha_{11}, \dots \frac{1}{2}\frac{\partial\omega}{\partial x_1} \\ \dots\dots\dots\dots\dots\dots \\ a_{16}+(\lambda_i+h)\alpha_{16}, \dots \frac{1}{2}\frac{\partial\omega}{\partial x_6} \\ y_1 \dots\dots\dots y_6 \quad 0 \end{vmatrix}, \quad \Phi_1=\begin{vmatrix} a_{11}+(\lambda_i+h)\alpha_{11}, \dots y_1 \\ \dots\dots\dots\dots\dots\dots \\ a_{16}+(\lambda_i+h)\alpha_{16}, \dots y_6 \\ y_1 \dots\dots\dots y_6 \quad 0 \end{vmatrix};$$

* This application was made by Klein, see "Ueber die Transformation der allgemeinen Gleichung 2. Grades zwischen Linien-Coordinaten auf eine canonische Form," Diss. Bonn (1868) and *Math. Ann.* XXIII.

so that $$\frac{B_{n,1}}{\sqrt{\Delta_0 \Phi_1}} = \Psi(\lambda_i + h) = \Psi(\lambda_i) + h\frac{\partial \Psi}{\partial \lambda_i} + \ldots;$$

while, since Φ_0 is zero when h is zero,

$$\left.\begin{aligned} &\Phi_1 = \{\sqrt{-A_{kk}}y_k\}^2, \text{ if the } A_{ik} \text{ are the first minors of } \Phi_0, \\ &B_{n,1} = \tfrac{1}{4}\Sigma \frac{\partial \omega}{\partial x_k}\frac{\partial \Phi_1}{\partial y_k} = \tfrac{1}{2}\Sigma\sqrt{-A_{kk}}y_k \,\Sigma\sqrt{-A_{kk}}\frac{\partial \omega}{\partial x_k}, \end{aligned}\right\} \text{for } h = 0.$$

So that $\Psi(\lambda_i) = \frac{1}{2}\frac{1}{\sqrt{\delta \Pi_j(\lambda_i - \lambda_j)}}\Sigma\sqrt{-A_{kk}}\frac{\partial \omega}{\partial x_k}$; where δ is the discriminant of ω, and $\lambda_j \neq \lambda_i$.

It follows that $$2\sqrt{\delta}\,.\,\xi_k = \frac{1}{\lfloor k-1}\left(\frac{\partial}{\partial \lambda_i}\right)^{k-1}\left\{\frac{\Sigma_k\sqrt{-A_{kk}}\,.\,\frac{\partial \omega}{\partial x_k}}{\sqrt{\Pi_j(\lambda_i - \lambda_j)}}\right\}.$$

If several elementary divisors are connected with the same root λ_i of the discriminant of $f + \lambda\omega$, a certain number of arbitrary constants are contained in the canonical form. If, for instance, ν elementary divisors relate to λ_i, the corresponding sets of variables being $\xi, \xi', \xi'', \ldots \xi^{(n)}$; then the forms of f and ω are unaltered if we substitute for $\xi_{e_p}, \xi'_{e_{p'}}, \ldots$ respectively

$$\begin{gathered} \xi_{e_p} + \alpha_1\xi_1 + \alpha_2\xi_1' + \ldots + \alpha_\nu\xi_1^{(n)} \\ \xi'_{e_{p'}} + \beta_1\xi_1 + \beta_2\xi_1' + \ldots + \beta_\nu\xi_1^{(n)}, \\ \cdots\cdots\cdots\cdots\cdots\cdots \end{gathered}$$

provided that

$$\begin{aligned} \xi_1\xi_{e_p} + \xi_1'\xi'_{e_p} + \ldots \equiv\ & \xi_1(\xi_{e_p} + \alpha_1\xi_1 + \alpha_2\xi_1' + \ldots + \alpha_\nu\xi_1^{(n)}) \\ & + \xi_1'(\xi'_{e_{p'}} + \beta_1\xi_1 + \ldots + \beta_\nu\xi_1^{(n)}) \\ & + \cdots\cdots\cdots\cdots \end{aligned}$$

This introduces ν^2 arbitrary constants $\alpha, \beta, \ldots$ between which exist, by virtue of the last condition, $\frac{\nu(\nu+1)}{2}$ equations, leaving $\frac{\nu(\nu-1)}{2}$ of these quantities arbitrary.

If this occurs μ_ν times the total number of arbitrary constants contained in the given canonical form is $\mu_\nu \frac{\nu(\nu-1)}{2}$*.

155. Complexes formed by linear congruences. When two numbers $e_p, e_{p'}$ are connected with the same root λ_i of $\Delta(\lambda) = 0$, the equation $f + \lambda_i\omega = 0$ involves only *four* variables, since the variables $\xi_{e_p}, \xi'_{e_{p'}}$ do not appear in the last equation. Hence in all

* Klein, "Transformation der Complexe 2. Grades," *Math. Ann.* Bd. XXIII.

types which involve the single bracket, the equation of the complex can be brought into a form which involves not more than four variables. Regarding for a moment these four variables as the coordinates of a *point*, the equation of the complex will represent a quadric; now every quadric can be brought into the form

$$XY - ZW = 0$$

where X, Y, Z, W are linear in the variables.

Hence the complex may also be brought to this form, and consists of a singly infinite number of linear congruences,

$$X = \mu Z,$$
$$\mu Y = W.$$

If P is any point of p a directrix of such a congruence, the other directrix being p', the pencil (P, p') belongs to the complex and hence the lines p, p' belong to the singular surface which is therefore *ruled*. The complex gives rise to a correspondence* among the generators of the singular surface.

156. Double Lines. The equations to determine the four pencils of complex lines to which a line x of the complex belongs, were seen to be in the case of the complex $f = 0$, $\omega = 0$, the following:

$$\left(\alpha \frac{\partial \omega}{\partial x}\right) = 0, \quad \left(\alpha \frac{\partial \omega}{\partial \beta}\right) = 0, \quad \left(\alpha \frac{\partial f}{\partial x}\right), \; f(\alpha) = 0, \text{ (Art. 75).}$$

If a line x is such that each point of it is singular, it is said to be a *double line* of the complex, and belongs to an infinite number of pencils; x will then belong to the singular surface. The condition for the existence of a double line is therefore that the preceding four equations should reduce to three, hence there is in the case of a double line x a quantity μ such that

$$\frac{\partial f}{\partial x_i} + \mu \frac{\partial \omega}{\partial x_i} = 0, \quad (i = 1, 2, \ldots 6).$$

Now taking f and ω as being composed of groups of variables ξ where corresponding portions of f and ω are

$$\lambda_i(\xi_1\xi_n + \ldots + \xi_n\xi_1) + \xi_1\xi_{n-1} + \ldots + \xi_{n-1}\xi_1,$$
$$-(\xi_1\xi_n + \ldots + \xi_n\xi_1),$$

* On the subject of this Article see Weiler, "Die Erzeugung von Complexen ersten und zweiten Grades aus linearen Congruenzen," *Zeitschrift* 1882 and 1884.

the parts of the immediately preceding six equations which arise from this group of variables are

$$(\lambda_i - \mu)\,\xi_n + \xi_{n-1}$$
$$(\lambda_i - \mu)\,\xi_{n-1} + \xi_{n-2}$$
$$\cdots\cdots\cdots\cdots\cdots\cdots$$
$$(\lambda_i - \mu)\,\xi_1.$$

The equations are therefore all satisfied by

$$\mu = \lambda_i, \quad \xi_1 = \xi_2 = \ldots = \xi_{n-1} = 0,$$

and all other variables except ξ_n zero; so that the line whose coordinates are all zero except ξ_n belongs to f, and is now seen to be a double line of the complex. Hence each group of variables ξ gives rise in general to one double line. In the type [111 111] in which each group consists of only one member there is no double line, in every other case a double line exists.

Every double line of the complex is a double line of the singular surface.

For taking the complex as $f(x) = 0$, $(x^2) = 0$, the tangents y of the singular surface are given by the equations

$$\frac{\partial f}{\partial x_i} + \mu x_i = \rho \,.\, y_i, \qquad (i = 1, \; 2, \ldots 6),$$

where x is a singular line of the complex; if x is also a double line, there is a value of μ for which the left-hand side of each of the preceding equations is zero, for which therefore ρ is zero, so that *any line* y which meets this double line x is a tangent line of the singular surface, hence x must be a double line of the singular surface.

In the case of any complex for which three numbers are enclosed in a single bracket, as [11(112)], the singular surface is a quadric *counted twice*; for if ξ, η, ζ, λ_i correspond to the enclosed numbers, the equations to determine the double lines are satisfied by equating to zero all the variables except the ξ, η, ζ with the highest suffixes, *i.e.* by the vanishing of three variables; this gives a regulus of double lines (which may become two pencils), each line of which is a double line of the singular surface.

157. The Cosingular Complexes. If a tangent linear complex of f is special, its directrix touches the singular surface of f. For if y is the directrix of such a complex we have

$$\frac{\partial f}{\partial x_i} + \lambda \,.\, \frac{\partial \omega}{\partial x_i} = \rho \,.\, \frac{\partial \omega}{\partial y_i} \;\cdots\cdots\cdots\cdots\cdots\cdots \text{(i)},$$

hence $$\Omega\left(\frac{\partial f}{\partial x}\right)+2\lambda\Sigma\frac{\partial f}{\partial x}\frac{\partial\Omega(X)}{\partial X_i}+4\lambda^2\Omega(X)=0,$$

where $$X_i=\frac{1}{2}\frac{\partial\omega}{\partial x_i};$$

also $$\frac{\partial\Omega(X)}{\partial X_i}=-2\Delta\,.\,x_i,\quad \Omega(X)=-\Delta\,.\,\omega(x);\quad \text{(Art. 21)},$$

therefore $\Omega\left(\frac{\partial f}{\partial x}\right)=0$, *i.e.* x is a singular line of f, (Art. 76).

It follows, since y meets x and all lines of f consecutive to x, that y must pass through the point of contact P of x and lie in the tangent plane π at P to the singular surface.

Now the equations (i) become, if we write

$$\frac{1}{2}\frac{\partial\omega}{\partial y_i}=Y_i,$$

$$\sum_k(a_{ik}+\lambda\alpha_{ik})x_k=\rho\,.\,Y_i,\quad (i=1,\ 2,\ldots 6);$$

solving for x we obtain

$$\Delta(\lambda)\,.\,x_k=\rho\,(Y_1A_{1k}+Y_2A_{2k}+\ldots+Y_6A_{6k})$$
$$=\frac{1}{2}\rho\frac{\partial Y}{\partial Y_k},$$

where $$Y\equiv\Sigma A_{ik}Y_iY_k;$$

moreover $$\rho\frac{\partial Y}{\partial y_i}=\rho\left(\frac{\partial Y}{\partial Y_1}\frac{\partial Y_1}{\partial y_i}+\ldots+\frac{\partial Y}{\partial Y_6}\frac{\partial Y_6}{\partial y_i}\right),$$
$$=\rho\left(\frac{\partial Y}{\partial Y_1}\alpha_{1i}+\ldots+\frac{\partial Y}{\partial Y_6}\alpha_{6i}\right),$$
$$=2\Delta(\lambda)(x_1\alpha_{1i}+\ldots+x_6\alpha_{6i}),$$
$$=\Delta(\lambda)\,.\,\frac{\partial\omega}{\partial x_i};$$

hence, $$\rho\,.\,\frac{\partial Y}{\partial y_i}=\Delta(\lambda)\,.\,\frac{\partial\omega}{\partial x_i},\quad (i=1,\ 2,\ldots 6).$$

From the last set of equations we conclude, as before, that y is a singular line of the complex $Y=0$, and that x is a tangent line to the singular surface of the latter complex at the point of contact of y. Hence the complexes $f=0$ and $Y=0$ are seen to be cosingular; the singly infinite number of complexes cosingular with f is obtained by giving all values to λ.

It was seen, (Art. 152), that for any quantities X_i,

$$\frac{-1}{\Delta(\lambda)}\begin{vmatrix} a_{11}+\lambda\alpha_{11}, \ldots a_{16}+\lambda\alpha_{16}, X_1 \\ \ldots\ldots\ldots\ldots \\ a_{16}+\lambda\alpha_{16}, \ldots a_{66}+\lambda\alpha_{66}, X_6 \\ X_1 \ldots\ldots\ldots X_6 \quad 0 \end{vmatrix} = \sum_{\lambda_i} \text{coeff.}\ \frac{1}{h}\ \text{in}\ \frac{-1}{\lambda-\lambda_i-h}\sum_{p=1}^{p=6}\frac{A^2_{6-p+1}}{\Phi_p\Phi_{p-1}},$$

where $A_{6-p+1} =$

$$\begin{vmatrix} a_{11}+(\lambda_i+h)\alpha_{11}, & \ldots\ldots\ldots & a_{16}+(\lambda_i+h)\alpha_{16}, & y_{11}\ldots y_{1,\,p-1} X_1 \\ a_{21}+(\lambda_i+h)\alpha_{21}, & \ldots\ldots\ldots & a_{26}+(\lambda_i+h)\alpha_{26}, & \ldots\ldots\ldots \\ \ldots\ldots & \ldots\ldots & \ldots\ldots & \ldots\ldots \\ a_{61}+(\lambda_i+h)\alpha_{61}, & \ldots\ldots\ldots & a_{66}+(\lambda_i+h)\alpha_{66}, & y_{61}\ldots y_{6,\,p-1} X_6 \\ y_{11} \ldots\ldots & & y_{61} & 0 \ldots 0 \quad 0 \\ \ldots\ldots & \ldots\ldots & \ldots\ldots & \ldots\ldots \\ y_{1p} \ldots\ldots & & y_{6p} & 0 \ldots 0 \quad 0 \end{vmatrix}.$$

If we now take X_i to be $\frac{1}{2}\frac{\partial\omega}{\partial y_i}$, the equation $|\, a_{ik}+\lambda\alpha_{ik}, X_i \,| = 0$ is the equation, $Y=0$, of the cosingular complexes; while A_{6-p+1} becomes the determinant formerly denoted by $B_{n,\,p}$, hence

$$\frac{1}{\lambda-\lambda_i-h}\cdot\frac{A^2_{6-p+1}}{\Phi_p\Phi_{p-1}} = \frac{1}{\lambda-\lambda_i-h}\cdot\frac{(\eta_1+\eta_2 h+\ldots)^2}{h^{e_p}};$$

the η_i being the same functions of the y_i as the ξ_i of the x_i.

Therefore

$$\frac{Y}{\Delta(\lambda)} \equiv -\Sigma\ \text{coeff. of}\ \frac{1}{h}\ \text{in}\ \frac{1}{\lambda-\lambda_i-h}\cdot\frac{(\eta_1+\eta_2 h+\ldots)^2}{h^{e_p}}$$

$$\equiv -\Sigma\ \text{coeff. of}\ \frac{1}{h}\ \text{in}\ \left[\left(\frac{1}{\lambda-\lambda_i}+\frac{h}{(\lambda-\lambda_i)^2}+\ldots\right)\times\right.$$

$$\left.\left(\ldots+\frac{\eta_1\eta_{e_p}+\ldots+\eta_{e_p}\eta_1}{h}+\frac{\eta_1\eta_{e_p-1}+\ldots+\eta_{e_p-1}\eta_1}{h^2}+\ldots\right)\right].$$

So that

$$\frac{Y}{\Delta(\lambda)} \equiv -\Sigma\left(\frac{\eta_1\eta_{e_p}+\ldots+\eta_{e_p}\eta_1}{\lambda-\lambda_i}+\frac{\eta_1\eta_{e_p-1}+\ldots+\eta_{e_p-1}\eta_1}{(\lambda-\lambda_i)^2}+\ldots\right) \quad \ldots\ldots\ldots (B);$$

where each elementary divisor $(\lambda-\lambda_i)^{e_p}$ contributes a set of terms on the right side of the last equation. This gives the expression of the cosingular complexes in terms of the variables η, &c. which enter into the expression of the canonical form of the given quadratic complex $f(x) \equiv \Sigma a_{ik}x_i x_k = 0$.

158. Correspondence between lines of cosingular complexes. It was shown in Chapter VIII. that a (1, 1) correspondence exists between the lines of $(\lambda x^2)=0$, $(x^2)=0$ and the lines of any one of the cosingular complexes. This result will now be established for the other varieties of the quadratic complex. The algebraical theorem which follows enables us to obtain the required result.

Let there be two sets of n variables $\xi_1 \ldots \xi_n$ and $\eta_1 \ldots \eta_n$, and let us denote by X_{n-r}, Y_{n-r} respectively the expressions

$$\xi_1\xi_{n-r}+\ldots+\xi_{n-r}\xi_1, \quad \eta_1\eta_{n-r}+\ldots+\eta_{n-r}\eta_1.$$

Further denote by A_{n-r} the expression $\alpha_1\alpha_{n-r}+\ldots+\alpha_{n-r}\alpha_1$, where the α's are constant quantities determined by the series of equations

$$A_1=\alpha_1^2=-1, \quad A_2=\alpha_1\alpha_2+\alpha_2\alpha_1=\frac{1}{\lambda},$$

$$A_3=\frac{-1}{\lambda^2}, \quad \ldots, \quad A_n=-\left(\frac{-1}{\lambda}\right)^{n-1}.$$

These equations determine uniquely the values of $\alpha_1 \ldots \alpha_n$; and it follows that

$$A_{n-r}+\frac{1}{\lambda}A_{n-r-1}=0, \qquad (r=0, 1, \ldots, n-2).$$

Now consider the equations

$$\left.\begin{aligned}\sqrt{\lambda}\,.\,\xi_1&=\alpha_1\eta_1,\\ \sqrt{\lambda}\,.\,\xi_2&=\alpha_2\eta_1+\alpha_1\eta_2,\\ \sqrt{\lambda}\,.\,\xi_3&=\alpha_3\eta_1+\alpha_2\eta_2+\alpha_1\eta_3,\\ &\ldots\ldots\ldots\ldots\ldots\\ \sqrt{\lambda}\,.\,\xi_n&=\alpha_n\eta_1+\alpha_{n-1}\eta_2+\ldots+\alpha_1\eta_n.\end{aligned}\right\}\ldots\ldots\ldots\ldots(\text{I}).$$

From these equations we deduce at once

$$\begin{aligned}\lambda X_n+{}_{,}X_{n-1}=\left(A_n+\frac{1}{\lambda}A_{n-1}\right)\eta_1^2+2\left(A_{n-1}+\frac{1}{\lambda}A_{n-2}\right)\eta_1\eta_2+\ldots\\ +2\left(A_2-\frac{1}{\lambda}\right)\eta_1\eta_{n-1}-2\eta_1\eta_n+\left(A_{n-2}+\frac{1}{\lambda}A_{n-3}\right)\eta_2^2+\ldots\\ -2\eta_2\eta_{n-1}+\ldots.\end{aligned}$$

From which we conclude that

$$\lambda X_n+X_{n-1}=-Y_n \ldots\ldots\ldots\ldots\ldots\ldots(\text{II});$$

similarly,

$$\lambda X_{n-1}+X_{n-2}=-Y_{n-1},$$

$$\ldots\ldots\ldots\ldots\ldots\ldots$$

$$\lambda X_1=-Y_1;$$

hence,

$$-X_n = \frac{1}{\lambda} Y_n - \frac{1}{\lambda^2} Y_{n-1} + \frac{1}{\lambda^3} Y_{n-2} \ldots + (-1)^{n-1} \left(\frac{1}{\lambda}\right)^n Y_1 \ldots \text{(III)}.$$

Equations (II) and (III) are thus a consequence of (I).

We shall now, in these equations, take $n = e_p$ and replace λ by $\lambda_i - \lambda$. Also we suppose that other sets of equations, similar to (I), are formed between variables $\xi'_1, \ldots, \xi'_{e_{p'}}$; $\eta'_1, \ldots, \eta'_{e_{p'}}$, &c., then by addition of all equations of the type (II) we have

$$\Sigma \{\lambda_i (\xi_1 \xi_{e_p} + \ldots + \xi_{e_p} \xi_1) + \xi_1 \xi_{e_p - 1} + \ldots + \xi_{e_p - 1} \xi_1\}$$
$$- \lambda \Sigma (\xi_1 \xi_{e_p} + \ldots + \xi_{e_p} \xi_1) \equiv - \Sigma (\eta_1 \eta_{e_p} + \ldots + \eta_{e_p} \eta_1).$$

Similarly by addition of equations of the type (III) we have

$$\Sigma \left\{ \frac{\eta_1 \eta_{e_p} + \ldots + \eta_{e_p} \eta_1}{\lambda_i - \lambda} - \frac{\eta_1 \eta_{e_p - 1} + \ldots + \eta_{e_p - 1} \eta_1}{(\lambda_i - \lambda)^2} + \ldots \right\}$$
$$\equiv - \Sigma (\xi_1 \xi_{e_p} + \ldots + \xi_{e_p} \xi_1).$$

These equations show that if the quantities

$$\xi_1, \ldots, \xi_{e_p};\ \xi'_1, \ldots, \xi'_{e_{p'}}, \text{ \&c.}$$

are the coordinates of a *line* ξ, and $\eta_1, \ldots, \eta_{e_p}$; $\eta'_1, \ldots, \eta'_{e_{p'}}$, &c., the coordinates of a *line* η, then ξ *belongs to* $f(x) = 0$ *and* η *to a cosingular complex* $Y = 0$.

It is therefore seen that by aid of the equations (I), a (1, 1) correspondence is established between the variables ξ of f and η of Y which arise from the elementary divisor $(\lambda - \lambda_i)^{e_p}$.

This holds for each elementary divisor; hence by aid of sets of equations (I), (whose number is that of the elementary divisors), a (1, 1) correspondence is established between the lines ξ of f and η of Y, such that if $f(x) = 0$ expressed in the canonical variables be denoted by $f(\xi) = 0$, and $Y = 0$ by $Y(\eta) = 0$,

$$f(\xi) + \lambda \omega (\xi) \equiv \omega (\eta),$$
$$Y(\eta) \equiv \omega (\xi).$$

All the results already deduced for $(\lambda x^2) = 0$ from the existence of this (1, 1) correspondence will therefore hold for any quadratic complex.

159. The singular surface of the complex. The equations of the complex are obtained by equating to zero the *aggregate* of such terms as

$$\lambda_i (\xi_1 \xi_n + \ldots + \xi_n \xi_1) + \xi_1 \xi_{n-1} + \ldots + \xi_{n-1} \xi_1,$$
$$\xi_1 \xi_n + \ldots + \xi_n \xi_1,$$

where other portions arise from each other group of variables η, ζ, &c. It has been seen, (i), Art. 157, that if the line $\xi', \eta', \zeta', \ldots$ is a tangent line at P of the singular surface and $\xi, \eta, \zeta, \ldots$ the singular line for P, we have

$$(\lambda_i+\sigma)\,\xi_n+\xi_{n-1}=\rho\,.\,\xi_n',$$
$$(\lambda_i+\sigma)\,\xi_{n-1}+\xi_{n-2}=\rho\,.\,\xi'_{n-1},$$
$$\cdots\cdots\cdots\cdots\cdots\cdots\cdots\cdots$$
$$(\lambda_i+\sigma)\,\xi_2+\xi_1=\rho\,.\,\xi_2',$$
$$(\lambda_i+\sigma)\,\xi_1=\rho\,.\,\xi_1';$$

with corresponding equations for the other sets of variables $\eta', \zeta', \ldots$ of the form

$$(\lambda_j+\sigma)\,\eta_m+\eta_{m-1}=\rho\,.\,\eta_m', \text{ \&c.}$$

Now we obtain the coordinates of a double line, (Art. 156), by taking as zero all the coordinates $\xi, \eta, \zeta, \ldots$ except ξ_n, and the special complex having this double line for directrix is $\xi_1=0$. Hence, for a tangent line of the singular surface which intersects this double line, we have $\xi_1'=0$, and therefore $\lambda_i+\sigma=0$.

Again from the equations connecting η and η' we easily find, if Y_m be written for $\eta_1\eta_m+\ldots+\eta_m\eta_1$ and Y'_m for the same function of the η',

$$(\lambda_j+\sigma)^2\,Y_m+2\,(\lambda_j+\sigma)\,Y_{m-1}+Y_{m-2}=\rho^2\,.\,Y'_m,$$

hence

$$\rho^2\left(\frac{Y'_m}{\lambda_j+\sigma}-\frac{Y'_{m-1}}{(\lambda_j+\sigma)^2}+\ldots+\frac{(-1)^{m-1}\,Y_1}{(\lambda_j+\sigma)^m}\right)=(\lambda_j+\sigma)\,Y_m+Y_{m-1};$$

in which, for a line $\xi', \eta', \zeta', \ldots$ which meets the double line, $\sigma=-\lambda_i$. But since $\xi, \eta, \zeta, \ldots$ belongs to f we have

$$\xi_1\xi_{n-1}+\ldots+\xi_{n-1}\xi_1+(\lambda_j-\lambda_i)\,Y_m+Y_{m-1}+\ldots=0;$$

hence, substituting in this equation for the $\xi, \eta, \zeta, \ldots$ their values in terms of $\xi', \eta', \zeta', \ldots$, we find

$$\xi_2'\xi_n'+\ldots+\xi_n'\xi_2'+\frac{Y_m'}{\lambda_j-\lambda_i}-\ldots+\frac{(-1)^{m-1}\,Y_1'}{(\lambda_j-\lambda_i)^m}+\ldots=0.$$

This last equation thus represents a quadratic complex to which belong those tangents of the singular surface which also intersect the given double line, *i.e.* which satisfy $\xi_1'=0$.

The singular surface must therefore be a Complex Surface of Plücker. In all cases, therefore, in which a *group of variables* ξ *occurs*, the singular surface is a complex surface.

If the ξ are the only *group* of variables, the terms of the last equation which follow the ξ' are composed of squares, *i.e.* are of the form

$$\frac{\eta^2}{\lambda_j-\lambda_i}+\frac{\zeta^2}{\lambda_k-\lambda_i}+\ldots.$$

160. Degree of a complex. The number of cosingular complexes which pass through any line is called the *degree** of the complex. For cases in which no two elementary divisors refer to the same root λ_i, (the eleven principal types), the degree of the complex is four, as may be seen from the preceding equation of the cosingular complexes. In other cases the degree is easily calculated, *e.g.* in [1113] the equation to determine λ having given $\xi, \eta, \zeta \ldots$ is of the fourth degree; but in [11(13)] the coefficient of λ^4 vanishes identically, so that the degree of the complex is three.

161. The varieties of the quadratic Complex. We shall now investigate the different varieties of the quadratic complex†; they consist as has been seen of eleven types or canonical forms, and each type contains a number of sub-cases.

First canonical form. [111 111].

$$\omega(x) \equiv x_1^2+x_2^2+x_3^2+x_4^2+x_5^2+x_6^2,$$

$$f(x) \equiv \lambda_1 x_1^2+\lambda_2 x_2^2+\lambda_3 x_3^2+\lambda_4 x_4^2+\lambda_5 x_5^2+\lambda_6 x_6^2.$$

This general form has been already considered in Chapter VI.

162. The sub-cases are

$$[1111(11)];\ \lambda_5=\lambda_6.$$

The complex is

$$f(x) \equiv (\lambda_1-\lambda_5)\,x_1^2+(\lambda_2-\lambda_5)\,x_2^2+(\lambda_3-\lambda_5)\,x_3^2+(\lambda_4-\lambda_5)\,x_4^2=0\,;$$

the lines which satisfy the equations $x_1=x_2=x_3=x_4=0$, are double lines of the complex, they are the edges A_1A_4, A_2A_3 of the tetrahedron of reference.

The singular lines satisfy the equations $f(x)=0$, $f_1(x)=0$, where

$$f_1(x) \equiv (\lambda_1-\lambda_5)^2\,x_1^2+(\lambda_2-\lambda_5)^2\,x_2^2+(\lambda_3-\lambda_5)^2\,x_3^2+(\lambda_4-\lambda_5)^2\,x_4^2.$$

* Segre.

† The classification which follows was given by Weiler, *Math. Ann.* VII., "Ueber die verschiedenen Gattungen der Complexe zweiten Grades." His memoir contains some inaccuracies which have been corrected by Segre, see "Note sur les complexes quadratiques dont la surface singulière est une surface du 2e dégré double," *Math. Ann.* XXIII. See also Segre's classification of the quadratic complex in the *Memorie della R. Accad. di Torino* (1883).

The singular surface is obtained by substituting $\lambda_5 = \lambda_6$ in the equation given in Art. 82, and is therefore

$$\begin{aligned}(\lambda_1 - \lambda_2)(\lambda_3 - \lambda_5)(\lambda_4 - \lambda_5)(y_1^2 y_2^2 + y_3^2 y_4^2)\\ + (\lambda_3 - \lambda_4)(\lambda_1 - \lambda_5)(\lambda_2 - \lambda_5)(y_1^2 y_3^2 + y_2^2 y_4^2)\\ + 2\{(\lambda_1 + \lambda_2 - 2\lambda_5)(\lambda_3 - \lambda_5)(\lambda_4 - \lambda_5)\\ - (\lambda_3 + \lambda_4 - 2\lambda_5)(\lambda_1 - \lambda_5)(\lambda_2 - \lambda_5)\}\, y_1 y_2 y_3 y_4 = 0;\end{aligned}$$

which is a ruled quartic, (Art. 155), possessing

$$y_1 = y_4 = 0, \quad y_2 = y_3 = 0,$$

as double directrices, and hence belonging to class I.

163. $[(111)111]$, $\lambda_1 = \lambda_2 = \lambda_3$.

$$f(x) \equiv (\lambda_4 - \lambda_1)\, x_4^2 + (\lambda_5 - \lambda_1)\, x_5^2 + (\lambda_6 - \lambda_1)\, x_6^2;$$

$$f_1(x) \equiv (\lambda_4 - \lambda_1)^2 x_4^2 + (\lambda_5 - \lambda_1)^2 x_5^2 + (\lambda_6 - \lambda_1)^2 x_6^2.$$

The double lines are those which satisfy the equations

$$x_4 = x_5 = x_6 = 0;$$

they are one set of generators of the quadric $y_1 y_3 - y_2 y_4 = 0$. This quadric, counted twice, constitutes the singular surface, (Art. 156), as may be seen by making $\lambda_1 = \lambda_2 = \lambda_3$ in the singular surface of $[(11)1111]$. The equations of the singular lines assume the form $\frac{x_4^2}{A} = \frac{x_5^2}{B} = \frac{x_6^2}{C}$, and therefore form four linear congruences.

164. The complex $[(111)111]$ is one of a series of five, the others being $[1(11)(111)]$, $[(111)12]$, $[(111)(12)]$, $[(111)3]$, which are formed by aid of an involution $[2]$ between the lines of a regulus.

The involution $[2]$ is defined by an equation of the form

$$Lz^2z'^2 + Mzz'(z + z') + N(z + z')^2 + Rzz' + S(z + z') + T = 0 \ldots\text{(I)}.$$

By making $z = z'$ it is seen that there are in general four elements of the involution each of which coincides with one of its corresponding elements; if these be called "double" elements, four special cases arise:

Case (i) two double elements coincide,
„ (ii) three „ „ „
„ (iii) four „ „ „
„ (iv) two pairs of double elements coincide.

Now we may, for clearness, regard the coordinates z, z' as defining points on a given line; to the four double elements will then correspond four points, say P, P'; Q, Q', on this line. There are two points, say A and B, which are harmonic with P and P' and also with Q and Q', *i.e. A and B are the double points of the ordinary involution determined by P, P'; Q, Q'.*

If α and β are the coordinates of A and B respectively, then substituting in (I) by aid of the equations

$$x = m\frac{z-\alpha}{z-\beta}, \quad x' = m\frac{z'-\alpha}{z'-\beta},$$

we obtain the involution [2] expressed in terms of x and x'. The coordinates of A and B are now seen to be zero and infinity respectively; hence the points P and P' harmonic with them must have coordinates of the form $\pm\gamma$; similarly Q and Q' have coordinates $\pm\delta$. So that the equation to determine the double elements must be of the form

$$Lx^4 + Kx^2 + T = 0;$$

where $K = 4N + R$: that is to say, the coefficients of the second and fifth terms in the equation defining the involution [2] must be zero.

The latter equation, therefore, is of the form

$$Lx^2x'^2 + N(x + x')^2 + Rxx' + T = 0;$$

which is therefore a form by which the general involution [2] may be defined.

By writing in the new equation $x = \left(\frac{T}{L}\right)^{\frac{1}{4}}.u$, $x' = \left(\frac{T}{L}\right)^{\frac{1}{4}}.u'$, the coefficients of the first and last terms are made equal.

165. It will now be shown that the complex [(111)111] is the locus of lines which intersect corresponding lines of a regulus in the general involution [2].

For the complex may be written

$$\lambda_4 x_4^2 + \lambda_5 x_5^2 + \lambda_6 x_6^2 = 0,$$

if we replace $\lambda_4 - \lambda_1$ by λ_4, etc.

Any line of this complex is therefore given by the equations

$$\sqrt{\lambda_4}\, x_4 : \sqrt{\lambda_5}\, x_5 : \sqrt{\lambda_6}\, x_6 = \mu^2 - 1 : i(\mu^2 + 1) : 2\mu;$$

which divide the complex into a singly infinite number of linear congruences.

Also any line of the regulus $x_1 = x_2 = x_3 = 0$ is given by

$$x_4 : x_5 : x_6 = \rho^2 - 1 : i(\rho^2 + 1) : 2\rho.$$

All lines of such a linear congruence meet lines of the regulus for values of ρ given by the equation

$$\frac{(\mu^2-1)(\rho^2-1)}{\sqrt{\lambda_4}}-\frac{(\mu^2+1)(\rho^2+1)}{\sqrt{\lambda_5}}+\frac{4\mu\rho}{\sqrt{\lambda_6}}=0.$$

If ρ_1 and ρ_2 are the roots of this quadratic we have

$$\rho_1+\rho_2=\frac{A\mu}{C\mu^2+D},\ \rho_1\rho_2=\frac{D\mu^2+C}{C\mu^2+D},$$

where $$A=\frac{4}{\sqrt{\lambda_6}},\ C=\frac{1}{\sqrt{\lambda_5}}-\frac{1}{\sqrt{\lambda_4}},\ D=\frac{1}{\sqrt{\lambda_5}}+\frac{1}{\sqrt{\lambda_4}}.$$

Hence between ρ_1 and ρ_2 the following equation exists

$$(\rho_1+\rho_2)^2=\frac{A^2}{(C^2-D^2)^2}(C-D\rho_1\rho_2)(C\rho_1\rho_2-D),$$

which defines a general involution [2].

The complex is therefore obtained by establishing a general involution [2] between the lines of a regulus, and taking all the lines which intersect each pair of corresponding lines.

166. If $\lambda_4=\lambda_5$, which gives the form $[(111)(11)1]$, $C=0$, and the involution becomes $(\rho_1+\rho_2)^2=\frac{A^2}{D^2}\rho_1\rho_2$, which is derived from the general case (i) (Art. 164) when $L=M=S=T=0$; here (i) has two pairs of coincident double elements (viz. two infinite and two zero), giving case (iv).

167. $[11(11)(11)]$, $\lambda_3=\lambda_4$, $\lambda_5=\lambda_6$.

In this case

$$f(x)\equiv(\lambda_1-\lambda_5)x_1^2+(\lambda_2-\lambda_5)x_2^2+(\lambda_3-\lambda_5)(x_3^2+x_4^2),$$
$$f_1(x)\equiv(\lambda_1-\lambda_5)^2x_1^2+(\lambda_2-\lambda_5)^2x_2^2+(\lambda_3-\lambda_5)^2(x_3^2+x_4^2).$$

There are two pairs of double lines, viz. A_1A_4, A_2A_3; A_1A_3, A_2A_4; which form a twisted quadrilateral.

If in the singular surface of $[1111(11)]$ we put $\lambda_3=\lambda_4$, it becomes of the form $y_1^2y_2^2+y_3^2y_4^2=Ky_1y_2y_3y_4$, and therefore consists of two quadrics which intersect in the four double lines of the complex.

From the equations of the singular lines we deduce that they satisfy the equation $f_1-(\lambda_3-\lambda_5)f=0$, which is of the form

$$(x_1+ax_2)(x_1-ax_2)=0,$$

hence the singular lines consist of two congruences (2, 2).

This is one of three complexes which have for singular surfaces a pair of quadrics, the others being [1 (11) (12)] and [(12) (12)]. Writing λ_1 for $\lambda_1 - \lambda_5$ it is seen that the complex consists of the singly infinite system of linear congruences

$$\sqrt{\lambda_1}(1+\kappa_1)x_1 + i\sqrt{\lambda_2}(1-\kappa_1)x_2 + \sqrt{\lambda_3}\left(\frac{-\kappa_1}{\mu}+\mu\right)x_3 + i\sqrt{\lambda_3}\left(\frac{\kappa_1}{\mu}+\mu\right)x_4 = 0,$$

$$\sqrt{\lambda_1}(1+\kappa_2)x_1 + i\sqrt{\lambda_2}(1-\kappa_2)x_2 + \sqrt{\lambda_3}\left(\frac{-\kappa_2}{\mu}+\mu\right)x_3 + i\sqrt{\lambda_3}\left(\frac{\kappa_2}{\mu}+\mu\right)x_4 = 0;$$

where κ_1, κ_2 are the roots of the equation

$$\kappa^2 + 1 + 2\kappa \, . \, \frac{\lambda_1 + \lambda_2 - 2\lambda_3}{\lambda_1 - \lambda_2} = 0.$$

The directrices p, p' of these (special) complexes have coordinates

	x_1	x_2	x_3	x_4	x_5	x_6
p	$\sqrt{\lambda_1}(1+\kappa_1)$,	$i\sqrt{\lambda_2}(1-\kappa_1)$,	$\sqrt{\lambda_3}\left(\frac{-\kappa_1}{\mu}+\mu\right)$,	$i\sqrt{\lambda_3}\left(\frac{\kappa_1}{\mu}+\mu\right)$,	0,	0
p'	$\sqrt{\lambda_1}(1+\kappa_2)$,	$i\sqrt{\lambda_2}(1-\kappa_2)$,	$\sqrt{\lambda_3}\left(\frac{-\kappa_2}{\mu}+\mu\right)$,	$i\sqrt{\lambda_3}\left(\frac{\kappa_2}{\mu}+\mu\right)$,	0,	0.

On varying μ, it is seen that p and p' describe two reguli, whose lines are in (1, 1) correspondence, and which have (for $\mu = 0$, $\mu = \infty$, respectively), the common self-corresponding lines A_1A_3, A_2A_4; while A_1A_4, A_2A_3 are seen to be common directrices of the two reguli; hence *the complex is the locus of lines which intersect corresponding pairs of lines of two reguli which are in* (1, 1) *correspondence, and which have two common self-corresponding lines.*

168. $[1(1\dot{1})(111)]$, $\lambda_2 = \lambda_3$, $\lambda_4 = \lambda_5 = \lambda_6$.

The complex is

$$f(x) \equiv (\lambda_1 - \lambda_4)x_1^2 + (\lambda_2 - \lambda_4)(x_2^2 + x_3^2) = 0,$$

while $$f_1(x) \equiv (\lambda_1 - \lambda_4)^2 x_1^2 + (\lambda_2 - \lambda_4)^2(x_2^2 + x_3^2) = 0.$$

Hence the singular lines form the congruences

$$x_1 = 0, \ x_2 + ix_3 = 0; \quad x_1 = 0, \ x_2 - ix_3 = 0.$$

The singular surface consists as in [111(111)] of a quadric taken doubly, while the double lines are one set of generators of this quadric together with the two lines

$$x_1 = x_4 = x_5 = x_6 = 0.$$

169. [(11)(11)(11)], $\lambda_1 = \lambda_2$, $\lambda_3 = \lambda_4$, $\lambda_5 = \lambda_6$.

This gives the Tetrahedral Complex

$$\lambda_1 (x_1^2 + x_2^2) + \lambda_3 (x_3^2 + x_4^2) + \lambda_5 (x_5^2 + x_6^2) = 0.$$

The complex possesses thirteen independent constants, viz. twelve from the tetrahedron and one from the constant double ratio.

170. [(111)(111)], $\lambda_1 = \lambda_2 = \lambda_3$, $\lambda_4 = \lambda_5 = \lambda_6$.

The equation of the complex has either of the forms

$$x_1^2 + x_2^2 + x_3^2 = 0, \quad x_4^2 + x_5^2 + x_6^2 = 0.$$

The generators of one system are $x_1 = x_2 = x_3 = 0$, and those of the other are $x_4 = x_5 = x_6 = 0$; hence any tangent line of the quadric will belong to the complex, which therefore consists of the tangents of a quadric.

171. *Second canonical form.* [11112].

$$\omega(x) \equiv x_1^2 + x_2^2 + x_3^2 + x_4^2 + 2x_5x_6,$$

$$f(x) \equiv \lambda_1 x_1^2 + \lambda_2 x_2^2 + \lambda_3 x_3^2 + \lambda_4 x_4^2 + 2\lambda_5 x_5 x_6 + x_5^2.$$

From the form of the identity $\omega(x) \equiv 0$ it is permissible to write

$$x_1 = p_{12} + p_{34}, \quad x_3 = p_{13} + p_{42}, \quad x_5 = p_{14},$$
$$ix_2 = p_{12} - p_{34}, \quad ix_4 = p_{13} - p_{42}, \quad x_6 = 2p_{23}.$$

The singular lines are those whose coordinates satisfy the equations

$$(\lambda_1 - \lambda_5) x_1^2 + (\lambda_2 - \lambda_5) x_2^2 + (\lambda_3 - \lambda_5) x_3^2 + (\lambda_4 - \lambda_5) x_4^2 + x_5^2 = 0,$$
$$(\lambda_1 - \lambda_5)^2 x_1^2 + (\lambda_2 - \lambda_5)^2 x_2^2 + (\lambda_3 - \lambda_5)^2 x_3^2 + (\lambda_4 - \lambda_5)^2 x_4^2 = 0.$$

The directrix of the special complex $x_5 = 0$ is seen to be a double line of the complex (Art. 156).

It is seen from Art. 159 that the singular surface is the Complex Surface of the congruence

$$x_5 = 0, \quad x_6^2 + \frac{x_1^2}{\lambda_1 - \lambda_5} + \frac{x_2^2}{\lambda_2 - \lambda_5} + \frac{x_3^2}{\lambda_3 - \lambda_5} + \frac{x_4^2}{\lambda_4 - \lambda_5} = 0;$$

and hence of the congruence

$$x_5'^2 + \frac{x_1^2}{\lambda_1 - \lambda_5} + \frac{x_2^2}{\lambda_2 - \lambda_5} + \frac{x_3^2}{\lambda_3 - \lambda_5} + \frac{x_4^2}{\lambda_4 - \lambda_5} = 0, \quad x'_5 + ix'_6 = 0,$$

where $x'_5 = x_6 + \frac{1}{2}x_5, \quad x'_6 = i(x_6 - \frac{1}{2}x_5),$

and therefore $\sum_1^4 x_i^2 + x'^2_5 + x'^2_6 = 0.$

Thus the Singular Surface is the Plücker surface for a general quadratic complex [111 111] and an edge of a fundamental tetrahedron.

It was seen (Art. 86) that the Plücker surface for the complex

$$(\lambda x^2) = 0, \quad (x^2) = 0 \text{ is } (\lambda u^2)(\lambda v^2) - (\lambda uv)^2 = 0,$$

where u is a line through A, v a line through B, A and B being any two points on the double line, and u, v meet in a point y of the Complex Surface. The double line being in this case the edge A_2A_3 of the tetrahedron of reference, we may take A as A_2 and B as A_3, which gives as the coordinates of u and v

	p_{12}	p_{13}	p_{14}	p_{23}	p_{34}	p_{42}
u	y_1	0	0	$-y_3$	0	y_4
v	0	y_1	0	y_2	$-y_4$	0,

whence

	x_1	x_2	x_3	x_4	x_5'	x_6'
u	y_1	$-iy_1$	y_4	iy_4	$-y_3$	$-iy_3$
v	$-y_4$	$-iy_4$	y_1	$-iy_1$	y_2	iy_2,

and the equation of the singular surface is

$$\left\{y_3^2 + \frac{y_1^2(\lambda_2-\lambda_1)}{(\lambda_1-\lambda_5)(\lambda_2-\lambda_5)} + \frac{y_4^2(\lambda_4-\lambda_3)}{(\lambda_3-\lambda_5)(\lambda_4-\lambda_5)}\right\} \times$$
$$\left\{y_2^2 + \frac{y_4^2(\lambda_2-\lambda_1)}{(\lambda_1-\lambda_5)(\lambda_2-\lambda_5)} + \frac{y_1^2(\lambda_4-\lambda_3)}{(\lambda_3-\lambda_5)(\lambda_4-\lambda_5)}\right\}$$
$$-\left\{-y_2y_3 + y_1y_4\left(-\frac{1}{\lambda_1-\lambda_5} - \frac{1}{\lambda_2-\lambda_5} + \frac{1}{\lambda_3-\lambda_5} + \frac{1}{\lambda_4-\lambda_5}\right)\right\}^2 = 0,$$

which reduces to the form

$$\begin{aligned}(\lambda_1-\lambda_2)(\lambda_3-\lambda_4)(y_1^4+y_4^4) &- (\lambda_3-\lambda_5)(\lambda_4-\lambda_5)(\lambda_1-\lambda_2)(y_1^2y_2^2+y_3^2y_4^2)\\ &- (\lambda_1-\lambda_5)(\lambda_2-\lambda_5)(\lambda_3-\lambda_4)(y_1^2y_3^2+y_2^2y_4^2)\\ &+ 2\left[(\lambda_1+\lambda_2-2\lambda_5)(\lambda_3+\lambda_4-2\lambda_5)\right.\\ &\left.- 2\{(\lambda_1-\lambda_5)(\lambda_2-\lambda_5)+(\lambda_3-\lambda_5)(\lambda_4-\lambda_5)\}\right] y_1^2y_4^2\\ &+ 2\{(\lambda_1-\lambda_5)(\lambda_2-\lambda_5)(\lambda_3+\lambda_4-2\lambda_5)\\ &- (\lambda_3-\lambda_5)(\lambda_4-\lambda_5)(\lambda_1+\lambda_2-2\lambda_5)\}\, y_1y_2y_3y_4 = 0.\end{aligned}$$

This equation might also have been obtained by finding the locus of points y for which the complex cones of [11112] become pairs of planes.

172. There are nine sub-cases, which are as follows:

$$[111(12)], \quad \lambda_4 = \lambda_5.$$

The singular surface becomes

$$(\lambda_1-\lambda_2)(\lambda_3-\lambda_4)(y_1^4+y_4^4)-(\lambda_1-\lambda_4)(\lambda_2-\lambda_4)(\lambda_3-\lambda_4)(y_1y_3-y_2y_4)^2$$
$$+2[(\lambda_3-\lambda_4)(\lambda_1+\lambda_2-2\lambda_4)-2(\lambda_1-\lambda_4)(\lambda_2-\lambda_4)]\,y_1^2y_4^2=0.$$

The line $y_1 = y_4 = 0$ is a double line of the surface, and any plane through it cuts the surface in two lines which meet on the double line; the surface is the ruled quartic of class II.

173. $$[11(11)2], \quad \lambda_3 = \lambda_4.$$

The complex possesses three double lines, viz. A_2A_3 and also A_1A_3, A_2A_4; any plane through A_1A_3 is seen to meet the singular surface in two lines intersecting on A_2A_4 and *vice versâ*, hence A_1A_3 and A_2A_4 are double directrices of the surface, while A_2A_3 is a double generator; the singular surface therefore belongs to class VII.

174. $$[11(112)], \quad \lambda_3 = \lambda_4 = \lambda_5.$$

The equation of the complex is

$$f(x) \equiv (\lambda_1-\lambda_3)\,x_1^2+(\lambda_2-\lambda_3)\,x_2^2+x_5^2=0,$$

the singular lines being given by $f(x) = 0$, and

$$f_1(x) \equiv (\lambda_1-\lambda_3)^2\,x_1^2+(\lambda_2-\lambda_3)^2\,x_2^2=0\,;$$

thus the congruence of singular lines consists of the four linear congruences

$$x_1 : x_2 : x_5 = i(\lambda_2-\lambda_3) : \pm(\lambda_1-\lambda_3) : \pm\sqrt{(\lambda_1-\lambda_3)(\lambda_2-\lambda_3)(\lambda_2-\lambda_1)}.$$

The complex is composed of the singly infinite number of linear congruences

$$x_1 + ix_2 = 2\rho \,.\, x_5, \quad x_1 - ix_2 = 2\sigma \,.\, x_5,$$

i.e. $$p_{12} = \rho \,.\, p_{14}, \quad p_{34} = \sigma \,.\, p_{14}\,;$$

where ρ and σ are connected by the equation

$$(\lambda_1-\lambda_3)(\rho+\sigma)^2-(\lambda_2-\lambda_3)(\rho-\sigma)^2+1=0.$$

Hence, *the complex is formed by lines which meet corresponding lines of two pencils* (A_3, α_1), (A_2, α_4) *which are in* (2, 2) *correspondence, and which have a common self-corresponding line* A_2A_3. *The* (2, 2) *correspondence has two of its four double elements in coincidence.*

The lines $x_1 = x_2 = x_5 = 0$, forming the two pencils (A_2, α_1), (A_3, α_4), are double lines of the complex. The singular surface is seen to consist of the centres and planes of these pencils.

175. $[(11)1(12)], \quad \lambda_1 = \lambda_2, \quad \lambda_4 = \lambda_5.$

The singular surface consists of the two quadrics

$$(\lambda_1 - \lambda_4)(\lambda_3 - \lambda_4)(y_1 y_3 - y_2 y_4)^2 - 4(\lambda_3 - \lambda_1) y_1^2 y_4^2 = 0,$$

which touch along their common generator $y_1 = y_4 = 0$, and which have also in common the lines $y_1 = y_2 = 0$, $y_3 = y_4 = 0$.

The complex is $\lambda_1(x_1^2 + x_2^2) + \lambda_3 x_3^2 + x_5^2 = 0$ (if λ_1 be written for $\lambda_1 - \lambda_4$), and consists of the congruences

$$\sqrt{\lambda_1}\, x_1\left(\mu + \frac{k_1}{\mu}\right) + i\sqrt{\lambda_1}\, x_2\left(\frac{k_1}{\mu} - \mu\right) + \sqrt{\lambda_3}\,(k_1 - 1)\, x_3 - i(k_1 + 1)\, x_5 = 0,$$

$$\sqrt{\lambda_1}\, x_1\left(\mu + \frac{k_2}{\mu}\right) + i\sqrt{\lambda_1}\, x_2\left(\frac{k_2}{\mu} - \mu\right) + \sqrt{\lambda_3}\,(k_2 - 1)\, x_3 - i(k_2 + 1)\, x_5 = 0,$$

where k_1, k_2 are the roots of the equation $\Omega(a) = 0$, *i.e.*

$$4\lambda_1 k + \lambda_3 (k - 1)^2 = 0.$$

Thus each of these complexes is special; the directrices describe (for different values of μ) two reguli which have A_1A_2, A_3A_4 in common.

Hence, *the complex consists of lines which meet corresponding pairs of lines of two reguli in* (1, 1) *correspondence which have two common self-corresponding lines, and two consecutive common directrices.*

176. $[(111)12], \quad \lambda_1 = \lambda_2 = \lambda_3.$

The complex is

$$(\lambda_4 - \lambda_1) x_4^2 + 2(\lambda_5 - \lambda_1) x_5 x_6 + x_5^2 = 0;$$

the double lines are those given by $x_4 = x_5 = x_6 = 0$, which form one set of generators of the quadric $y_1 y_3 - y_2 y_4 = 0$. The singular surface consists of this quadric taken doubly. The singular lines are given by the equations

$$\frac{x_4^2}{(\lambda_5 - \lambda_1)^2} = \frac{2x_5 x_6}{(\lambda_4 - \lambda_1)(\lambda_1 + \lambda_4 - 2\lambda_5)} = \frac{x_5^2}{(\lambda_4 - \lambda_1)(\lambda_5 - \lambda_1)(\lambda_5 - \lambda_4)},$$

they therefore form the special linear congruence $x_4 = x_5 = 0$, and two general linear congruences.

If in the equation of the complex λ_4 be written for $\lambda_4 - \lambda_1$ &c., it is easily seen that any line of the complex is given by the equation

$$x_4 : x_5 : x_6 = \sqrt{\lambda_5}\,(2\mu\sqrt{\lambda_5} - i) : \sqrt{\lambda_4\lambda_5} : -2\mu\sqrt{\lambda_4}\,(\mu\sqrt{\lambda_5} - i),$$

giving ∞^1 linear congruences; the line $(0, 0, 0, 2\rho, 2, -\rho^2)$ is any line of the regulus $x_1 = x_2 = x_3 = 0$, and is met by the lines of the preceding congruence provided that

$$\sqrt{\lambda_4\lambda_5}\rho^2 + 4\sqrt{\lambda_4}\mu(\mu\sqrt{\lambda_5} - i) - 2\rho\sqrt{\lambda_5}(2\mu\sqrt{\lambda_5} - i) = 0.$$

If ρ_1 and ρ_2 are the roots of this quadratic it is seen that

$$(\rho_1 + \rho_2)^2 + \frac{4}{\lambda_4} = 4\frac{\lambda_5}{\lambda_4}\rho_1\rho_2.$$

This equation defines an involution [2] *which has two coincident double elements, and is case* (i) *previously mentioned*, (Art. 164).

If $\lambda_4 = \lambda_5$, giving [(111)(12)], *the involution has all its double elements coincident*, since it is given by the equation $(\rho_1 - \rho_2)^2 + \frac{\lambda_4}{4} = 0$; *this is case* (iii) *of the involution* [2].

177. [(11)(11)2], $\lambda_1 = \lambda_2$, $\lambda_3 = \lambda_4$.

The singular surface as derived from that of [11112] is seen to consist of the planes $y_1 = 0$, $y_4 = 0$ and a quadric whose equation is $y_1y_4 - \kappa y_2y_3 = 0$. If the equation of the singular surface for [11112] be formed in plane-coordinates and if in it we put $\lambda_1 = \lambda_2$, $\lambda_3 = \lambda_4$, it is seen that v_2v_3 is a factor of the equation, which shows that the singular surface is completed by the points A_2, A_3 which raise its class to four.

If λ_3 be written for $\lambda_3 - \lambda_1$ &c., the equation of the complex is

$$4\lambda_3 p_{13}p_{42} + 4\lambda_5 p_{14}p_{23} + p^2_{14} = 0.$$

It consists therefore of the linear congruences

$$2\sqrt{\lambda_3}p_{13} - \mu p_{14} = 0,$$

$$2\sqrt{\lambda_3}(\lambda_3 - \lambda_5)\mu^2 p_{42} + 4\lambda_5(\lambda_3 - \lambda_5)\mu p_{23} + \mu\lambda_3 p_{14} - 2\sqrt{\lambda_3}\lambda_5 p_{13} = 0.$$

The directrices of the first of these linear complexes form the pencil (A_2, α_1), the directrices of the second form a regulus; the lines of the pencil and of the regulus are in (1, 1) correspondence, and have a common self-corresponding line A_2A_4; hence, *the complex consists of the lines which meet paired lines of a pencil and regulus in* (1, 1) *correspondence having a common self-corresponding line.*

178. [(11)(112)], $\lambda_1 = \lambda_2$, $\lambda_3 = \lambda_4 = \lambda_5$.

In this case

$$f(x) \equiv (\lambda_1 - \lambda_5)(x_1^2 + x_2^2) + x_5^2 = 4(\lambda_1 - \lambda_5)p_{12}p_{34} + p^2_{14}.$$

The complex consists of the singly infinite series of congruences

$$2\sqrt{\lambda_1 - \lambda_5}\,.\,p_{12} = \mu p_{14},\quad 2\sqrt{\lambda_1 - \lambda_5}\,.\,p_{34} = -\frac{1}{\mu}p_{14}.$$

Each of these complexes is special and the directrices form the respective pencils (A_3, α_1), (A_2, α_4) which are in (1, 1) correspondence, *hence the complex is the locus of lines which meet corresponding lines of two projective pencils where the plane of each pencil passes through the centre of the other.*

From consideration of the complex [11(112)] the double lines are seen to be the pencils (A_2, α_1), (A_3, α_4), and the singular surface consists of the centres and planes of these pencils.

179. $[(111)(12)], \quad \lambda_1 = \lambda_2 = \lambda_3, \quad \lambda_4 = \lambda_5.$

The complex is

$$(\lambda_4 - \lambda_1)(x_4^2 + 2x_5x_6) + x_5^2 = 0.$$

The singular surface, obtained from that of [111(12)], is $(y_1y_3 - y_2y_4)^2 = 0$. By comparison with [111(12)] and [(111)12] the double lines are seen to consist of one set of generators of the singular surface together with the generator $y_1 = y_4 = 0$.

180. *Third canonical form.* [1113].

$$\omega(x) \equiv x_1^2 + x_2^2 + x_3^2 + x_5^2 + 2x_4x_6,$$

$$f(x) \equiv \lambda_1x_1^2 + \lambda_2x_2^2 + \lambda_3x_3^2 + \lambda_4(x_5^2 + 2x_4x_6) + 2x_4x_5.$$

From the form of $\omega(x)$ it is permissible to write

$$x_1 = p_{12} + p_{34}, \quad x_3 = p_{13} + p_{42}, \quad x_4 = p_{14},$$

$$ix_2 = p_{12} - p_{34}, \quad ix_3 = p_{13} - p_{42}, \quad x_6 = 2p_{23}.$$

The singular surface is (Art. 159) the complex surface for

$$\phi(x) \equiv \frac{x_1^2}{\lambda_1 - \lambda_4} + \frac{x_2^2}{\lambda_2 - \lambda_4} + \frac{x_3^2}{\lambda_3 - \lambda_4} + 2x_5x_6 = 0, \quad x_4 = 0;$$

and is therefore $\phi(u)\,\phi(v) - \left(\frac{1}{2}u_i\frac{\partial\phi}{\partial v_i}\right)^2 = 0$, where the lines u and v have the same coordinates p_{ik} as in [11112]; hence, finding the x coordinates of u and v from Art. 171, the singular surface is seen to be

$$\left\{\frac{y_1^2(\lambda_2 - \lambda_1)}{(\lambda_1 - \lambda_4)(\lambda_2 - \lambda_4)} - \frac{y_4^2}{\lambda_3 - \lambda_4} - 4y_3y_4\right\}$$

$$\times \left\{\frac{y_4^2(\lambda_2 - \lambda_1)}{(\lambda_1 - \lambda_4)(\lambda_2 - \lambda_4)} - \frac{y_1^2}{\lambda_3 - \lambda_4} + 4y_1y_2\right\}$$

$$- \left\{y_1y_4\left(-\frac{1}{\lambda_1 - \lambda_4} - \frac{1}{\lambda_2 - \lambda_4} + \frac{1}{\lambda_3 - \lambda_4}\right) + 2(y_2y_4 - y_1y_3)\right\}^2 = 0;$$

which reduces to the form

$$(\lambda_1-\lambda_2)(y_1^4+y_4^4)-4(\lambda_3-\lambda_4)(\lambda_1-\lambda_2)(y_1^3y_2-y_3y_4^3)$$
$$-4(\lambda_1-\lambda_4)(\lambda_2-\lambda_4)(\lambda_3-\lambda_4)(y_1y_3+y_2y_4)^2$$
$$-4\{(\lambda_3-\lambda_4)(\lambda_1+\lambda_2-2\lambda_4)$$
$$-2(\lambda_1-\lambda_4)(\lambda_2-\lambda_4)\}\,y_1y_4(y_1y_3-y_2y_4)$$
$$+2(\lambda_1+\lambda_2-2\lambda_3)\,y_1^2y_4^2=0.$$

It is clear that the double line $y_1=y_4=0$, for which all the coordinates except x_6 are zero (Art. 156), belongs to the complex $\phi(x)=0$. Along this double line the tangent planes to the surface coincide with those of $y_1y_3+y_2y_4=0$.

181. $[11(13)]$, $\lambda_3=\lambda_4$.

The singular surface is

$$(\lambda_1-\lambda_2)(y_1^4+y_2^4)+8(\lambda_1-\lambda_3)(\lambda_2-\lambda_3)\,y_1y_4(y_1y_3-y_2y_4)$$
$$+2(\lambda_1+\lambda_2-2\lambda_3)\,y_1^2y_4^2=0.$$

The line $y_1=y_4=0$ is a triple line of this surface; and any plane through $y_1=0$, $y_4=0$, cuts out one line from the surface, which therefore is a ruled quartic of class XII.

182. $[(11)13]$, $\lambda_1=\lambda_2$.

Here there are two double lines A_1A_2, A_3A_4 in addition to the cuspidal double line A_2A_3; the singular surface is therefore a special case of class VII.

183. $[1(113)]$, $\lambda_2=\lambda_3=\lambda_4$.

The singular surface consists of the planes

$$y_1+iy_4=0,\quad y_1-iy_4=0,$$

counted twice, together with two points on their line of intersection. The double lines are those which belong to the three complexes $x_1=x_4=x_5=0$, forming two pencils which have A_2A_3 as common line.

The complex is

$$(\lambda_1-\lambda_2)\,x_1^2+2x_4x_5=0;$$

the lines of the complex belong to the singly infinite number of linear congruences

$$\sqrt{\lambda_1-\lambda_2}\,(x_5+ix_1)+2\mu x_4(\sqrt{\lambda_1-\lambda_2}\,\mu-i)=0,$$
$$\sqrt{\lambda_1-\lambda_2}\,(x_5-ix_1)+2\mu x_4(\sqrt{\lambda_1-\lambda_2}\,\mu+i)=0.$$

Each of these complexes is special; the coordinates of their directrices p, p' are

	x_1	x_2	x_3	x_4	x_5	x_6
p	$i\sqrt{\lambda_1-\lambda_2}$	0	0	0	$\sqrt{\lambda_1-\lambda_2}$	$2\mu(\sqrt{\lambda_1-\lambda_2}\,\mu-i)$
p'	$-i\sqrt{\lambda_1-\lambda_2}$	0	0	0	$\sqrt{\lambda_1-\lambda_2}$	$2\mu(\sqrt{\lambda_1-\lambda_2}\,\mu+i)$.

Hence p, p' are corresponding lines of two pencils in (2, 2) correspondence, and which have A_2A_3 as common self-corresponding line. Corresponding lines are of the form $\alpha_i+\rho\beta_i$, $\alpha'_i+\rho'\beta_i$ (where β_i is A_2A_3). The connexion between ρ and ρ' is given by the equation

$$\sqrt{\lambda_1-\lambda_2}(\rho-\rho')^2+2(\rho+\rho')=0.$$

In this involution [2] three double elements coincide (they are infinite).

184. $[(11)(13)]$, $\lambda_1=\lambda_2$, $\lambda_3=\lambda_4$.

Putting $\lambda_1=\lambda_2$ in $[11(13)]$ the singular surface is seen to consist of the planes $y_1y_4=0$, together with the quadric

$$2(\lambda_1-\lambda_3)(y_1y_3-y_2y_4)+y_1y_4=0.$$

The complex has three double lines.

The equation of the complex is

$$4\lambda_1p_{12}p_{34}+2p_{14}(p_{13}+p_{42})=0,$$

writing λ_1 for $\lambda_1-\lambda_4$. It is formed by the linear congruences

$$\sqrt{\lambda_1}p_{12}+\mu p_{14}=0,$$

$$\sqrt{\lambda_1}\,p_{12}+\mu p_{14}-2\mu\lambda_1(2\mu\sqrt{\lambda_1}\,p_{34}-p_{13}-p_{42})=0.$$

The directrices of the first complex, for different values of μ, form the pencil (A_3, α_1); those of the second form a regulus; *the lines of the pencil and regulus are in* (1, 1) *correspondence and have A_3A_4 as common self-corresponding line, while the line A_3A_2 of the pencil is a directrix of the regulus.* This gives that case of the correspondence in $[(11)(11)2]$ in which the pencil contains a directrix of the regulus.

185. $[(111)3]$, $\lambda_1=\lambda_2=\lambda_3$.

The singular surface reduces to $(y_1y_3+y_2y_4)^2=0$; the double lines are its generators $y_1=\mu y_4$, $y_3=-\dfrac{1}{\mu}y_2$.

The singular lines form the special congruence $x_4=x_5=0$, together with a general linear congruence.

The equation of the complex is

$$\lambda_4 (x_5^2 + 2x_4 x_6) + 2x_4 x_5 = 0,$$

if λ_4 be written for $\lambda_4 - \lambda_1$. The lines of the complex are given by the equations

$$x_4 : x_5 : x_6 = \lambda_4^{\frac{3}{2}} : 2\mu\lambda_4 : -2\mu (1 + \mu \sqrt{\lambda_4}),$$

forming ∞^1 linear congruences; the lines of such a congruence meet the line $(0, 0, 0, 2\rho^2, 2\rho, -1)$ of the regulus $x_1 = x_2 = x_3 = 0$, provided that

$$4\mu\rho^2 (1 + \mu \sqrt{\lambda_4}) - 4\mu\lambda_4\rho + \lambda_4^{\frac{3}{2}} = 0.$$

If the roots of this equation are ρ_1 and ρ_2, it is seen that

$$(\rho_1 - \rho_2)^2 + \frac{4}{\lambda_4} \rho_1\rho_2 (\rho_1 + \rho_2) = 0.$$

This is an involution [2] in which three double elements coincide, giving case (ii) (Art. 164).

186. *Fourth canonical form.* [1122].

$$\omega (x) \equiv x_1^2 + x_2^2 + 2x_3x_4 + 2x_5x_6,$$

$$f(x) \equiv \lambda_1 x_1^2 + \lambda_2 x_2^2 + 2\lambda_3 x_3 x_4 + 2\lambda_4 x_5 x_6 + x_3^2 + x_5^2.$$

We may write

$$x_1 = p_{12} + p_{34},\quad x_3 = p_{13},\quad x_5 = p_{14},$$

$$ix_2 = p_{12} - p_{34},\quad x_4 = 2p_{42},\quad x_6 = 2p_{23}.$$

The singular surface is the complex surface for

$$x_5 = 0,\quad x_6^2 + \frac{x_1^2}{\lambda_1 - \lambda_4} + \frac{x_2^2}{\lambda_2 - \lambda_4} + \frac{2x_3x_4}{\lambda_3 - \lambda_4} - \frac{x_3^2}{(\lambda_3 - \lambda_4)^2} = 0.$$

Repeating the process previously adopted, the equation of the singular surface is seen to be (if λ_1 be written for $\lambda_1 - \lambda_4$, &c.),

$$\begin{aligned}(\lambda_1 - \lambda_2) y_1^4 - 4\lambda_3^2 (\lambda_1 - \lambda_2) (y_1^2 y_2^2 + y_3^2 y_4^2) - 4\lambda_1\lambda_2 y_1^2 y_3^2 \\ + 4 (\lambda_2\lambda_3 + \lambda_3\lambda_1 - \lambda_1\lambda_2 - \lambda_3^2) y_1^2 y_4^2 \\ - 8\lambda_3 \{\lambda_3 (\lambda_1 + \lambda_2) - 2\lambda_1\lambda_2\} y_1 y_2 y_3 y_4 = 0.\end{aligned}$$

This is a Plücker surface with the two intersecting double lines A_2A_3, A_2A_4. There are six special cases.

187. [11(22)].

The singular surface is obtained from the general case by putting $\lambda_3 = 0$, which gives

$$y_1^2 \{(\lambda_1 - \lambda_2) y_1^2 - 4\lambda_1\lambda_2 (y_3^2 + y_4^2)\} = 0;$$

while if we find the envelope of the singular planes we obtain

$$v_2^2 \{(\lambda_1 - \lambda_2) v_2^2 - 4\lambda_1\lambda_2 (v_3^2 + v_4^2)\} = 0.$$

Hence the singular surface consists of a quadric cone and a conic whose plane passes through the vertex of the cone.

The double lines are given by $x_1 = x_2 = x_3 = x_5 = 0$, and consist of the pencil (A_2, α_1).

188. $[12(12)]$, $\lambda_2 = \lambda_4$.

The singular surface is

$$y_1^2 \{\lambda_1 y_1^2 + 4\lambda_3 (\lambda_1 - \lambda_3) y_4^2\} - 4\lambda_1 \lambda_3^2 (y_1 y_2 + y_3 y_4)^2 = 0.$$

Any plane through A_2A_3 cuts out two lines from the surface which intersect on A_2A_3, while A_2A_4 is a double generator; this is the case VIII. of ruled quartics. It gives a case of $[11(11)2]$. There are three double lines, viz. A_2A_4 and two lines coinciding with A_2A_3.

189. $[1(122)]$, $\lambda_2 = \lambda_3 = \lambda_4$.

If in the equations of the cone and conic which form the singular surface of $[11(22)]$ we make $\lambda_2 = 0$, the cone becomes a plane (counted twice), and the conic a point (counted twice).

The singular surface thus consists of a plane and a point in it taken four times.

This complex is a special case of $[11(112)]$; the double lines are those given by $x_1 = x_3 = x_5 = 0$, and therefore form the pencil (A_2, α_1); they are to be taken twice as being derived from the two pencils of $[11(112)]$.

190. $[(11)22]$, $\lambda_1 = \lambda_2$.

The singular surface is seen to reduce to a ruled cubic with A_1A_2 as *double* and A_3A_4 as *simple* directrix, together with the plane $y_1 = 0$, and the point $v_2 = 0$.

The double lines are

$$A_1A_2,\ A_3A_4,\ A_2A_4,\ A_2A_3.$$

191. $[(112)2]$, $\lambda_1 = \lambda_2 = \lambda_4$.

The equation of the complex may be put in the form

$$2(\lambda_3 - \lambda_4) x_3 x_4 + x_3^2 + x_5^2 = 0,$$

i.e.

$$4(\lambda_3 - \lambda_4) p_{13} p_{42} + p_{13}^2 + p_{14}^2 = 0.$$

The singular surface reduces to $y_1^2 y_4^2 = 0$, $v_2^2 v_3^2 = 0$; the double lines are A_2A_4 together with those given by $x_3 = x_4 = x_5 = 0$, *i.e.* the pencils (A_2, α_4), (A_3, α_1).

The complex is formed by lines which belong to the congruence

$$\mu x_3 + x_5 = 0,\ 2\mu(\lambda_3 - \lambda_4) x_4 - (\mu^2 + 1) x_5 = 0;$$

and is therefore *the locus of lines which intersect paired lines of two pencils in* (1, 2) *correspondence, which have a common self-corresponding line.*

192. $[(11)(22)],\ \lambda_1 = \lambda_2,\ \lambda_3 = \lambda_4.$

The equation of the singular surface is $y_1^2(y_3^2 + y_4^2) = 0$, together with $v_2^2(v_3^2 + v_4^2) = 0$.

The double lines are those which satisfy the equations

$$x_1 = x_2 = x_3 = x_5 = 0,$$

i.e. the pencil (A_2, α_1), together with A_1A_2, A_3A_4.

The complex is

$$(\lambda_1 - \lambda_3)(x_1^2 + x_2^2) + x_3^2 + x_5^2 = 0,$$

i.e.
$$4(\lambda_1 - \lambda_3) p_{12}p_{34} + p_{13}^2 + p_{14}^2 = 0;$$

and is formed by the lines which belong to the complexes

$$2\sqrt{\lambda_1 - \lambda_3}\, p_{12} - \mu(p_{13} + ip_{14}) = 0,$$

$$2\sqrt{\lambda_1 - \lambda_3}\, p_{34} + \frac{1}{\mu}(p_{13} - ip_{14}) = 0.$$

Each of these complexes is special, the directrices p, p' having the coordinates given by

	p_{12}	p_{13}	p_{14}	p_{23}	p_{34}	p_{42}
p	0	0	0	$-i\mu$	$2\sqrt{\lambda_1 - \lambda_3}$	$-\mu$
p'	$2\sqrt{\lambda_1 - \lambda_3}\,\mu$	0	0	$-i$	0	1 ;

hence p and p' form two projective pencils, the centre of the former being a point O on A_3A_4 and its plane α_1, the latter having A_2 for centre and A_1A_2O for its plane; hence, *the complex is the locus of lines which intersect corresponding lines of two projective pencils in which the plane of one pencil passes through the centre of the other.*

193. $[(12)(12)],\ \lambda_1 = \lambda_3,\ \lambda_2 = \lambda_4.$

The singular surface consists of the quadrics

$$y_1^2 = \pm 2\lambda_1 (y_1y_2 + y_3y_4),$$

i.e. two quadrics touching along A_2A_4, A_2A_3. The line A_2A_3 is a "doubled" double line, as also is A_2A_4. The complex is (writing λ_2 for $\lambda_2 - \lambda_1$),

$$\lambda_2(x_2^2 + 2x_5x_6) + x_3^2 + x_5^2 = 0.$$

It consists of the linear congruences

$$2\sqrt{\lambda_2}(1-i\mu)x_2-2\mu x_3+(\mu-\frac{1}{\mu}+2i)x_5+2\lambda_2\mu x_6=0,$$

$$2\sqrt{\lambda_2}(1+i\mu)x_2+2\mu x_3+(\mu-\frac{1}{\mu}-2i)x_5+2\lambda_2\mu x_6=0.$$

The directrices p, p' of these (special) complexes, as μ varies, describe two reguli. The line A_2A_4 is seen to be a directrix and A_2A_3 a common line of each regulus. Hence, *the complex consists of the lines which meet conjugate lines of two reguli in* (1, 1) *correspondence, the reguli having two common consecutive lines and two common consecutive directrices.*

194. *Fifth canonical form* [114].

$$\omega(x)\equiv x_1^2+x_2^2+2x_3x_6+2x_4x_5,$$

$$f(x)\equiv\lambda_1x_1^2+\lambda_2x_2^2+2\lambda_3(x_3x_6+x_4x_5)+2x_3x_5+x_4^2.$$

We may write

$$x_1=p_{12}+p_{34},\quad x_3=p_{14},\quad x_5=p_{42},$$
$$ix_2=p_{12}-p_{34},\quad x_4=2p_{13},\quad x_6=2p_{23}.$$

The singular surface is the complex surface for

$$x_3=0,\ 2x_4x_6+x_5^2+\frac{x_1^2}{\lambda_1-\lambda_3}+\frac{x_2^2}{\lambda_2-\lambda_3}=0.$$

The double line of $f(x)$, which is A_2A_3, is a singular line of the last complex.

We obtain, as before, for the equation of the singular surface

$$y_4^4(\lambda_1-\lambda_2)+16\lambda_1\lambda_2y_1^2y_3^2+8(\lambda_1+\lambda_2)y_1^2y_3y_4+4y_1^2y_4^2$$
$$-8\lambda_1\lambda_2y_1y_2y_4^2+8(\lambda_1-\lambda_2)y_1^3y_2=0,$$

where λ_1, λ_2 are written for $\lambda_1-\lambda_3$, $\lambda_2-\lambda_3$ respectively.

This is a Plücker surface for a quadratic complex and one of its singular lines. There are three special forms.

195. [1(14)], $\lambda_2=\lambda_3$.

Putting $\lambda_2=0$ in the equation of the singular surface gives

$$y_4^2(\lambda_1y_4^2+4y_1^2)+8\lambda_1y_1^2(y_3y_4+y_1y_2)=0.$$

Any plane through $y_1=0$, $y_4=0$ meets the surface in this line together with one other; the line A_2A_3 is therefore a double generator and simple directrix of the surface, hence we have a case of class XII. In the singular surface of [11(13)], each of the planes $y_1=0$, $y_4=0$, meets the surface in the line A_2A_3 merely, *i.e.* there are two "stationary" tangent planes; in the present case both stationary planes have come into coincidence with $y_1=0$.

196. [(11)4].

Putting $\lambda_1=\lambda_2$, we obtain as the singular surface

$$y_1(4\lambda_1^2y_1y_3^2+4\lambda_1y_1y_3y_4+y_1y_4^2-2\lambda_1^2y_2y_4^2)=0.$$

Hence the singular surface consists of a ruled cubic together with the plane $y_1=0$.

The line $y_1=0$, $y_2=0$ is the simple directrix; $y_3=0$, $y_4=0$ the double directrix. The plane $y=0$ meets the surface in A_3A_4 and in the two *coincident* generators A_3A_2; hence $y_1=0$ is a cuspidal tangent plane of the surface. Using plane coordinates we find v_2 as a factor of the left side of the equation of the singular surface. The cuspidal point A_2, therefore, completes the singular surface.

197. [(114)].

The equation of the complex may be put in the form

$$2x_3x_5+x_4^2=0,$$

and hence consists of the singly infinite number of congruences

$$2x_3=\mu x_4,\quad 2x_3+\mu^2x_5=0;$$

that is

$$p_{14}=\mu p_{13},\quad \mu^2p_{42}+2p_{14}=0.$$

The directrices of these special complexes form the pencils (A_2, α_1), (A_3, α_4), which are thus in (2, 1) correspondence and have A_2A_3 as common self-corresponding line.

In a (1, 2) correspondence, which is given by an equation of the form

$$x(ay^2+by+c)+a'y^2+b'y+c'=0,$$

the two values of y which correspond to any value of x form an involution; in the present case, the involution formed in the pencil (A_2, α_1) has A_2A_3 as a double line.

The complex is therefore formed as follows: in two pencils having a common line a, connect linearly the pairs of lines p_1, p_2 of an involution in one pencil with the lines p of the other pencil so that a is a double line of the involution and a self-corresponding line; *the lines which intersect p and p_1, p and p_2 form the complex.*

The planes of the pencils and their centres, each taken doubly, form the singular surface.

198. *Sixth canonical form* [123].

$$\omega(x) \equiv x_1^2 + 2x_2x_3 + 2x_4x_6 + x_5^2,$$

$$f(x) \equiv \lambda_1 x_1^2 + 2\lambda_2 x_2 x_3 + x_2^2 + \lambda_3(2x_4x_6 + x_5^2) + 2x_4x_5.$$

We may write

$$x_5 = p_{12} + p_{34}, \quad x_2 = 2p_{13}, \quad x_4 = p_{14},$$

$$ix_1 = p_{12} - p_{34}, \quad x_3 = p_{42}, \quad x_6 = 2p_{23}.$$

The singular surface is the Complex Surface for

$$x_4 = 0, \ 2x_5x_6 + \frac{x_1^2}{\lambda_1 - \lambda_3} + \frac{2x_2x_3}{\lambda_2 - \lambda_3} - \frac{x_2^2}{(\lambda_2 - \lambda_3)^2} = 0.$$

This gives as the equation of the singular surface (writing λ_1, λ_2 for $\lambda_1 - \lambda_3, \lambda_2 - \lambda_3$),

$$y_1^4 - \lambda_1\lambda_2^2(y_1y_2 - y_3y_4)^2 + 4\lambda_1 y_1^3 y_3 - (\lambda_1 - \lambda_2) y_1^2 y_4^2 - 2\lambda_2(\lambda_1 - \lambda_2) y_1 y_4 (y_1y_2 + y_3y_4) = 0.$$

The line $y_1 = y_3 = 0$, is double; $y_1 = y_4 = 0$, is cuspidal.

There are four special cases.

199. [1(23)], $\lambda_2 = \lambda_3$.

Putting $\lambda_2 = 0$ in the last equation, we derive as the equation of the singular surface in point coordinates

$$y_1^2(y_1^2 + 4\lambda_1 y_1 y_3 - \lambda_1 y_4^2) = 0,$$

in plane coordinates

$$v_2^2(v_2^2 + 4\lambda_1 v_2 v_4 - \lambda_1 v_3^2) = 0;$$

thus giving a cone and a conic whose plane touches the cone, while the vertex of the cone lies upon the conic.

200. [2(13)], $\lambda_1 = \lambda_3$.

The singular surface is

$$y_1\{y_1^3 + \lambda_2 y_1 y_4^2 + 2\lambda_2^2 y_4(y_1y_2 + y_3y_4)\} = 0.$$

This is a ruled cubic (Cayley's) together with a plane of its bitangent developable and a point upon it.

201. [(12)3], $\lambda_1 = \lambda_2$.

The singular surface is

$$y_1^4 - \lambda_1^3(y_1y_2 - y_3y_4)^2 + 4\lambda_1 y_1^3 y_3 = 0.$$

Any plane through A_2A_4 intersects the surface in two lines which meet on A_2A_4, while A_2A_3 is a double generator along which the tangent planes of the surface coincide; hence the surface belongs to class VIII. with a cuspidal generator.

202. [(123)].

The complex has as its equation

$$x_2^2 + 2x_4x_5 = 0.$$

The singular surface consists of a plane and a point in it, each counted four times.

203. *Seventh canonical form* [222].

$$\omega(x) \equiv x_1x_2 + x_3x_4 + x_5x_6,$$

$$f(x) \equiv 2\lambda_1x_1x_2 + 2\lambda_2x_3x_4 + 2\lambda_3x_5x_6 + x_1^2 + x_3^2 + x_5^2.$$

There are three double lines, viz. (010000), (000100), (000001), which meet each other; when these double lines are *coplanar* we may write

$$x_1 = p_{12}, \quad x_3 = p_{13}, \quad x_5 = p_{14},$$

$$x_2 = p_{34}, \quad x_4 = p_{42}, \quad x_6 = p_{23}.$$

The singular surface is seen to be*

$$y_1\{y_1^2 - (\lambda_2-\lambda_3)^2y_2^2 - (\lambda_1-\lambda_3)^2y_3^2 - (\lambda_1-\lambda_2)^2y_4^2\} - 2(\lambda_1-\lambda_2)(\lambda_2-\lambda_3)(\lambda_3-\lambda_1)y_2y_3y_4 = 0,$$

which is Cayley's cubic surface of the fourth class, together with the plane $y_1 = 0$ through the three double lines.

Secondly, when the double lines are *concurrent*, we may write

$$x_1 = p_{34}, \quad x_3 = p_{42}, \quad x_5 = p_{23},$$

$$x_2 = p_{12}, \quad x_4 = p_{13}, \quad x_6 = p_{14}.$$

The singular surface is in this case, (writing λ_1, λ_2 for $\lambda_1 - \lambda_3$, $\lambda_2 - \lambda_3$),

$$y_4^2\left(\frac{y_3^2}{\lambda_1^2} + \frac{y_2^2}{\lambda_2^2}\right) + y_2^2y_3^2\frac{(\lambda_1-\lambda_2)^2}{\lambda_1^2\lambda_2^2} - 2y_1y_2y_3y_4\frac{\lambda_1-\lambda_2}{\lambda_1\lambda_2} = 0;$$

i.e. an equation of the form

$$y_3^2y_4^2 + y_2^2y_4^2 + y_2^2y_3^2 = 2y_1y_2y_3y_4.$$

The latter equation is one of the forms to which Steiner's quartic surface of the third class can be brought. The singular surface is completed by the point of intersection of the three double lines.

* This equation is most easily derived as follows: Let y_i, y_i' be two points on a line of the complex, where y_i' lies in a_1, so that

$$p_{12} = -y_2' \cdot y_1, \quad p_{13} = -y_3' \cdot y_1, \quad p_{14} = -y_4' \cdot y_1, \text{ \&c.};$$

then for any point y_i of the singular surface the locus of y_i' is a pair of lines; from which the given equation follows at once.

If for y_i in the second singular surface, plane coordinates v_i be written, we obtain the equation of the first singular surface in plane coordinates.

There are two special cases.

204. $[2(22)],\ \lambda_2 = \lambda_3.$

In the first case the singular surface consists of a quadric cone and a pair of points, reciprocally in the second case we have a conic and a pair of planes.

205. $[(222)].$

In the first case of Art. 203

$$f(x) = p_{12}^2 + p_{13}^2 + p_{14}^2,$$

and the complex consists of the lines which meet the conic

$$y_1 = 0,\ y_2^2 + y_3^2 + y_4^2 = 0.$$

In the second case

$$f(x) \equiv p_{34}^2 + p_{42}^2 + p_{23}^2,$$

and the complex consists of the tangents to the cone

$$v_1 = 0,\ v_2^2 + v_3^2 + v_4^2 = 0.$$

206. *Eighth canonical form* [15].

$$\omega(x) \equiv x_1^2 + 2x_2x_6 + 2x_3x_5 + x_4^2.$$

$$f(x) \equiv \lambda_1 x_1^2 + \lambda_2 (2x_2x_6 + 2x_3x_5 + x_4^2) + 2x_2x_5 + 2x_3x_4.$$

Here we may write

$$ix_1 = p_{12} + p_{34},\quad x_2 = p_{14},\quad x_3 = p_{13},$$
$$x_4 = p_{12} - p_{34},\quad x_6 = 2p_{23},\quad x_5 = 2p_{42}.$$

The singular surface is the complex surface for

$$x_2 = 0,\quad \phi \equiv x_2 X + 2x_3x_6 + 2x_4x_5 + \frac{x_1^2}{\lambda_1 - \lambda_2} = 0;$$

$X = 0$ being any linear complex.

It is easy to see that the double line of the complex surface (all of whose coordinates are zero except x_6) is a singular line of ϕ, and also that each tangent line y of the singular surface of ϕ given by the equations $\frac{\partial \phi}{\partial x_i} + \mu \frac{\partial \omega}{\partial x_i} = \rho \cdot \frac{\partial \omega}{\partial y_i}$, (Art. 157), belongs to ϕ, if x is the double line. Hence the singular surface of $f(x) = 0$ is the complex surface for the general quadratic complex, in which the double line is a singular line of the second order.

The equation of the singular surface is

$$(\lambda_1 - \lambda_2)\{(y_4^2 - y_1y_3)^2 - 4y_1^2y_2y_4\} + 2y_1y_4^3 + y_1^3y_2 - y_1^2y_3y_4 = 0.$$

The line $y_1 = y_4 = 0$ is the double line, the point

$$y_1 = y_3 = y_4 = 0$$

is a triple point.

207. $[(15)], \quad \lambda_1 = \lambda_2.$

The singular surface is seen to be Cayley's ruled cubic, together with a cuspidal plane and a point, as in [(11)4].

208. *Ninth canonical form* [24].

$$\omega(x) \equiv x_1 x_2 + x_3 x_6 + x_4 x_5.$$

$$f(x) \equiv 2\lambda_1 x_1 x_2 + x_1^2 + \lambda_2 (2x_3 x_6 + 2x_4 x_5) + 2x_3 x_5 + x_4^2.$$

As in the form [222] there are two reciprocal cases.

Case (i). $x_1 = p_{12}, \quad x_3 = p_{14}, \quad x_4 = p_{13},$

$x_2 = p_{34}, \quad x_6 = p_{23}, \quad x_5 = p_{42}.$

The singular surface is, writing λ_1 for $\lambda_1 - \lambda_2$,

$$y_1 (\lambda_1 y_3 + y_4)^2 + 2y_2 (y_1^2 - \lambda_1^2 y_4^2) = 0,$$

together with $y_1 = 0$. This is a special case of Cayley's surface, and belongs to the species VIII. of Schläfli*, (a cubic of the 6th class with three proper nodes).

Case (ii). $x_1 = p_{34}, \quad x_3 = p_{23}, \quad x_4 = p_{42},$

$x_2 = p_{12}, \quad x_6 = p_{14}, \quad x_5 = p_{13}.$

The singular surface is one of the third class and fourth degree, a special case of Steiner's surface, together with the point of intersection of the double lines.

209. [(24)].

If $\lambda_1 = 0$, the equation of the singular surface is in case (i) $y_1^2 (y_4^2 + 2y_1 y_2) = 0$, in point-coordinates, and $v_3^2 (v_2^2 + v_3^2) = 0$ in plane-coordinates; hence, the singular surface consists of a cone and a pair of points on a generator of the cone: in case (ii) we have, reciprocally, a conic and a pair of planes whose line of intersection touches the conic. The complex is a special case of [(22)11].

In each case the complex has $A_2 A_3$, $A_3 A_4$ as double lines.

210. *Tenth canonical form* [33].

$$\omega(x) \equiv 2x_1 x_3 + x_2^2 + 2x_4 x_6 + x_5^2,$$

$$f(x) \equiv \lambda_1 (2x_1 x_3 + x_2^2) + 2x_1 x_2 + \lambda_2 (2x_4 x_6 + x_5^2) + 2x_4 x_5.$$

* See Schläfli "On Surfaces of the third order," *Phil. Trans.* 1863; also Cayley "A Memoir on cubic surfaces," *Phil. Trans.* 1869.

Here we may write

$$x_4 = ip_{14}, \quad x_2 = p_{12} + p_{34}, \quad x_1 = p_{13},$$
$$x_6 = -2ip_{23}, \quad ix_5 = p_{12} - p_{34}, \quad x_3 = 2p_{42}.$$

The equation of the singular surface is found by the usual process to be

$$y_1^3(y_3 + y_4 + \lambda_1 y_2) - 3\lambda_1 y_1^2 y_3 y_4 + \lambda_1^3 (y_1 y_2 + y_3 y_4)^2 = 0.$$

It possesses the two cuspidal lines $y_1 = 0$, $y_3 = 0$; $y_1 = 0$, $y_4 = 0$.

211. [(33)].

If we take λ_1 as zero in the previous equation, it is seen that the singular surface consists of the plane $y_1 = 0$ triply, and the plane $y_3 + y_4 = 0$, together with the point A_2 triply, and one other point in $y_1 = 0$.

The complex is

$$x_1 x_2 + x_4 x_5 = 0, \text{ or, } p_{13}(p_{12} + p_{34}) + p_{14}(p_{12} - p_{34}) = 0.$$

It consists therefore of the lines of the ∞^1 congruences

$$p_{13} = \mu(p_{12} - p_{34}), \quad p_{14} = -\mu(p_{12} + p_{34});$$

i.e.
$$p_{13} + p_{14} + 2\mu p_{34} = 0, \quad p_{13} - p_{14} - 2\mu p_{12} = 0.$$

These complexes are special, their directrices p and p' having the coordinates

	p_{12}	p_{13}	p_{14}	p_{23}	p_{34}	p_{42}
p	2μ	0	0	1	0	1
p'	0	0	0	-1	-2μ	1.

Thus p and p' form two projective pencils, the plane α_1 of the latter passing through A_2 the centre of the former, while O the centre of the latter lies upon A_3A_4, and the line OA_2 corresponds to the line of intersection of the planes of the pencils. *The complex is therefore the locus of lines which intersect corresponding lines of two projective pencils, in which the plane of one passes through the centre of the other, and the intersection of the planes corresponds to the line joining the centres.*

212. *Eleventh canonical form* [6].

$$\omega(x) \equiv 2x_1x_6 + 2x_2x_5 + 2x_3x_4,$$
$$f(x) = \lambda(2x_1x_6 + 2x_2x_5 + 2x_3x_4) + 2x_1x_5 + 2x_2x_4 + x_3^2.$$

There are two reciprocal cases:

Case (i).
$$x_1 = p_{14}, \quad x_2 = p_{12}, \quad x_3 = p_{13},$$
$$x_6 = p_{23}, \quad x_5 = p_{34}, \quad x_4 = p_{42}.$$

The equation of the singular surface is

$$y_1\{y_1y_2^2+2y_4^3-2y_1y_3y_4\}=0,$$

and therefore consists of the plane $y_1=0$ together with a surface of the third degree and fourth class.

This surface is the complex surface for the congruence

$$x_1=0,\quad \phi\equiv x_1X+2x_2x_6+2x_3x_5+x_4^2=0,$$

where $X\equiv\Sigma a_ix_i$ is any linear complex; and it is easy to see, if x is the double line of the complex surface, that the members of the

pencil
$$\frac{\partial\phi}{\partial x_i}+\mu\frac{\partial\omega}{\partial x_i},$$

i.e. the lines for which $y_5=1$, $y_6=a_6+\mu$, $y_1=y_2=y_3=y_4=0$, are all singular lines of ϕ.

Hence this double line is a singular line of ϕ of the third order. The singular surface is the species XIX. of Schläfli.

Case (ii).
$$x_1=p_{23},\quad x_2=p_{34},\quad x_3=p_{42},$$
$$x_6=p_{14},\quad x_5=p_{12},\quad x_4=p_{13}.$$

The singular surface is

$$(y_2^2-y_3y_4)^2-2y_1y_3^3=0,$$

together with the triple point $y_2=y_3=y_4=0$, of this surface. As before, the surface whose equation has just been given is the Plücker surface for a general quadratic complex with regard to a singular line of the third order.

In each case the complex has A_2A_3 as double line.

213. Number of constants in a canonical form. The number of independent constants in the general complex [111111] is 19; for each case of equality of roots of the discriminant of $f+\lambda\omega$ one constant is lost; while for each case of ν elementary divisors relating to the same root λ_i of this discriminant, $\dfrac{\nu(\nu-1)}{2}$ arbitrary constants are introduced. This enables us to find the number of independent constants in all cases.

Thus consider [(11)22], there are three pairs of equal roots which reduces the number of constants to 16, while there is one arbitrary constant contained in the canonical form, so that the number of independent constants is 15.

214. The Table which follows exhibits the characteristics of the different varieties of the quadratic complex.

	Number of Constants	Degree	Singular Surface	
[111 111]	19	4	Kummer Surface	
[2 1111]	18	4	Complex Surface for a general quadratic complex ϕ and	any line,
[3 111]	17	4		a line of ϕ,
[4 11]	16	4		a singular line of ϕ,
[5 1]	15	4		.... of 2nd order.
[22 11]	17	4	Complex Surface for a general quadratic complex ϕ and a tangent of its Kummer Surface. The complex surface has two double lines which are	both double,
[123]	16	4		one cuspidal,
[33]	15	4		two cuspidal.
[222]	16	4	Cayley's surface of 3rd degree and 4th class and the plane through the three double lines, or reciprocally, Steiner's Roman Surface and the point of intersection of the double lines.	There are three double lines which are coplanar or concurrent; in [42] two coincide; in [6] all three coincide.
[4 2]	15	4		
[6]	14	4	Complex Surface for a general quadratic complex ϕ relative to a singular line of the 3rd order.	
[(11)1111]	17	3	Class I. of ruled quartics (two double directrices).	
[(11)211]	16	3	Class VII. of ruled quartics (double directrix).	
[(11)31]	15	3	The same with a cuspidal generator.	
[(11)22]	15	3	Ruled cubic with two directrices together with a point on the double directrix, and a plane through the single directrix.	
[(11)4]	14	3	The same when the point and plane are cuspidal.	
[(21)111]	16	3	Class II. of ruled quartics.	
[(21)21]	15	3	Class VIII. of ruled quartics.	
[(21)3]	14	3	The same with a cuspidal generator.	
[(31)11]	13	3	Class XII. of ruled quartics.	
[(31)2]	14	3	Ruled cubic (Cayley's) with a point and a plane as in [(11)22].	
[(41)1]	14	3	A case of Class XII., the stationary tangent planes coincide.	

	Number of Constants	Degree	Singular Surface
[(51)]	13	3	Cayley's ruled cubic and a point and plane as in [(11)4].
[(22)11]	14	2	A quadric cone and a conic.
[1(23)]	13	2	Quadric cone and conic through its vertex, the plane of the conic touches the cone.
[(22)2]	13	2	Quadric cone and pair of points or reciprocally a conic and a pair of planes.
[(42)]	12	2	Two planes and a conic touching their intersection or reciprocally a cone and two points on one of its generators.
[(33)]	11	1	A triple plane and a triple point together with another plane and a point on it. (The complex is the locus of lines which meet corresponding lines of two projective pencils, in which the plane of one passes through the centre of the other, and the line joining the centres corresponds to the intersection of the planes.)
[(11)(11)11]	15	2	Two quadrics meeting in a twisted quadrilateral.
[(11)(11)2]	14	2	One quadric becomes two tangent planes of the other.
[(21)(11)1]	14	2	Two quadrics touching along a generator and having in common two generators of the other system.
[(12)(12)]	13	2	Two quadrics touching along two intersecting generators.
[(22)(11)]	12	1	Two planes with a third plane taken twice together with two points and a third point taken twice. (The complex is the locus of lines which meet corresponding lines of two projective pencils in which the plane of one pencil passes through the centre of the other.)
[(31)(11)]	13	2	As in [(11)(11)2].
[(11)(11)(11)]	13	1	Tetrahedron.
[(111)111]	14	2	A quadric taken twice (the complex is the locus of lines meeting paired lines of a regulus in involution [2]).
[(111)(11)1]	12	1	As in the last case. (The involution has two pairs of coincident double elements.)
[(111)21]	13	2	As above. (The involution has two coincident double elements.)
[(111)(21)]	11	1	As above. (The involution has all its double elements coincident.)
[(111)3]	12	2	As above. (The involution has three of its double elements in coincidence.)

	Number of Constants	Degree	Singular Surface
[(211)11]	13	2	Two planes and on their intersection two points, all taken doubly. (Formed by aid of two pencils in (2, 2) correspondence having a self-corresponding line.)
[(211)(11)]	11	1	As before. (The complex is locus of lines which meet corresponding lines of two projective pencils having a common line.)
[(211)2]	12	2	As before. (The pencils are in (1, 2) correspondence.)
[(311)1]	12	2	Two planes and on their intersection two points all taken doubly. (The complex is formed by the lines which meet corresponding lines of 2 pencils in (2, 2) correspondence and having three coincident double elements.)
[(411)]	11	2	As in [(211)2]. (The common line of the planes is double in the involution of one pencil.)
[(221)1]	11	1	A plane and a point on it each counted four times.
[(123)]	10	1	As in [(221)1].
[(222)]	8	0	Complex consists of tangents of a cone or of lines which meet a conic.
[(111)(111)]	9	0	Complex consists of the tangents to a quadric.

There are thus 49 distinct species of quadratic complexes, if the reciprocal cases which occur in [222], [42], [6] and their sub-cases, be considered as of the same species; if considered as forming different species, we obtain 55 species of quadratic complexes.

CHAPTER XII.

CONNEXION OF LINE GEOMETRY WITH SPHERE GEOMETRY.

215. THE fact that a line and a sphere in space of three dimensions have the same number of coordinates suggests the existence of a connexion between the geometry of the line and that of the sphere. Such a connexion was discovered by Lie*, and is discussed in the present chapter.

The equation of any sphere in Cartesian coordinates is

$$-2\alpha x - 2\beta y - 2\gamma z + x^2 + y^2 + z^2 + C = 0,$$

where $$C = \alpha^2 + \beta^2 + \gamma^2 - R^2 \quad \text{.....................(i)},$$

R being the radius of the sphere.

The quantities α, β, γ, C regarded as its coordinates, determine the sphere, and if a fifth coordinate R be employed, the equation (i) holds between these five coordinates.

The last equation may be written

$$C = (\alpha + \beta i)(\alpha - \beta i) - (R + \gamma)(R - \gamma) \quad \text{.........(ii)},$$

by comparison with the equation which holds for the five coordinates r, s, ρ, σ, η of any line (Art. 2), which is

$$\eta = r\sigma - s\rho \quad \text{.........................(iii)}:$$

the last two equations become identical if we assume

$$\left.\begin{array}{lll} C = -\eta, & \alpha + \beta i = s, & R + \gamma = r \\ & \alpha - \beta i = \rho, & R - \gamma = \sigma \end{array}\right\} \quad \text{............(iv)}.$$

A correspondence is thus established between the lines and spheres of space of three dimensions. It will be convenient for clearness to suppose the lines and spheres connected in this manner to belong to two separate spaces Λ and Σ, though each space is, of course, identical with ordinary space of three dimensions.

* See his celebrated memoir "Ueber Complexe, insbesondere Linien- und Kugel-Complexe mit Anwendung auf die Theorie partieller Differential-Gleichungen," *Math. Ann.* v.

It should be observed that if the line of equations (iv) is real, then one of the coordinates of the centre of the corresponding sphere is imaginary. If the line is such that $r+\sigma$ is a positive quantity we obtain a positive value for R, if $r+\sigma$ is negative, a *negative* value for R; such a geometrical form will still be termed a *sphere.*

It is clear that if the line r, s, ρ, σ is given, the corresponding sphere of Σ is uniquely determined by equations (iv), but if a sphere is given, *i.e.* if α, β, γ, and C are given, the equation (i) gives two values for R, viz.

$$\pm\sqrt{\alpha^2+\beta^2+\gamma^2-C},$$

and we therefore obtain *two lines*; to a point of Σ, *i.e.* to a sphere of zero radius, *one* line corresponds in Λ, which from (iv) is seen to belong to the complex $r+\sigma=0$.

216. Intersection of lines corresponds to contact of spheres. The equation (iii) is the form taken by the fundamental relation for the special coordinates r, s, ρ, σ, η: the result of the "polar process" equated to zero, *i.e.* $\omega(x\,|\,x')=0$, expresses, (Art. 8), the condition of intersection of the lines x and x'; with the coordinates r, s, ρ, σ, η the result of the polar process applied to equation (iii) gives

$$\eta+\eta'+s'\rho+s\rho'-r'\sigma-r\sigma'=0 \quad\text{..............(v)}$$

and is therefore the condition of intersection of the lines $r\ldots\eta$, $r'\ldots\eta'$. Now the condition of contact of the two corresponding spheres of Σ is

$$(\alpha-\alpha')^2+(\beta-\beta')^2+(\gamma-\gamma')^2=(R-R')^2,$$

i.e.
$$C+C'-2\alpha\alpha'-2\beta\beta'-2\gamma\gamma'+2RR'=0,$$

and is the result of the application of the polar process to equation (i).

The last equation may be written

$$C+C'-(\alpha'+\beta'i)(\alpha-\beta i)-(\alpha+\beta i)(\alpha'-\beta'i)$$
$$+(R+\gamma)(R'-\gamma')+(R'+\gamma')(R-\gamma)=0,$$

and by aid of equations (iv) is seen to be identical with (v). Hence, *if two lines intersect, the corresponding spheres touch.* The two lines (r', s', ρ', σ'), $(-\sigma', s', \rho', -r')$, which give the same values for the coordinates α, β, γ, C of the corresponding sphere in Λ, are *polar lines for the complex* $r+\sigma=0$; since *any* line (r, s, ρ, σ) of this complex which meets the first line satisfies the condition

$$\eta+\eta'+s'\rho+s\rho'-r'\sigma-r\sigma'=0,$$

and this is also the condition that (r, s, ρ, σ) should meet $(-\sigma', s', \rho', -r')$, since $r+\sigma=0$.

It is now seen that, by aid of equations (iv), a connexion between the spaces Λ and Σ is established, whose nature is expressed in the following table:

Space Λ	Space Σ
any line,	one sphere,
two lines polar with regard to the complex $r+\sigma=0$,	one sphere,
two intersecting lines,	two spheres which touch,
a line of the complex $r+\sigma=0$,	a point,
a line meeting a given line of this complex.	a sphere through a given point.

217. Points of Λ correspond to minimal lines of Σ. If (xyz) is any given point P of Λ, the coordinates of the lines through P of the complex $r+\sigma=0$, satisfy the equations

$$x = rz + \rho,$$
$$y = sz - r.$$

Hence the coordinates (α, β, γ) of the points of Σ which correspond to these lines satisfy the equations

$$\left.\begin{aligned} x &= \gamma z + \alpha - \beta i \\ y &= (\alpha + \beta i) z - \gamma \end{aligned}\right\} \quad \text{.....................(vi).}$$

These points, therefore, lie in a line of Σ which meets the sphere-circle of Σ, *i.e.* a "*minimal*" line. We are thus led back to the equations of Art. 101. *To a point P of Λ corresponds a minimal line q in Σ, and to the lines of $r+\sigma=0$ through P correspond the points of q.*

Since any line through P meets all the lines of $r+\sigma=0$ which pass through P, to the sheaf of lines through P in Λ correspond ∞^2 spheres of Σ of which each contains all the points of this minimal line q. To any one line p through P corresponds a sphere which passes through q. Through each point of p there pass ∞^1 lines of $r+\sigma=0$, to which correspond points of a minimal line of this sphere. Thus to the points of a line p of Λ correspond one set of minimal lines of a sphere; the other set of minimal lines will correspond to the points of p' the polar line of p for $r+\sigma=0$.

If in the equation of any linear complex of lines the substitution (iv) be effected, we obtain a *sphere complex* of the form

$$a\alpha + b\beta + c\gamma + dR + C + f = 0.$$

But the condition that any sphere $(\alpha\beta\gamma C)$ should intersect a given sphere $(\alpha'\beta'\gamma' C')$ at a constant angle ϕ is of the same form, viz.

$$C + C' - 2\alpha\alpha' - 2\beta\beta' - 2\gamma\gamma' + 2RR' \cos\phi = 0;$$

hence, *the spheres of a linear sphere-complex intersect a fixed sphere at a constant angle.*

Again, since all the lines which belong to two given linear complexes intersect two given lines, it follows that the spheres which belong to two linear sphere complexes touch two fixed spheres. To the table of correspondence lately given must therefore be added the following:

Space Λ	Space Σ
any point,	a minimal line,
a point and a line of $r+\sigma=0$ through it,	a minimal line and a point on it,
the points of a given line p,	one set of minimal lines of a given sphere,
the points of p' the polar of p,	the other set of minimal lines of the sphere,
a linear line complex,	a sphere complex, composed of spheres intersecting a given sphere at a constant angle,
a linear line congruence,	the spheres touching two given spheres,
a regulus.	the spheres touching three given spheres.

218. Surface Element. The association with any point, of an indefinitely small area containing the point, gives rise to the idea of a "*surface element*"; with each point of space are connected ∞^2 surface elements, and in space there are ∞^5 surface elements. With any surface are connected ∞^2 surface elements, the plane of each surface element being a tangent plane of the surface. If with each point P of any given surface S we associate in some definite way a plane π through P, and for the lines p of the pencil (P, π) take the polar lines p' with regard to a given linear complex C, the lines p' form a pencil whose plane π' and centre P' are respectively the polar plane of P, and the pole of π, for C.

If now π is the tangent plane of S at P, then, by Art. 44, the locus of P' is the *polar surface* S' of S, and the tangent plane to S' at P' is π'.

219. Corresponding surfaces in Λ and Σ. In each pencil of tangent lines of S there is one line which belongs to the complex C, or $r+\sigma=0$; thus to each point of S one line of C is assigned, and thereby one point Q is determined in Σ. Through each point of S there passes one complex curve k of C, and on S there are ∞^1 such curves k. To a point P of such a curve k and its tangent at P, correspond a minimal line q and a point Q on q. Thus to the locus of tangents and points of k correspond a minimal curve the locus of Q, and its tangents; to S, the locus of the curves k, corresponds therefore a surface T formed by ∞^1 minimal curves.

Again consider the surface S' which is the polar of S with regard to C, the line PP' touches at P' one of ∞^1 complex curves k' of C on S', to these curves k' correspond the other set of ∞^1 minimal curves on T. If p is any tangent line at P to S, and p' its polar line, which therefore touches S' at P', to p and p' will correspond the *same sphere* in Σ; this sphere contains the minimal lines corresponding to P and P', and these lines meet in the point Q which corresponds to the line PP' of C; hence *this sphere touches* T *at* Q. To the pencil of tangent lines to S at P (or to S' at P') correspond the ∞^1 spheres which touch T at Q. Thus *each surface element of* S *determines one surface element of* T.

The connexion between surface elements of S, S' and T admits of simple analytical expression. For the polar plane of (xyz), or P, in the complex $r+\sigma=0$, is

$$x-\xi+z\eta-y\zeta=0,$$

where $(\xi\eta\zeta)$ are current coordinates.

Since this is the tangent plane to S' at P', if the surface element of P' be

$$(x', y', z'; -1, m', n'),$$

where the quantities $-1, m', n'$ are proportional to the direction cosines of the normal at P', it is clear that

$$m'=z, \quad n'=-y,$$

and, from the symmetry of the relation between S and S',

$$m=z', \quad n=-y',$$

while since the polar plane of P passes through P'

$$x'=x-nz-my;$$

thus the surface element at P' is determined.

To determine the surface element of T at Q, we observe that the coordinates α, β, γ of Q satisfy the equations (vi), also $\gamma=r$, where r is a coordinate of the line PP', hence

$$\gamma=r=\frac{x-x'}{z-z'}=\frac{my+nz}{z-m} \quad \text{.........................(vii).}$$

Again since the direction cosines of the two minimal lines through Q are from (vi) proportional to $1-z^2, -i(1+z^2), 2z$; $1-z'^2, -i(1+z'^2), 2z'$, respectively, the direction cosines of a line perpendicular to their plane are proportional to $1-zz', -i(zz'+1), z+z'$, hence if the surface element of T at Q is $(\alpha, \beta, \gamma; -1, m_1, n_1)$ we obtain

$$m_1=\frac{i(zz'+1)}{1-zz'}, \quad n_1=\frac{z+z'}{zz'-1},$$

i.e.
$$m_1=\frac{i(zm+1)}{1-zm}, \quad n_1=\frac{z+m}{zm-1} \quad \text{.....................(viii).}$$

The equations (vi), (vii), (viii) completely determine the surface element at Q.

220. Principal tangents and principal spheres. It has been seen that to a tangent line p of S at P there corresponds a sphere touching T at Q, *i.e.* having with T a common surface element at Q; if the tangent to S is a *principal* tangent, *i.e.* if a consecutive surface element of S passes through p, the corresponding sphere must have in common with T a consecutive surface element, *i.e.* it touches T at a consecutive point as well as at P, and is therefore a *principal* sphere at P; hence, *to a principal tangent curve on S corresponds a line of curvature on T.*

If S is a ruled surface, any one of its generators p has ∞^1 surface elements in common with S, thus T is touched along a line of curvature by the sphere corresponding to p. *This line of curvature k is a circle*; for let Q be a point on k, and Q' a consecutive point on the second line of curvature through Q, through Q' there passes a consecutive line of curvature k', and the tangent plane at Q' passes through Q; it follows that Q and Q' are equidistant from the centre of the sphere which touches T along k', therefore k lies upon this consecutive sphere, and the two consecutive spheres intersect in k, which is therefore a circle: T is the envelope of ∞^1 spheres.

To a ruled surface, therefore, corresponds a surface which is the envelope of ∞^1 spheres and of which one set of lines of curvature are circles.

To a quadric corresponds a surface which is the envelope of two sets of ∞^1 spheres, and of which all the lines of curvature are circles, *i.e.* a *Dupin's Cyclide*; each sphere of one set touches each sphere of the other set.

221. Pentaspherical Coordinates. The analogy between line geometry and point geometry with pentaspherical coordinates will now be investigated*.

* For full discussion of pentaspherical coordinates see Darboux, *La Théorie générale des surfaces*, prem. partie, p. 213.

The equation of any sphere may be written in the form

$$2\alpha x + 2\beta y + 2\gamma z + \delta\frac{x^2+y^2+z^2-R^2}{R} + i\epsilon\frac{x^2+y^2+z^2+R^2}{R} = 0.$$

We find for the coordinates x_0, y_0, z_0 of the centre, the following expressions

$$x_0 = \frac{-\alpha R}{\delta + i\epsilon}, \quad y_0 = \frac{-\beta R}{\delta + i\epsilon}, \quad z_0 = \frac{-\gamma R}{\delta + i\epsilon};$$

the radius ρ is equal to $\dfrac{R\sqrt{\alpha^2+\beta^2+\gamma^2+\delta^2+\epsilon^2}}{\delta + i\epsilon}$. If the sphere is not a point we may take $\alpha^2+\beta^2+\gamma^2+\delta^2+\epsilon^2 = 1$, and then $\rho = \dfrac{R}{\delta + i\epsilon}$. Taking a second sphere $(x_0' y_0' z_0' \rho')$ we easily find that

$$(x_0 - x_0')^2 + (y_0 - y_0')^2 + (z_0 - z_0')^2 - \rho^2 - \rho'^2$$
$$= -2R^2\frac{(\alpha\alpha' + \beta\beta' + \gamma\gamma' + \delta\delta' + \epsilon\epsilon')}{(\delta + i\epsilon)(\delta' + i\epsilon')},$$

and hence that the spheres are orthogonal if

$$\alpha\alpha' + \beta\beta' + \gamma\gamma' + \delta\delta' + \epsilon\epsilon' = 0.$$

Now consider five mutually orthogonal spheres

$$S_1 = 0, \ldots, S_5 = 0,$$

of radii $\rho_1, \ldots, \rho_5$; the equation of any one of them is

$$2\alpha_k x + 2\beta_k y + 2\gamma_k z + \delta_k\frac{x^2+y^2+z^2-R^2}{R}$$
$$+ i\epsilon_k\frac{x^2+y^2+z^2+R^2}{R} = 0 \quad \ldots\ldots\ldots (a),$$

where we have by hypothesis

$$\left.\begin{aligned} \alpha_k^2 + \beta_k^2 + \gamma_k^2 + \delta_k^2 + \epsilon_k^2 &= 1, \\ \text{and} \qquad \alpha_k\alpha_{k'} + \beta_k\beta_{k'} + \gamma_k\gamma_{k'} + \delta_k\delta_{k'} + \epsilon_k\epsilon_{k'} &= 0 \end{aligned}\right\} \ldots\ldots\ldots\ldots (b);$$

these two groups of equations are those of linear orthogonal substitutions in five variables and we infer therefore that

$$\left.\begin{aligned} \alpha_1^2 + \alpha_2^2 + \alpha_3^2 + \alpha_4^2 + \alpha_5^2 &= 1, \\ \alpha_1\beta_1 + \alpha_2\beta_2 + \alpha_3\beta_3 + \alpha_4\beta_4 + \alpha_5\beta_5 &= 0, \ \&\text{c.} \end{aligned}\right\} \ldots\ldots\ldots (c).$$

Now if S_k is the "*power*" of any point in reference to the sphere $S_k = 0$, $\dfrac{S_k}{\rho_k}$ is the first member of (a); denoting it by x_k we see from (c) that

$$\sum_1^5 x_k^2 = \left(\frac{x^2+y^2+z^2-R^2}{R}\right)^2 - \left(\frac{x^2+y^2+z^2+R^2}{R}\right)^2 + 4(x^2+y^2+z^2) = 0.$$

The quantities x_k are called the pentaspherical coordinates of the point with reference to the given five mutually orthogonal spheres, and the fundamental relation between them has been shown to be $\Sigma x_k^2 \equiv 0$.

The following results proceed from this definition of the quantities x_k:

(i) Any linear equation in x, as $\Sigma a_k x_k = 0$, represents a sphere.

(ii) Two spheres $\Sigma a_k x_k = 0$, $\Sigma a'_k x_k = 0$, are orthogonal if $\sum_1^5 a_k a'_k = 0$; for $\Sigma a_k x_k \equiv 2x\Sigma a_k \alpha_k + 2y\Sigma a_k \beta_k + 2z\Sigma a_k \gamma_k$

$$+ \frac{x^2 + y^2 + z^2 - R^2}{R} \Sigma a_k \delta_k + i \frac{x^2 + y^2 + z^2 + R^2}{R} \Sigma a_k \epsilon_k,$$

and this sphere is orthogonal to $\Sigma a'_k x_k = 0$ if

$$\Sigma a_k \alpha_k \Sigma a'_k \alpha_k + \dots\dots\dots = 0;$$

i.e., by (*b*), if $\qquad \Sigma a_k a'_k = 0.$

(iii) The radius of $\Sigma a_k x_k = 0$ is seen to be

$$\frac{R\sqrt{\Sigma a_k^2}}{\Sigma a_k \delta_k + i\Sigma a_k \epsilon_k} = \frac{\sqrt{\Sigma a_k^2}}{\Sigma \dfrac{a_k}{\rho_k}}.$$

If therefore $\Sigma a_k^2 = 0$ we have a point, if $\Sigma \dfrac{a_k}{\rho_k} = 0$ a plane; in the former case the quantities a_k are the pentaspherical coordinates of this point $(x'y'z')$, and

$$\Sigma a_k x_k \equiv \sigma \{\overline{x - x'}^2 + \overline{y - y'}^2 + \overline{z - z'}^2\}.$$

(iv) If x_k and x'_k are any two points, the condition $\Sigma x_k x'_k = 0$ states that each of the points lies on the sphere whose centre is the other point and which has zero radius*; this may be expressed by saying that *each point lies upon the null-sphere of the other.*

(v) If $\Sigma a_k x_k = 0$ is any sphere and x_k any point, the coordinates of its inverse point with regard to this sphere are x'_k where

$$\sigma x'_k = x_k \Sigma a_k^2 - 2a_k \Sigma a_k x_k;$$

for if P and P' are inverse points with reference to any sphere, and Q is any point on the sphere, we have

$$\overline{PQ}^2 = m\overline{P'Q}^2,$$

where m is a constant.

* This does not involve the coincidence of the points.

Hence each point Q of the given sphere which lies on the null-sphere of P will also lie on the null-sphere of P'; so that if P is x_k and P' is x_k', then any point ξ_k which satisfies the equation $\Sigma a_k \xi_k = 0$ and *either* of the equations $\Sigma \xi_k x_k = 0$, $\Sigma \xi_k x_k' = 0$, will satisfy the other; hence

$$\sigma x_k' = \kappa x_k + \lambda a_k, \ldots\ldots\ldots (k = 1, \ldots\ldots\ldots, 5);$$

from which, by squaring each side and adding, the result follows.

The connexions between line geometry and point geometry with pentaspherical coordinates, derived from these results, are set forth in the following Table.

Line Geometry	Point Geometry with pentaspherical coordinates
Six fundamental complexes, in mutual involution, $x_1=0, \ldots, x_6=0$.	Five spheres mutually orthogonal, $x_1=0, \ldots, x_5=0$.
Fundamental relation between coordinates of a line, $\sum_1^6 x_k^2=0$.	Fundamental relation between coordinates of a point, $\sum_1^5 x_k^2=0$.
Linear Complex $\sum_1^6 a_k x_k=0$.	Sphere $\sum_1^5 a_k x_k=0$.
Special Complex, if $\sum_1^6 a_k^2=0$.	Point Sphere, if $\sum_1^5 a_k^2=0$.
Two linear complexes $(ax)=0$, $(a'x)=0$ are in Involution if $\sum_1^6 a_k a_k'=0$.	Two spheres are orthogonal if $\sum_1^5 a_k a_k'=0$.
Two lines x_k, x_k' intersect if $\sum_1^6 x_k x_k'=0$.	Each of two points is on the null-sphere of the other if $\sum_1^5 x_k x_k'=0$.
Two lines x_k, x_k' are polar with regard to a linear complex $\Sigma a_k x_k=0$ if $x_k=\lambda a_k+\mu x_k', \ldots (k=1, \ldots 6)$.	Two points are inverse with regard to the sphere $\Sigma a_k x_k=0$ if $x_k=\lambda a_k+\mu x_k', \ldots (k=1, \ldots 5)$.

At any point x_i of the surface $f(x_1 \ldots x_5) = 0$ there are ∞^1 tangent spheres, whose equation is

$$\sum_1^5 \left(\frac{\partial f}{\partial x_i} + \mu x_i\right) y_i = 0.$$

This is shown just as in the case of the tangent linear complexes of $f(x_1 \ldots x_6) = 0$. The similarity in form of the equations of the ∞^1 tangent spheres at the point x_i and the ∞^1 tangent linear complexes of the line x_i, suggests one of the most important connexions of line- and sphere-geometry. For as has been seen (Arts. 74, 76, 115), when a tangent linear complex is special its directrix touches the singular surface; when it is special and bitangent (Art. 131), its directrix is a double tangent of the singular surface; while, for the quadratic complex $\sum_1^6 \lambda_i x_i^2 = 0$, these bitangents form the six congruences (Art. 83)

$$y_i = 0, \quad \sum_k \frac{y_k^2}{\lambda_k - \lambda_i} = 0, \quad (k \neq i), \quad (i = 1, \ldots 6).$$

But a special linear complex corresponds to a point-sphere, and the centre of a point-sphere which is bitangent to a surface is a *focus**, hence, *to a double-tangent of the singular surface of a complex there corresponds a focus of a surface.*

The five focal curves of the surface $\sum_1^5 \lambda_i x_i^2 = 0$ are thus seen to be given by the equations

$$y_i = 0, \quad \sum \frac{y_k^2}{\lambda_k - \lambda_i} = 0, \quad (k \neq i);$$

if to i the values $1, \ldots 5$ are successively attributed.

As in the case of the congruences of bitangents of the singular surface of $\sum_1^6 \lambda_i x_i^2 = 0$, these curves are not affected by the substitution of $\frac{1}{\lambda_i + \mu}$ for λ_i; *i.e.* the ∞^1 surfaces

$$\sum_1^5 \frac{x_i^2}{\lambda_i + \mu} = 0$$

are confocal.

Hence we have the result that *confocal cyclides correspond to cosingular quadratic complexes.*

For the purpose of comparison we place side by side corresponding characteristics of cosingular complexes and confocal cyclides.

∞^1 cosingular complexes,	∞^1 confocal cyclides,
a bitangent of the singular surface,	a focus of the cyclides,
six bitangent congruences of the singular surface,	five focal curves,
four complexes of the system through any line l,	three cyclides of the system through any point P,
the four tangent linear complexes of l with regard to these complexes are mutually in involution.	the three tangent spheres at P of these surfaces are mutually orthogonal, *i.e.*, confocal cyclides cut at right angles.

* Salmon, *Geom. of three dimensions*, Third edition, p. 108.

The quantities a_i in the equation of a sphere $\sum_1^5 a_i x_i = 0$, completely define the sphere: we may introduce a sixth coordinate a_6, where $ia_6 = \sqrt{\sum_1^5 a_i^2}$, in which case $\sum_1^6 a_i^2 = 0$. The condition of (internal) contact of the spheres

$$(x_0,\ y_0,\ z_0;\ \rho) \text{ and } (x_0',\ y_0',\ z_0';\ \rho')$$

being $(x_0 - x_0')^2 + (y_0 - y_0')^2 + (z_0 - z_0')^2 - \rho^2 - \rho'^2 + 2\rho\rho' = 0$,
is seen from the preceding to be

$$\sum_1^5 a_i a_i' - \sum_1^5 a_i^2 \sum_1^5 a_i'^2 = 0,$$

i.e.

$$\sum_1^6 a_i a_i' = 0.$$

A complete correspondence is therefore now established between the geometry of the line in Klein coordinates and that of the sphere in the coordinates a_i; in fact we have returned to the sphere-geometry of Lie which has been discussed in the preceding Articles of the present chapter, in which the *intersection* of lines corresponds to the *contact* of spheres.

CHAPTER XIII.

CONNEXION OF LINE GEOMETRY WITH HYPERGEOMETRY.

222. For five variable quantities $X_1 \dots X_5$, there are ∞^4 sets of values of their ratios: each such set of values may, from the analogy of space of three dimensions, be said to define a "point," and the totality of such points to constitute a "space" S_4 of four dimensions, which thus contains ∞^4 points. If A and B are any two points of S_4 the locus of the ∞^1 points $A_i + \lambda B_i$ will be called a *line* of S_4.

Any linear equation of the form $\sum_1^5 a_i X_i = 0$ singles out ∞^3 points from S_4, which will then form a space of three dimensions; the locus of these ∞^3 points will be called a *hyperplane*; there are clearly ∞^4 hyperplanes in S_4. Any line which does not lie in a given hyperplane will obviously meet it in one point only; if two points of a line lie in a hyperplane the line will lie altogether in that hyperplane.

Two hyperplanes intersect in a plane; for any line in one of them meets the other in one point; a plane in S_4 is therefore defined by two linear equations in the quantities X; there are ∞^6 planes in S_4.

Three hyperplanes which have not a plane in common, intersect in one line, thus a line is determined by three linear equations; there are ∞^6 lines in S_4.

Four hyperplanes have in general one point in common, *i.e. any two planes of S_4 meet in one point.*

Three points not in the same line define a plane of S_4, and four points not in the same plane determine a hyperplane; hence two lines which do not intersect determine a hyperplane.

A line p in one hyperplane will not in general meet a plane α in another hyperplane; hence through any line p there pass ∞^2

planes obtained by constructing the planes which pass through p and the points of α.

223. Equations connecting lines of Λ and points of S_4*. Any six linear complexes $x_1 = 0, \ldots, x_6 = 0$ being taken as those of reference, and $\omega \equiv \Sigma a_{ik} x_i x_k = 0$ being the fundamental relation, then, since

$$-\Omega(a) = A_{11} a_1^2 + \ldots + 2A_{rs} a_r a_s + \ldots,$$

$$-\Omega(a|b) = A_{11} a_1 b_1 + \ldots + A_{rs}(a_r b_s + a_s b_r) + \ldots,$$

it follows, that if any one of the complexes, say $x_i = 0$, is *special*, $A_{ii} = 0$, and if $x_i = 0$, $x_k = 0$ are in involution, $A_{ik} = 0$.

If now $x_1 = 0$, $x_2 = 0$, $x_3 = 0$, $x_4 = 0$ are any four linear complexes in mutual involution, and $x_5 = 0$, $x_6 = 0$ the two special complexes whose directrices are the two lines common to the first four complexes, then

$$-\Omega(a) = A_{11} a_1^2 + A_{22} a_2^2 + A_{33} a_3^2 + A_{44} a_4^2 + 2A_{56} a_5 a_6;$$

all the other coefficients A being zero in this case.

Now since the A_{ik} are the first minors of the discriminant of $\Sigma a_{ik} x_i x_k$, the a_{ik} are proportional to the first minors of the discriminant of $\Omega(a)$, and therefore in the present case,

$$a_{ik} = \frac{\rho}{A_{ik}};$$

and the fundamental relation assumes the form

$$\frac{x_1^2}{A_{11}} + \frac{x_2^2}{A_{22}} + \frac{x_3^2}{A_{33}} + \frac{x_4^2}{A_{44}} + \frac{2x_5 x_6}{A_{56}} = 0.$$

But we may take

$$A_{11} = A_{22} = A_{33} = A_{44} = \frac{-A_{56}}{2} = 1,$$

in which case the fundamental relation becomes

$$x_1^2 + x_2^2 + x_3^2 + x_4^2 - x_5 x_6 = 0.$$

If we now write

$$\left.\begin{aligned} \rho \,.\, x_1 &= X_1 X_5, \quad \rho \,.\, x_5 = X_5^2, \\ \rho \,.\, x_2 &= X_2 X_5, \quad \rho \,.\, x_6 = \sum_1^4 X_i^2 \\ \rho \,.\, x_3 &= X_3 X_5, \\ \rho \,.\, x_4 &= X_4 X_5, \end{aligned}\right\} \ldots\ldots\ldots\ldots\ldots \text{(i)},$$

* The theory here given is due to Klein, see "Ueber Liniengeometrie und metrische Geometrie," *Math. Ann.* v.

a (1, 1) correspondence is established between the lines of the three dimensional space Λ and the points X of S_4. Exceptional elements, however, occur in this correspondence; for denoting by s the directrix of the special complex x_5, if $x_5=0$ then is also $X_5=0$, and therefore one solution is $x_1=x_2=x_3=x_4=0$, which gives s; so that s corresponds to *any* point of the hyperplane $X_5=0$. If, however, in addition to $X_5=0$ we have $\sum_1^4 X_i^2=0$, then must $\rho=0$, and the lines corresponding to a given point X whose coordinates satisfy the last two equations, are determined by

$$\frac{x_1}{X_1}=\frac{x_2}{X_2}=\frac{x_3}{X_3}=\frac{x_4}{X_4}, \qquad x_5=0, \ldots\ldots\ldots\ldots \text{(ii)}.$$

The locus $X_5=0$, $\sum_1^4 X_i^2=0$ is a quadric surface contained in the hyperplane $X_5=0$, it will be denoted by Φ; thus to any point X of Φ the ∞^1 lines given by equation (ii) correspond; *these lines are those of a pencil which includes* s; for they are determined by the equations

$$x_i=\rho X_i, \ (i=1, 2, 3, 4),$$
$$x_5=0;$$

hence if a is any line, the equation $\sum_1^6 x_i \frac{\partial \omega}{\partial a_i}=0$ gives one value for $\frac{x_6}{\rho}$, *i.e., one* of these lines meets any given line.

Hence the lines determined by (ii) pass through a point P of s and lie in a plane π through s; all the lines of such a pencil correspond to the same point of Φ. To the ∞^3 lines of the complex $x_5=0$ there correspond the ∞^2 points of Φ.

If two lines x, x', corresponding to different points X, X' of Φ, *intersect*, $\sum_1^4 x_i x_i'=0$; hence it follows from (ii) that $\sum_1^4 X_i X_i'=0$, which, together with the equations $\sum_1^4 X_i^2=0$, $\sum_1^4 X_i'^2=0$, shows that each point of the line $X_i+\lambda X_i'$ lies on Φ, XX' is therefore a generator of Φ: hence it follows, *firstly*, to each of the ∞^1 pencils through s of centre P corresponds a point on a generator σ_1 of Φ; *secondly*, to each of the ∞^1 pencils through s whose plane is π corresponds a point on a generator σ_2 of Φ. Now these two sets of pencils have one pencil in common, viz. (P, π); hence σ_1 and σ_2 intersect each other and therefore belong to *different systems.* Thus with each point of s is associated one generator σ_1 of Φ, and with each plane through s is associated one generator σ_2; so that with each pencil containing s is associated one point of Φ.

The nature of the correspondence expressed by equations (i) is therefore the following:

Space Λ	Space S_4
any line,	a point,
exceptionally the line s,	any point of $X_5=0$,
the lines of a pencil whose centre and plane are united to s,	one point on Φ,
a linear complex through s,	a hyperplane,
a linear congruence through s,	a plane,
a regulus through s.	a line.

224. Correlation of Schumacher*. The foregoing correspondence may be obtained as a special case of a more general (1, 1) correspondence between the lines of Λ and the points of S_4. Having given two lines p_1 and p_2 of S_4 which do not intersect, we can establish a (1, 1) correspondence between the ∞^2 planes of S_4 which pass through p_1 and the points of any plane α_1 in Λ, and similarly a (1, 1) correspondence between the ∞^2 planes of S_4 through p_2 and the points of any plane α_2 in Λ; provided that Λ is not the hyperplane Σ determined by p_1 and p_2. Any point P of S_4 determines one plane through p_1 and one plane through p_2, and hence one point Q_1 in α_1 and one point Q_2 in α_2, thus P corresponds to the line Q_1Q_2; conversely, Q_1Q_2 determines a point in α_1 and a point in α_2, and hence two planes through p_1 and p_2 respectively, and therefore one point P in S_4.

The ∞^2 planes through p_1 are

$$X_1 = \lambda X_5,\ X_2 = \mu X_5 \quad\text{.................... (i),}$$

where $X_5 = 0$ is the hyperplane containing p_1 and p_2, and $X_1 = 0$, $X_2 = 0$ are any two hyperplanes through p_1.

In the collineation of the planes through p_1 and the points of α_1, a plane given by (i) corresponds to the point Q_2, or x_i, of α_1, where

$$\kappa\,.\,x_i = C_i + \lambda B_i + \mu A_i \quad\text{.................. (ii),}$$

provided that the point C is the correlative of the plane (X_1, X_2), the point B of (X_5, X_2), and the point A of (X_5, X_1).

* "Classification der algebraischen Strahlensysteme," *Math. Ann.* 37.

Any hyperplane through p_1 will have an equation of the form

$$\alpha X_1 + \beta X_2 + \gamma X_5 = 0 \quad \text{.................(iii)};$$

and on substitution in the last equation from (i) we obtain

$$\alpha\lambda + \beta\mu + \gamma = 0.$$

This equation connects the quantities λ, μ of the planes through p_1 in the hyperplane (iii). It shows that the locus of the points Q_1, corresponding to these ∞^1 planes, is a line of α_1; hence, *each hyperplane through* p_1 *determines a line of* α_1, *and vice versâ.*

In the general case the hyperplane X_5 will not correspond to s (the line of intersection of α_1 and α_2).

Similarly the ∞^2 planes through p_2 are

$$X_3 = \rho X_5, \; X_4 = \sigma X_5 \quad \text{......................(iv)},$$

where $X_3 = 0$, $X_4 = 0$ are any two hyperplanes through p_2; and the point Q_2, or y_i, which corresponds to this plane is given by the equations

$$\tau . y_i = C_i' + \rho B_i' + \sigma A_i' \quad \text{..................(v)},$$

where C' is the correlative of (X_3, X_4), B' of (X_5, X_4), A' of (X_5, X_3).

If p is the line Q_1Q_2, it follows from (ii) and (v) that

$$\nu . p_{ik} = (cc')_{ik} + \rho\,(cb')_{ik} + \sigma\,(ca')_{ik} + \lambda\,(bc')_{ik} + \mu\,(ac')_{ik} + \lambda\rho\,(bb')_{ik}$$
$$+ \mu\sigma\,(aa')_{ik} + \lambda\sigma\,(ba')_{ik} + \mu\rho\,(ab')_{ik} \quad \text{.............(vi)},$$

where $(cc')_{ik}$ is a Plücker coordinate of the line CC', &c.

If we now assume the collineation between the planes through p_1 and the points of α_1 to be such that the line s corresponds to $X_5 = 0$, both A and B will lie on s. Similarly if also in the second collineation the line s corresponds to $X_5 = 0$, the points A' and B' will lie on s. The quantities $(bb')_{ik}$, $(aa')_{ik}$, $(ba')_{ik}$, $(ab')_{ik}$ are now coordinates of the same line s; hence taking s as one edge of the tetrahedron of reference in Λ, the four quadratic terms disappear from *five* of the equations (vi).

Finally, if Σ_3, or $\sum_1^5 a_i X_i = 0$, is any hyperplane whatever and if we substitute in its equation for the X_i by means of (i) and (iv), we obtain the equation

$$a_1\lambda + a_2\mu + a_3\rho + a_4\sigma + a_5 = 0 \quad \text{...........(vii)},$$

connecting the quantities λ, μ, ρ, σ.

Now eliminating the latter quantities between (vii) and the previous five equations of (vi), we obtain a linear complex, which passes through s, and whose lines correspond to the points of Σ_3.

Hence, *if the line corresponding to* $X_5=0$ *is for both collineations the line* s, *to the points of any hyperplane will correspond the lines of a linear complex through* s.

We observe from equations (vi) that, *in the general case*, to a linear complex of Q_1Q_2 corresponds a locus of points in S_4 of the second degree.

It was seen that to the points of any given line in α_1 there correspond the planes through p_1 in a given hyperplane; thus to the lines which intersect any two given lines of α_1 and α_2 respectively, correspond the points of a plane in S_4 which meets p_1 and p_2; in particular, to the lines of a *plane system* in Λ correspond the points of such a plane in S_4.

When the collineations satisfy the condition that to s the hyperplane (p_1, p_2) corresponds in each of them, the equations of Art. 223 may be regained by taking as coordinate complexes in Λ the special complex $x_5=0$ whose directrix is s, four complexes $x_1=0$, $x_2=0$, $x_3=0$, $x_4=0$, in mutual involution to each of which s belongs, and the special complex $x_6=0$ whose directrix is the second line common to $x_1=0$, $x_2=0$, $x_3=0$, $x_4=0$. For let the hyperplanes which correspond respectively to the four complexes in involution be $X_1=0$, $X_2=0$, $X_3=0$, $X_4=0$, and let $X_5=0$ be the hyperplane which corresponds to s, then we have from (vi)

$$x_1=\sigma X_1,\quad x_2=\sigma X_2,\quad x_3=\sigma X_3,\quad x_4=\sigma X_4,\quad x_5=\sigma X_5,$$

while x_6 is a quadratic function of $X_1 \ldots X_5$; but since

$$\omega(x) \equiv x_1^2+x_2^2+x_3^2+x_4^2-x_5x_6$$

it follows that $$x_6=(X_1^2+X_2^2+X_3^2+X_4^2)\frac{\sigma}{X_5},$$

and writing $\sigma=\rho X_5$, we obtain the original equations.

225. Correlatives of the lines of any plane system and sheaf of Λ. When $\omega(x)$ has the above form, the Invariant $-\Omega(a)$ of any complex $\Sigma a_i x_i=0$ is $a_1^2+a_2^2+a_3^2+a_4^2-4a_5a_6$; if the complex is special and contains s, then $a_6=0$, $\sum_1^4 a_i^2=0$, and the coordinates of its directrix are $(a_1, a_2, a_3, a_4, 0, -2a_5)$; to the lines of this special complex correspond the points of the hyperplane $\sum_1^5 a_i X_i=0$.

This hyperplane 'touches' Φ at the point $(a_1, a_2, a_3, a_4, 0)$, since it passes through this point and every point on Φ consecutive to it; the hyperplane therefore contains the generators σ_1, σ_2 of Φ through the point. If a_5 vary, the directrix describes a pencil (P, π) containing s (Art. 223), where P is the point corresponding to σ_1 and π the plane corresponding to σ_2.

If $\sum_1^5 b_i x_i = 0$ is another special complex which contains s, and of which the directrix is either *concurrent* with s and the directrix of $\sum_1^5 a_i x_i = 0$, or *coplanar* with them, then

$$\sum_1^4 a_i^2 = \sum_1^4 a_i b_i = \sum_1^4 b_i^2 = 0 \quad\text{.....................}\ \text{(i).}$$

Hence to the lines of the *plane system*, or of the *sheaf*, determined by

$$\sum_1^5 a_i x_i = \sum_1^5 b_i x_i = 0,$$

correspond the points of the plane

$$\sum_1^5 a_i X_i = \sum_1^5 b_i X_i = 0:$$

this plane passes through the line

$$(a_1 + \lambda b_1,\ a_2 + \lambda b_2,\ a_3 + \lambda b_3,\ a_4 + \lambda b_4,\ 0)$$

of S_4, which from (i) is seen to be a generator of Φ.

Hence *to the lines of a plane system correspond the points of a plane through a generator σ_1 of Φ, to the lines of a sheaf correspond the points of a plane through a generator σ_2 of Φ.*

If the centre of any sheaf is P, and the plane π of any plane system meets s in P', then the line PP' of the sheaf and the intersection of π with the plane (s, PP') correspond to the same point of Φ (Art. 223); hence the generators σ_1 and σ_2 have one point in common and therefore *belong to different systems.*

To sheaves whose centres lie in the same plane π through s correspond planes through the same generator σ_2, to plane systems whose planes pass through the same point P of s correspond planes through the same generator σ_1.

If a and b are any two intersecting lines, to the lines of the pencil $a_i + \lambda b_i$ correspond the points of a line $A_i + \lambda B_i$, and since any pencil in Λ contains one line which meets s, the corresponding line will meet Φ, *i.e., to the lines of a pencil in Λ correspond the points of a line which meets Φ; to the ∞^2 pencils which contain the line $(a_1, a_2, a_3, a_4, 0, -2a_5)$ correspond the ∞^2 lines through the point $(a_1, a_2, a_3, a_4, 0)$ in the hyperplane $\sum_1^5 a_i X_i = 0$.*

226. Metrical Geometry. The locus of points in S_4 of which the equation is $X_1^2+X_2^2+X_3^2+X_4^2=0$, and which corresponds to the lines of $x_6=0$, forms a three-dimensional space which is met by any line of S_4 in two points; such a space is therefore denoted by S_3^2. The points common to S_3^2 and any hyperplane Σ_3 form a two-dimensional space which is met by any line of Σ_3 in two points and is therefore a quadric in Σ_3. In geometry of three dimensions the properties connected with the sphere-circle and the plane at infinity are called *metrical*, and a quadric through the intersection of $x^2+y^2+z^2=0$, and the plane at infinity is a sphere; by analogy, in four-dimensional space, the three-dimensional quadric spaces through the intersection of S_3^2 and $X_5=0$, *i.e.* which contain Φ, may be termed '*hyperspheres.*'

To a linear complex $\sum_1^6 a_i x_i=0$ corresponds therefore the hypersphere

$$X_5 \sum_1^5 a_i X_i + a_6 \sum_1^4 X_i^2 = 0.$$

In the equations of connexion of Λ and S_4 we may, to complete the analogy, take X_5 as being unity, we then have

$$X_1=\frac{x_1}{x_5},\quad X_2=\frac{x_2}{x_5},\quad X_3=\frac{x_3}{x_5},\quad X_4=\frac{x_4}{x_5},\quad \sum_1^4 X_i^2=\frac{x_6}{x_5}.$$

The equation of a hypersphere is then

$$\sum_1^4 a_i X_i + a_5 + a_6 \sum_1^4 X_i^2 = 0.$$

To complete the parallel with metrical geometry of three dimensions, the expression $\sqrt{\sum_1^4 (X_i - X_i')^2}$ will be called the 'distance' between the points X and X', and the equation of a hypersphere may be written

$$\sum_1^4 \left(X_i + \frac{a_i}{2a_6}\right)^2 = \frac{\sum_1^4 a_i^2 - 4a_5 a_6}{4a_6^2}$$

If the quantity $\dfrac{\sqrt{\sum_1^4 a_i^2 - 4a_5 a_6}}{2a_6}$ be called the *radius* of this hypersphere, it follows that to a special linear complex corresponds a hypersphere of zero radius, or a *null-hypersphere.*

Two hyperspheres will be termed orthogonal if

$$\sum_1^4\left(\frac{a_i}{2a_6}-\frac{a_i'}{2a_6'}\right)^2=\frac{\sum\limits_1^4 a_i^2-4a_5a_6}{4a_6^2}+\frac{\sum\limits_1^4 a_i'^2-4a_5'a_6'}{4a_6'^2},$$

which reduces to

$$\sum_1^4 a_ia_i'-2a_5a_6'-2a_5'a_6=0,$$

in which case the corresponding linear complexes are in Involution.

To two intersecting lines x, x' correspond two points X, X' whose coordinates must satisfy the condition

$$2\sum_1^4 X_iX_i'-\sum_1^4 X_i^2-\sum_1^4 X_i'^2=0,$$

i.e., each of the points X, X' lies on the null-hypersphere whose centre is the other point.

Inverse points in S_4. The equation of any linear complex can be brought to the form $x_6-k^2x_5=0$, and to this complex corresponds the hypersphere $\sum\limits_1^4 X_i^2=k^2$; the points Y, Y' may be called '*inverse*' with regard to this hypersphere if $Y_i'=\dfrac{k^2Y_i}{\sum\limits_1^4 Y_i^2}$.

If any hypersphere $\sum\limits_1^4 a_iX_i+a_5+a_6\sum\limits_1^4 X_i^2=0$ pass through both Y and Y', we have

$$\sum_1^4 a_iY_i+a_5+a_6\sum_1^4 Y_i^2=0,$$

$$\sum_1^4 a_iY_i'+a_5+a_6\sum_1^4 Y_i'^2=0;$$

now the last equation may be written

$$k^2\sum_1^4 a_iY_i+a_5\sum_1^4 Y_i^2+a_6k^4=0,$$

which, from the first equation, requires that $a_5-k^2a_6=0$; but this is the condition that the two hyperspheres should cut orthogonally; hence, *any hypersphere passing through two points inverse with regard to another hypersphere cuts the latter orthogonally*: conversely, if every hypersphere through two given points cuts a given hypersphere orthogonally, these points are inverse with regard to the given hypersphere. Since every linear complex through two lines p and p', which are polar with regard to a complex C, is in involution with C (Art. 25), it follows that *to two polar lines of a linear complex C correspond two points which are inverse with regard to the hypersphere which corresponds to C.*

To the lines of a linear congruence correspond the points of intersection of two hyperspheres, *i.e.* of a hyperplane and a hypersphere; to its directrices the two null-hyperspheres which pass through the intersection of the hyperspheres. To the lines of a regulus correspond the points common to three hyperspheres, *i.e.* the intersection of two hyperplanes and a hypersphere, or a *conic.*

227. Automorphic transformations in Λ correspond to anallagmatic transformations of S_4. Automorphic transformations in Λ are those for which $\omega(x) \equiv \omega(x')$; they involve either a *collineation* or a *reciprocity*, in Λ, (Art. 40).

It will now be shown that the corresponding transformations in S_4 are *anallagmatic, i.e.* change hyperspheres into hyperspheres.

For any such transformation in Λ being

$$x_1 = a_1'x_1' + \ldots\ldots + a_6'x_6',$$
$$\ldots\ldots\ldots\ldots\ldots\ldots\ldots\ldots$$
$$x_6 = f_1'x_1' + \ldots\ldots + f_6'x_6';$$

the connexion between the corresponding points X and X' is given by the equations

$$X_1 = \frac{a_1'X_1' + \ldots\ldots + a_5' + a_6'\sum_1^4 X_i'^2}{e_1'X_1' + \ldots\ldots + e_5' + e_6'\sum_1^4 X_i'^2},$$
$$\ldots\ldots\ldots\ldots\ldots\ldots\ldots\ldots$$
$$X_4 = \frac{d_1'X_1' + \ldots\ldots + d_5' + d_6'\sum_1^4 X_i'^2}{e_1'X_1' + \ldots\ldots + e_5' + e_6'\sum X_i'^2},$$
$$\sum_1^4 X_i^2 = \frac{f_1'X_1' + \ldots\ldots + f_5' + f_6'\sum_1^4 X_i'^2}{e_1'X_1' + \ldots\ldots + e_5' + e_6'\sum_1^4 X_i'^2},$$

with, of course, equations of similar form having the accents only on the left. Thus in S_4 *a hypersphere is changed into a hypersphere.* All anallagmatic transformations are equivalent to one or a finite number of combinations of the following kind*

* See Koenigs, *La Géométrie réglée*, p. 125.

(i) *translations*

$$X_1' = X_1 + h_1, \quad X_2' = X_2 + h_2, \quad X_3' = X_3 + h_3, \quad X_4' = X_4 + h_4;$$

(ii) *similar transformations*

$$X_1' = mX_1, \quad X_2' = mX_2, \quad X_3' = mX_3, \quad X_4' = mX_4;$$

(iii) *reflexions*

$$X_1' = \pm X_1, \quad X_2' = \pm X_2, \quad X_3' = \pm X_3, \quad X_4' = \pm X_4';$$

(iv) *inversions*

$$X_1' = \frac{k^2 X_1}{\Sigma X^2}, \quad X_2' = \frac{k^2 X_2}{\Sigma X^2}, \quad X_3' = \frac{k^2 X_3}{\Sigma X^2}, \quad X_4' = \frac{k^2 X_4}{\Sigma X^2}.$$

228. Principal Surfaces of Λ and Lines of Curvature of S_4. The Principal Surfaces of a complex, (Art. 132), are chiefly of interest as being the analogues of the *Lines of Curvature* of a hypersurface. For, taking the hypersurface $F(X_1, X_2, X_3, X_4, 1) = 0$ obtained from the line complex $f(x_1 \dots x_6) = 0$, to the tangent linear complexes of the latter there correspond the tangent hyperspheres of $F = 0$, viz.

$$\sum_1^4 (Y_i - X_i)\frac{\partial F}{\partial X_i} + \mu\left(2\sum_1^4 Y_i X_i - \sum_1^4 X_i^2 - \sum_1^4 Y_i^2\right) = 0.$$

It was seen, (Art. 132), that for any line x of f there are three tangent linear complexes which touch f in x and also in three respective lines consecutive to x; we have therefore in S_4 the result that *for each point of F there are three tangent hyperspheres which also touch F in three respective consecutive points.*

From the analogy with three-dimensional space, these hyperspheres are called Principal Spheres; and we infer that there are for each point P of a hypersurface three Principal Spheres, the lines joining P to the three consecutive points of contact P', P'', P''' being called the Lines of Curvature of F at P.

The three lines consecutive to x, just mentioned, are in an *involutory position* with regard to each other, (Art. 133); correspondingly, the lines PP', PP'', PP''' are *mutually orthogonal**.

As another instance of the analogy between Λ and S_4, the theorem of Art. 65 leads to the following result in S_4:—of any *six* hyperplanes in S_4, any *four* pass through one point, and by means of five of the hyperplanes we obtain *five* such points; through each such set of five points one hypersphere passes, and hence six hyperspheres are obtained, corresponding to the six sets of five hyperplanes; our theorem is then that these six hyperplanes have one common point.

* For taking $x_5 = 1$, from the equation $\omega(dx' \mid dx'') = 0$, (Art. 133), we derive

$$\sum_1^4 dX_i' dX_i'' = 0, \text{ \&c.}$$

229. Line Geometry is point geometry of an S_4^2 in an S_5. Any six quantities $x_1 \dots x_6$ may be regarded from another point of view as being point coordinates of a 'space' of five dimensions S_5, and when they are coordinates of a line in Λ they satisfy $\omega(x)=0$, *i.e.* they are coordinates of a point in a 'space' of four dimensions contained in S_5; moreover understanding by a 'line' of S_5 a locus of points which satisfy four linear equations

$$\sum_1^6 a_i x_i = 0, \quad \sum_1^6 b_i x_i = 0, \quad \sum_1^6 c_i x_i = 0, \quad \sum_1^6 d_i x_i = 0,$$

the space represented by $\omega(x)=0$ is met by any line of S_5 in two points, and is therefore denoted by S_4^2; thus line geometry may be regarded as point geometry of an S_4^2 in S_5. To a linear complex $\sum_1^6 a_i x_i = 0$, $\omega(x)=0$, corresponds the intersection of a hyperplane of S_5 with S_4^2. The complex is special if $\Omega(a)=0$, *i.e.* if the hyperplane *touches* the S_4^2; if two linear complexes are in involution, the corresponding hyperplanes are *conjugate* with regard to the quadric S_4^2.

230. Line Geometry in Klein coordinates is point geometry of S_4 with hexaspherical coordinates. If

$$\xi_1 = 0, \ \dots, \ \xi_6 = 0$$

are six complexes in mutual involution, then if

$$\Omega(\xi_1) = \dots = \Omega(\xi_6),$$

we know that
$$\sum_1^6 \xi_i^2 \equiv 0.$$

Let
$$\xi_1 \equiv a_1 x_1 + \dots + a_5 x_5 + a_6 x_6,$$
where
$$x_1^2 + x_2^2 + x_3^2 + x_4^2 - x_5 x_6 \equiv 0,$$
then
$$\Omega(\xi_1) \equiv a_1^2 + a_2^2 + a_3^2 + a_4^2 - 4a_5 a_6,$$
and to the complex $\xi_1 = 0$ corresponds the hypersphere

$$\sum_1^4 a_i X_i + a_5 + a_6 \sum_1^4 X_i^2 = 0.$$

It was seen, (Art. 226), that if r_a is the radius of this hypersphere,

$$r_a^2 = \frac{\Omega(\xi_1)}{4a_6^2};$$

and to the six complexes $\xi_i = 0$ correspond six hyperspheres which are mutually orthogonal; hence, extending the word 'power' as used in geometry of three dimensions to the space S_4, if P_a is the power of any point in S_4 with regard to the hypersphere of radius r_a,

$$\frac{\xi_1}{x_5} = a_6 P_a = \frac{P_a}{r_a} \cdot \frac{\sqrt{\Omega(\xi_1)}}{2},$$

and since $\sum_1^6 \xi_i^2 \equiv 0$, it follows that $\sum_1^6 \left(\frac{P_i}{r_i}\right)^2 \equiv 0$; hence *the quantities* ξ_i *are hexaspherical coordinates of any point in* S_4 *with regard to six fundamental hyperspheres which are mutually orthogonal.*

231. Congruences of the mth order and nth class. A set of ∞^2 lines such that through every point there pass m lines and in every plane lie n lines, is said to form a congruence of order m and class n. The ∞^1 lines of the congruence which meet any given line l form a ruled surface of degree $m+n$; since if P is the point of intersection of any line l' with l, and π the plane (l, l'), the lines of the congruence which meet l and l' are the m lines through P and the n lines in π: any linear congruence contains $m+n$ lines of the given congruence.

In particular, to the congruence which is the intersection of two linear complexes each of which contains s, belong $m+n$ lines of the given congruence; hence, applying the transformation of Art. 223, we see that in any plane of S_4 there are $m+n$ points corresponding to these lines. Again in a plane π of Λ there are n lines of the congruence, and if this plane meets s in P there are m lines of the congruence through P; to these $m+n$ lines there correspond in S_4 $m+n$ points in a plane ϵ_1 of S_4 which passes through a generator σ_1 of Φ (Art. 225); and m of these $m+n$ points lie on σ_1, (Art. 223).

Similarly in any plane ϵ_2 through a generator σ_2 there lie $m+n$ points, corresponding to the lines of the congruence which belong to any sheaf of centre P and the plane system (P, s); of these points n lie in σ_2.

Thus in S_4 we have ∞^2 points, corresponding to the lines of the congruence, and such that $m+n$ of them lie in any plane of S_4; these points therefore form a 'surface' in S_4 which may be denoted by F_{m+n}.

Since any line meets a hyperplane in one point only unless it is wholly contained in it, if any point O of S_4 be joined to all the points of F_{m+n}, the joining lines will *project* these points into ∞^2 points of any given hyperplane Σ, which does not contain O, to form a surface f_{m+n}: *the degree of this surface is* $m+n$; for, since in each plane of S_4 there are $m+n$ points of F_{m+n}, the plane through O and any line p of Σ contains $m+n$ points of F_{m+n}, which are projected into $m+n$ points of f_{m+n} lying on p; *i.e.* p meets f_{m+n} in $m+n$ points. The surface f_{m+n} will, in general, possess a double curve, for *any* hyperplane through O contains a curve of points of F_{m+n}, and this curve will have a certain number of "apparent double points" which are projected into *coincident* points of f_{m+n}. Through O will therefore pass a cone of chords of F_{m+n}.

An exception occurs when F_{m+n} is entirely contained in the same hyperplane.

232. Rank of a congruence. Any point O on Φ being taken as the centre of projection, then if $(a_1, a_2, a_3, a_4, 0)$ are the coordinates of O, it was seen (Art. 225) that the hyperplane $\sum_1^5 a_i X_i = 0$ touches Φ and corresponds to a special linear complex whose directrix d belongs to the pencil (P, π) associated with O (Art. 223). To the points of any line through O in this hyperplane there correspond the lines of a pencil containing d.

Denoting this tangent hyperplane by Σ, we observe that Σ meets the cone of chords of F_{m+n} through O in h chords, if h is the degree of the cone; now Σ contains the two generators σ_1 and σ_2 of Φ at O, while σ_1 contains $\frac{1}{2}m(m-1)$ pairs of points of F_{m+n}, and σ_2 contains $\frac{1}{2}n(n-1)$ pairs of points of F_{m+n}; the number of chords common to Σ and the cone, exclusive of σ_1 and σ_2, is therefore

$$h - \tfrac{1}{2}m(m-1) - \tfrac{1}{2}n(n-1) = r.$$

This number r is called the "rank" of the congruence, and has the following meaning for the space Λ; *if P and π are the point and plane of s which correspond to σ_1 and σ_2, and d is the line of the pencil (P, π) which is the directrix of the special complex which corresponds to Σ in Λ, then d lies in a pencil with two lines of the congruence r times.*

Since now with any given congruence the coordinate systems employed are quite arbitrary, it is seen that in general two lines of the congruence will lie in one pencil with any given line d a definite number r of times.

CHAPTER XIV.

CONGRUENCES OF LINES.

233. Order and class of a congruence. A set of ∞^2 lines, such that any two given conditions determine a definite finite number of lines of the set, is said to constitute a *congruence*. The locus of lines which belong to each of two complexes is one instance of a congruence. The requirement that a line of the system should pass through a given point is equivalent to two conditions, and the number of lines of the congruence which pass through any given point is called its *order*; similarly the number of lines of the system which lie in any given plane is called its *class**. When the congruence is the complete intersection of two complexes, its order and also its class is equal to the product of the degrees of the complexes. It is convenient to designate a line of a given congruence by the term *ray*. A congruence whose order is m and class n is termed a congruence (m, n).

A given congruence (m, n) establishes on any planes π and π' a correspondence, such that to each point P of π correspond m points of π' (where the rays through P meet π'), and to each point of π' correspond m points of π; while, if P is any point on the line (π, π'), the pencils (P, π), (P, π') are so related that on any line p of one pencil and on any line p' of the other pencil there are n pairs of respectively corresponding points, determined by the rays of the plane (p, p').

The most general congruence can be constructed as follows: it was seen (Art. 231) that the degree of a ruled surface, whose generators are those rays of a given congruence which meet any given line, is $m+n$; in any plane π take any pencil of lines and construct the equation in point-coordinates of the most general ruled surface of degree $m+n$ which has a line p of this pencil as m-fold directrix (*i.e.* such that through each point of p there

* See Art. 231.

pass m generators), and which has any plane through p as n-fold tangent plane (*i.e.* any plane through p contains n generators); then taking λ as the parameter upon whose variation p depends, express each coefficient of the equation of this surface as a polynomial of degree m in λ; the ∞^2 generators of the ∞^1 ruled surfaces form the required congruence (m, n).

The ruled surface of degree $m+n$ formed by the lines of a given congruence which meet any given line l will be denoted by (l).

234. Halphen's Theorem. *Two congruences* $(m, n), (m', n')$ *have* $mm'+nn'$ *rays in common;* this theorem was shown by Halphen, the following proof is due to Schubert. The *class* of a surface (l) of the first congruence is $m+n$, for the tangent planes to (l), through any line l' which meets it in the $m+n$ generators $p, p' \ldots$, are the planes through $p, p' \ldots$ and l; hence their number is $m+n$*. Similarly the class of (l) for the second congruence is $m'+n'$.

The class of the developable common to the two surfaces (l) is therefore $(m+n)(m'+n')$, which is the number of planes through any point A passing through a generator of each surface; but nn' of these planes coincide with the plane (A, l), hence there remain $mm'+mn'+m'n$ planes through any point which contain a ray p of (m, n) and a ray p' of (m', n'), where p, p', l are *concurrent.* We now seek the surface which is the locus of a line p of (m, n) such that the point of intersection P of p with a line p' of (m', n') lies in a given plane α, while the plane π of p and p' passes through a given point A.

By taking the line l in α, it is seen, from what has just been proved, that for any line l there are $mm'+mn'+m'n$ planes π through A, hence the locus of P is a curve k of degree

$$mm'+mn'+m'n;$$

reciprocally, the envelope of the planes π is a cone of vertex A whose class is $nn'+nm'+n'm$.

The degree of the required surface is therefore equal to $mm'+m'n+mn'$ together with the number of pairs p, p' for which p lies in α; and since there are n rays p in α which are paired with the n' rays of (m', n') in the planes (A, p), the degree of the required surface is seen to be

$$mm'+m'n+mn'+nn'=(m+n)(m'+n').$$

This is also the degree of the locus of the corresponding rays p'.

* The degree and class of a ruled surface are in general the same.

If (B, β) is any arbitrary pencil, such a ray p determines one line of the pencil, while it meets the locus of p' in $(m+n)(m'+n')$ points the generators at which determine just as many other lines of the pencil; thus in the pencil a $\{(m+n)(m'+n'), (m+n)(m'+n')\}$ correspondence is established. The number of coincidences is therefore $2(m+n)(m'+n')$, (Introd. xv.).

Such coincidences may arise

(i) from rays common to the two systems;

(ii) from rays p, p' whose plane passes through B; this occurs $nn'+mn'+m'n$ times, since this number is the class of the cone vertex A;

(iii) from rays p, p' whose point of intersection lies on β; this occurs $mm'+mn'+m'n$ times, since this is the degree of the curve in α;

therefore the required number of common rays is

$$2(m+n)(m'+n')-(nn'+mn'+m'n)-(mm'+m'n+mn')=mm'+nn'.$$

By taking $m'=n'=k$ it is seen that the number of rays of (m, n) which belong to a complex of degree k and meet any line is $k(m+n)$; hence, *a complex of degree k has in common with (m, n) a ruled surface of degree $k(m+n)$.*

235. Characteristic numbers of a congruence. In addition to the order and class of a congruence, it was seen in the last chapter that there is a third characteristic number, viz. the rank* r, which is the number of times two rays belong to the same pencil with any given line l. If the system is the intersection of two complexes of which one is linear, it is seen that $r=0$, since l does not in general belong to the linear complex.

Two loci are of special importance in the present theory†: (i) the locus of points of intersection of rays in a plane π which turns about a line l, is a curve which will be denoted by $|l|$; r points of this curve lie on l, and in any plane π there are $\frac{1}{2}n(n-1)$ points of the curve which do not lie on l, hence the order of $|l|$ is $\frac{1}{2}n(n-1)+r$:

(ii) when π describes a sheaf of centre P, the locus of such points of intersection is a surface which will be denoted by (P). This surface is seen to be the locus of points on the lines l of this sheaf at which two rays belong to the same pencil with l, it is also the locus of the ∞^1 curves $|l|$ when l describes a plane

* Schumacher, to whom the introduction of this characteristic is due, used the term *Art.*

† See Art. 250.

pencil of centre P. Now since there are $\frac{1}{2}m(m-1)$ pairs of rays through P, there will be that number of these curves through P, which is therefore a point of multiplicity $\frac{1}{2}m(m-1)$ on (P), and since any line through P meets this surface in r points distinct from P, it is seen that the degree of (P) is $r+\frac{1}{2}m(m-1)$.

236. Focal points, planes and surface. The ruled surface formed by rays which meet any given ray l is still of degree $m+n$; for l' being *any* line, l meets (l') in $m+n$ points, hence (l') and (l) have $m+n$ rays in common.

Any line p meets $m+n$ generators of the surface; if p meets l in P, then p meets (l) where it meets (i) the $n-1$ rays of the plane (l, p) distinct from l, (ii) the $m-1$ rays of the point P distinct from l; hence the intersection of l with p must be counted twice to complete the number $m+n$ of intersections of p and (l). It follows that l is a double generator of (l), and there are two points upon any ray for each of which there are only $m-2$ rays distinct from it; hence, *each ray is intersected by two consecutive rays*. These two points on a ray are called *Focal Points*, (Art. 119): the locus of the ∞^2 Focal Points is called the Focal Surface of the congruence.

If l and l_1 are consecutive rays meeting in F_1 and l_1' is consecutive to l_1 and meets it in F_1', then F_1 and F_1' are ultimately focal points, thus l_1 (and hence any ray) touches the focal surface at each focal point.

Let the focal points on any ray l be F_1 and F_2, and on l_1, the consecutive ray through F_1, be F_1' and F_2', then F_2 being consecutive to F_2', the plane (l, l_1) touches the focal surface at F_2, since it contains two tangent lines of the surface at F_2, viz. l and F_2F_2'. Thus if l_1 and l_2 be the two rays consecutive to l which meet it, the plane (l, l_1) touches the focal surface at F_2 and the plane (l, l_2) touches it at F_1. The planes (l, l_1), (l, l_2) are called Focal Planes. All the rays of a congruence touch the focal surface twice, but all bitangents of the surface are not rays of the given congruence.

237. Degree and Class of the Focal Surface. Upon any arbitrary pencil of planes whose axis is l the congruence effects an involutory correspondence $[n(m-1)]$; for, any plane π through l contains n rays which determine n points P on l through each of which pass $m-1$ other rays not in π, and regarding the $n(m-1)$ planes through l and these other rays as corresponding to π, we have determined an involutory correspondence $[n(m-1)]$.

The number of coincidences is $2n(m-1)$ which may arise

(i) from the coincidence of two rays through a point P of l; such a point P belongs to the focal surface;

(ii) from the r points on l through each of which two rays lie in the same pencil with l; in this case the plane π through such a pair of rays p and p' coincides with *two* of its corresponding planes, since one of the planes determined by p is (l, p'), *i.e.* π, and one of the planes determined by p' is (l, p), *i.e.* π.

But in an involutory correspondence in which an element coincides with two of its corresponding elements, such an element counts for *two* of the coincidences (Introd. xv.); hence if m_1 is the degree of the Focal Surface, *i.e.* the number of points in which l meets the Focal Surface,

$$m_1 = 2n(m-1) - 2r.$$

Reciprocally, we obtain an involutory correspondence $[m(n-1)]$ of points on l in which two points correspond which are the intersections with l of two rays in one plane through l; and it follows by similar reasoning that if n_1 is the class of the Focal Surface, *i.e.* the number of tangent planes to the Focal Surface through l,

$$n_1 = 2m(n-1) - 2r.$$

238. Singular Points. If more than m rays pass through a point S, then ∞^1 rays will pass through S, which is called a *singular* point of the congruence. For taking any line l through S, $m+n$ of the rays which meet l also meet any line l', but if a finite number m', greater than m, of rays pass through S, then any line l' through S, such that the plane (l, l') does not contain any of these rays, will meet $m'+n$ rays of (l), which is impossible. Thus the rays through S form a cone, say of degree h, and (l) breaks up into this cone, which will be denoted by (S_h), and a ruled surface of degree $m+n-h$. Similarly, if more than n rays lie in a plane there will be ∞^1 rays in that plane which envelope a curve.

It is clear that each singular point S_h lies on each surface (l), and since (S_h) meets l in h points, therefore S_h is an h-fold point of (l).

Each surface (P) passes through each singular point. In the case of a singular point S_1, each line of the pencil of rays whose centre is S_1 touches the focal surface; the plane of the pencil is therefore a singular tangent plane of the focal surface, the curve of contact being of degree $n(m-1)-r$.

239. Expression of the coordinates of a ray in terms of two variables*. The equations of Art. 223 establish a (1, 1) correspondence which connects any line of the space Λ with a point of four-dimensional space S_4; so that to each line of the given congruence (m, n) there corresponds one of the points of S_4, and *vice versâ*; the latter points being projected from a point of S_4 upon any given hyperplane, give rise to a surface f_{m+n}. Hence a (1, 1) correspondence is determined between the lines of the congruence and the points of f_{m+n}. If u and v are the variables in terms of which the coordinates of any point of f_{m+n} may be expressed, and x_i is any line of the congruence, this correspondence may be set forth by six equations of the form

$$x_i = f_i(u, v),$$

so that the coordinates of any line of the congruence may be expressed in terms of two variables.

The foregoing mode of expression enables us to deduce important properties of a congruence: it was seen, (Art. 9), that if two consecutive lines x and $x + dx$ intersect, the equation $(dx^2) = 0$ must hold; if each of these lines belongs to the congruence, we have

$$E du^2 + 2F du dv + G dv^2 = 0,$$

where $$E = \sum_1^6 \left(\frac{\partial x_i}{\partial u}\right)^2, \quad F = \sum_1^6 \frac{\partial x_i}{\partial u}\frac{\partial x_i}{\partial v}, \quad G = \sum_1^6 \left(\frac{\partial x_i}{\partial v}\right)^2;$$

we infer, therefore, that *any line of a congruence is intersected by two consecutive lines of the congruence*, viz. those which correspond to the two values of $\dfrac{du}{dv}$ determined by the last equation, (see also Art. 236).

The equations of a linear congruence are of the form

$$y_1 = \sum_3^6 a_i y_i, \quad y_2 = \sum_3^6 b_i y_i;$$

if the eight constants a_i, b_i satisfy the six equations

$$x_1 = \sum_3^6 a_i x_i, \quad \frac{\partial x_1}{\partial u} = \sum_3^6 a_i \frac{\partial x_i}{\partial u}, \quad \frac{\partial x_1}{\partial v} = \sum_3^6 a_i \frac{\partial x_i}{\partial v},$$

$$x_2 = \sum_3^6 b_i x_i, \quad \frac{\partial x_2}{\partial u} = \sum_3^6 b_i \frac{\partial x_i}{\partial u}, \quad \frac{\partial x_2}{\partial v} = \sum_3^6 b_i \frac{\partial x_i}{\partial v},$$

* This mode of treatment of the congruence belongs to what is known as Differential Geometry. The most celebrated memoir in this field is that of Kummer, *Crelle's Journal*, Bd. 57 (1860), "Allgemeine Theorie der geradlinigen Strahlensysteme."

this linear congruence contains x and any consecutive line

$$x+\frac{\partial x}{\partial u}du+\frac{\partial x}{\partial v}dv$$

of the given congruence; hence, *there are ∞^2 linear congruences which contain any line x of a given congruence and all lines of the latter congruence consecutive to x.*

The constants a_i, b_i are determined if the additional condition be imposed that the linear congruence should pass through any given line x'; hence, *through a given line and any line x of a given congruence K, one linear congruence can be constructed which passes through all lines of K consecutive to x.*

It is easily seen that there are ∞^2 linear complexes which pass through any line x of K and all lines of K consecutive to x.

240. As an instance of the expression of the coordinates of the lines of a congruence in terms of two variables, consider the case in which the functions f_i are quadratic expressions; *i.e.*

$$\rho\,.\,x_i=a_iu^2+2h_iuv+b_iv^2+2g_iu+2f_iv+c_i.$$

The identical equation $(x^2)=0$ gives rise to a number of equations between the coefficients, among which are the following

$$(a^2)=0,\ (ah)=0,\ (b^2)=0,\ 2(h^2)+(ab)=0.$$

The two equations which determine the rays which meet any two lines are quadratics in u and v; hence there are four rays which meet any two lines, *i.e.* the sum of the order and class of the congruence is *four* (Art. 231).

On making the substitution

$$u=\frac{nu'}{u'+1},\quad v=\frac{v'}{u'+1}$$

the coordinates x_i become proportional to quadratic expressions in u' and v'; in these expressions, the coefficient h_i' of $2u'v'$ is easily seen to be nh_i+f_i. If therefore n be taken as either of the roots n_1, n_2 of the equation

$$n^2(h^2)+2n(hf)+(f^2)=0,$$

the h_i' are coordinates of a line; while, as before, we have

$$(a'^2)=0,\ (a'h')=0,\ (b'^2)=0,\ 2(h'^2)+(a'b')=0;$$

hence $(a'b')=0$, and the quantities a_i', h_i', b_i' are the coordinates of three mutually intersecting lines. These lines may be either *concurrent* or *coplanar*; in the former case from consideration of the equations

$$\sigma\,.\,x_i=a_i'u'^2+2h_i'u'v'+b_i'v'^2+2g_i'u'+2f_i'v'+c_i',$$

we observe that the values $u'=\infty$, $v'=\infty$, give ∞^1 rays, corresponding to the values of $\frac{u'}{v'}$, *which are concurrent*; hence in the original equations $u=n_1$ and $u=n_2$ each give a ray-cone of the congruence. If p be any line through the vertex of one of these cones, the ruled quartic formed by rays which intersect p breaks up into this ray-cone and a regulus; through any point of p will pass *one* ray, viz. the generator at this point of the regulus: therefore, *the order of the congruence is unity, and the class is three*, any point of the curve of intersection of the two ray-cones must therefore be a singular point of the congruence as having two rays through it; and the congruence possesses a curve of singular points which is a twisted cubic, since the vertex of each ray-cone lies on the other. The congruence consists of the chords of this twisted cubic (Introd. xii.).

Similarly if a_i', h_i' and b_i' are coplanar, the congruence is of the third order and first class, and consists of the lines of intersection of tangent planes of a developable; three such planes pass through any point.

A special case arises when the congruence is contained in a linear complex $\Sigma a_i x_i = 0$. If this condition is satisfied we must have

$$\Sigma a_i a_i = 0, \quad \Sigma a_i h_i = 0, \quad \Sigma a_i b_i = 0, \quad \Sigma a_i g_i = 0, \quad \Sigma a_i f_i = 0, \quad \Sigma a_i c_i = 0.$$

The complex must be special, as containing *either* a ray-cone *or* the tangents of a conic.

If the directrix of this special complex be taken as the edge $A_1 A_2$ of the tetrahedron of reference, we have for the rays

$$p_{34} = 0, \quad \rho \cdot p_{ik} = a_{ik} u^2 + 2h_{ik} uv + b_{ik} v^2 + 2g_{ik} u + 2f_{ik} v + c_{ik}.$$

To determine the rays which pass through the point $(a_1, a_2, 0, 0)$, we have the equations (Art. 3),

$$p_{42} a_1 + p_{14} a_2 = 0, \quad p_{23} a_1 - p_{13} a_2 = 0.$$

These equations, since $p_{34} = 0$, are seen to be equivalent to one only; hence each point of $A_1 A_2$ is a singular point of the congruence.

The locus of singular points breaks up into $A_1 A_2$ and a conic c^2, in the case where a_i', h_i' and b_i' are concurrent; and $A_1 A_2$ must meet c^2, for if not, then through any point P there would pass *two* rays, viz. the lines joining P to the intersections of the plane $(P, A_1 A_2)$ with c^2.

Similarly if a_i', h_i', b_i' are coplanar, the congruence consists of the lines of intersection of the planes through $A_1 A_2$ and the tangent planes of a cone which is touched by $A_1 A_2$.

The equations which give the points ξ_i of the above twisted cubic may be taken as

$$\rho \cdot \xi_1 = t^3, \quad \rho \cdot \xi_2 = t^2, \quad \rho \cdot \xi_3 = t, \quad \rho \cdot \xi_4 = 1;$$

the equations giving the coordinates p_{ik} of its chords are then

$$\sigma \cdot p_{12} = t^2 \tau^2, \quad \sigma \cdot p_{13} = t\tau (t + \tau), \quad \sigma \cdot p_{14} = (t + \tau)^2 - t\tau,$$
$$\sigma \cdot p_{23} = t\tau, \quad \sigma \cdot p_{34} = 1, \quad -\sigma \cdot p_{42} = t + \tau.$$

On writing $t\tau = u$, $t + \tau = v$, these equations become

$$\sigma \cdot p_{12} = u^2, \quad \sigma \cdot p_{13} = uv, \quad \sigma \cdot p_{14} = v^2 - u, \quad \sigma \cdot p_{23} = u, \quad \sigma \cdot p_{34} = 1, \quad -\sigma \cdot p_{42} = v.$$

241. Schumacher's method*. In the last chapter a process was investigated by which the rays of a given congruence are made to correspond to the points of a locus of ∞^2 points in four dimensions; this locus, denoted by F_{m+n}, is projected from any point P of S_4 upon any hyperplane not containing P into an *algebraic* surface denoted by f_{m+n}, whose degree is $m + n$.

* On a first reading it would be well to pass at once to Chapter XV.

Since *one* linear congruence can be described through the line s, connected with the equations of transformation, (Art. 223), to contain any given ray and every ray consecutive to it, (Art. 239), it follows that there is one plane of S_4 which contains any point Q of F_{m+n} and every point of this locus consecutive to Q; *i.e.* the points of F_{m+n} consecutive to Q lie in the same plane.

It was also seen that through P there pass ∞^1 lines each of which contains two points of F_{m+n}; if Q_1, Q_2 are such a pair they are projected into *one* point Q' of f, and thus f has at Q' *two* tangent planes, viz. the projections of the tangent planes at Q_1 and Q_2; the locus of Q' is therefore a *double curve* on f.

Since to three points on a line of S_4 correspond three generators of a regulus of Λ which also contains s, it follows from the foregoing that there are ∞^1 reguli in Λ each of which contains s, the line corresponding to P, and two rays of a given congruence.

Taking $2h$ as the degree of the locus of the points Q_1, Q_2, *i.e.* the number of its points in any hyperplane, then any hyperplane through P will contain h lines of the cone of P, each of these lines containing two points Q. If Σ is the hyperplane upon which the points of F_{m+n} are projected, and Σ' any hyperplane through P, the points of Σ' are projected into points of the plane (Σ, Σ'); hence the $2h$ points Q of the curve which lie in Σ' are projected into h points of the double curve in the plane (Σ, Σ'), *i.e. the order of the double curve is h.*

242. Tangents to F_{m+n}. Certain of the points Q_1, Q_2 of the cone of P are consecutive, we thus obtain *tangents* from P to F_{m+n}; for such points the two tangent planes at the corresponding point Q' coincide, and we have a "cuspidal point" of the double curve.

Denoting by π the number of such tangents, it is shown by Schumacher that $\pi = 2h - 2r$.

For this purpose he takes the projection upon Σ of the cone of P from any point P' of S_4 and thus obtains a cone κ of Σ, of which the degree is h; this cone will in general contain a certain number σ of double edges; for, as before, there is a cone through P' of ∞^1 lines which are chords of the cone of P; but having given one chord $P'RR'$, ∞^1 others are thereby determined, viz. the lines of the pencil of centre P' and plane $(P, P'RR')$; thus the ∞^1 chords through P' form a certain finite number σ of plane pencils, and each such pencil gives rise to a double edge of κ.

Again the cone of P may contain a multiple line which meets F_{m+n} in α points and is therefore an $\frac{\alpha(\alpha-1)}{2}$-fold chord; the projection κ of the cone will then contain an $\frac{\alpha(\alpha-1)}{2}$-fold edge. Thus on κ we have a curve C, the projection of the locus of Q, while κ has σ double edges and an edge which contains α points of C.

Since the cone of P has an $\frac{\alpha(\alpha-1)}{2}$-fold edge, any hyperplane through this edge meets the cone in h' other edges, where

$$h=\frac{\alpha(\alpha-1)}{2}+h',$$

on these h' edges are $2h'$ points Q, hence each point Q on the multiple edge must be an $\overline{\alpha-1}$-fold point of the locus of Q. It follows that each point of the intersection of the multiple edge of κ with C is an $\overline{\alpha-1}$-fold point of C.

The curve C is of degree $2h$, for any hyperplane (Σ') through P' contains $2h$ points Q which are projected into $2h$ points of C in the plane (Σ, Σ').

Now take any plane projection of C in Σ, which therefore gives a curve of degree $2h$, and let D be the total number of its double points (counting $\frac{r(r-1)}{2}$ double points for an r-fold point); it is clear that the number of tangent planes to C through any line is equal to the class of this projection, or $2h(2h-1)-2D$. But the number of such tangent planes, if the line passes through the vertex of κ, is equal to the number of tangents to C from the vertex, *i.e.* π, together with *twice* the number of tangent planes through the line to κ; hence

$$2h(2h-1)-2D$$
$$=\pi+2\left\{h(h-1)-2\sigma-2\frac{\frac{\alpha(\alpha-1)}{2}\left(\frac{\alpha(\alpha-1)}{2}-1\right)}{2}\right\}.$$

To determine D, we observe that it is the number of chords of C which pass through any point; and the ruled surface formed by chords of C which meet any line p through the vertex of κ consists of κ together with a ruled surface G; thus p is a D-fold directrix of G, so that D is the number of generators of G which also belong to κ, *i.e.* which pass through the vertex of κ. Now each tangent plane through p to κ gives one such generator; others arise from the σ double edges and the $\frac{\alpha(\alpha-1)}{2}$-fold edge of κ; each double edge (which contains four points of C) counts as six chords of which two belong to κ and therefore four to G; the multiple edge contains $\alpha(\alpha-1)$ points of C and hence counts as $\frac{\alpha(\alpha-1)\{\alpha(\alpha-1)-1\}}{2}$ chords, of which $\frac{\alpha(\alpha-1)}{2}$ belong to κ and therefore $\frac{\alpha(\alpha-1)}{2}\{\alpha(\alpha-1)-2\}$ to G, so that

$$D=\left\{h(h-1)-2\sigma-2\frac{\frac{\alpha(\alpha-1)}{2}\left\{\frac{\alpha(\alpha-1)}{2}-1\right\}}{2}\right\}$$
$$+4\sigma+\frac{\alpha(\alpha-1)\{\alpha(\alpha-1)-2\}}{2},$$

i.e. $$D=h(h-1)+2\sigma+\frac{\alpha(\alpha-1)\{\alpha(\alpha-1)-2\}}{4}.$$

It follows by insertion in the previous equation that $\pi=2h$. Should one or more additional multiple edges occur the previous method clearly applies. The resulting number of tangents is thus independent of multiple edges of κ.

These tangents however are not all proper tangents; it will be shown that $2r$ of them are due to the existence of r *double points* on F_{m+n}. For there are r pairs of rays in Λ which belong to a pencil with s, and each ray of such a pair p, p' corresponds to the same point on Φ (Art. 223). Now if we construct the ∞^1 ruled surfaces formed by the rays which meet the lines of any given pencil (P, π), all the rays are included; corresponding to these surfaces we obtain ∞^1 curves on F_{m+n}, and since in general each ray determines a line of (P, π), and hence the ruled surface on which it lies, so each point on F_{m+n} determines the curve which passes through it, and two curves do not in general intersect (except in the m points which correspond to the rays through P and the n points corresponding to the rays in π, through all of which each curve passes); but at the point of Φ which corresponds to each of the lines p, p' we have two intersecting curves, hence such a point is a double point on F_{m+n}, (since the tangent plane is there indeterminate).

The cone of chords for every point of S_4 passes through these r double points which lie on Φ, and such a double point is projected into a double point of C.

Hence the number of tangents proper from any point of S_4 to F_{m+n} is

$$2h - 2r = m(m-1) + n(n-1).$$

Applying to the congruence, we obtain these results:

(i) through any two lines of Λ there are ∞^1 reguli each of which contains two rays of the congruence;

(ii) $m(m-1)+n(n-1)$ of these reguli contain two consecutive rays.

243. Triple secants of F. Of the chords of F_{m+n} through any point P, there is in general a finite number t of chords which meet F_{m+n} three times, therefore there is a finite number of triplets of rays which belong to the same regulus with any two given lines. This gives t triple points on the double curve. In terms of the four numbers m, n, r, t, all the characteristics of the focal surface may be calculated.

244. The Focal Surface. Since to the lines of a plane system there correspond the points of a plane ϵ_1 of S_4, (Art. 231), if such a plane touches F, *i.e.* contains two consecutive points of F, the corresponding plane system contains two consecutive rays, and its plane is therefore a Focal Plane; similarly to a plane ϵ_2 which touches F corresponds a Focal Point: it will now be shown by Schumacher's method that the locus of focal points is identical with the envelope of focal planes.

Observe, in the first place, that the ∞^1 planes through any line l of S_4 which meet F in two consecutive points, are projected from any point of l upon any hyperplane Σ into the tangent lines through O to f_{m+n}, O being the point in which l meets Σ; and since any hyperplane Σ' through l is projected into the plane (Σ, Σ'), such of these tangent planes as lie in Σ' are projected into tangents from O to the section of f_{m+n} by (Σ, Σ'). Now let X, X' be two consecutive points of F and Σ' in such a plane, *then if* Σ' *contains another point* X'' *of* F, *consecutive to* X, *which does not lie in the plane* (XX', l), Σ' *will contain two consecutive planes through* l *each of which meets* F *in two consecutive points;* for since the plane $XX'X''$ is projected into (Σ, Σ'), the latter must touch f; thus through O in (Σ, Σ') there pass two (consecutive) generators of the tangent cone from O to f, which are the projections of two (consecutive) planes through l in Σ', each of which meets F_{m+n} in two consecutive points.

Again through any plane, and hence through the plane $XX'X''$, there pass two tangent hyperplanes of Φ; for if the plane be

$$\sum_1^5 a_i X_i = 0, \quad \sum_1^5 b_i X_i = 0,$$

the hyperplane
$$\sum_1^5 (a_i + \lambda b_i) X_i = 0$$

touches Φ, if
$$\sum_1^4 (a_i + \lambda b_i)^2 = 0;$$

let now the points X', X'' be so chosen that both XX' and XX'' meet Φ, and l be taken as the generator σ_1 through the point in which XX' meets Φ, (XX', XX'' are thus the two lines through X in the tangent plane at X to F_{m+n} which meet Φ); it follows that one of these two tangent hyperplanes contains σ_1, and taking it as Σ', *in* Σ' *there pass two consecutive planes through* σ_1, *each of which meets* F *in two consecutive points.*

To Σ' corresponds in Λ a special linear complex whose directrix p meets s (Art. 225); to this complex belong the ray x and the rays x', x'' consecutive to x and meeting it, since XX', XX'' both meet Φ; thus p passes through the intersection of x with one of these lines and meets the other, *e.g.* p passes through (x, x') and meets x''. Then, by what has been proved, *it follows that on the line* p *there will intersect two rays consecutive to* x, x'; hence p is a tangent to the locus of focal points, and the plane (x, p), *i.e.* the plane (x, x''), is the tangent plane at the point (x, p) to this locus (since it contains two tangents to the locus at the point, viz. x and p), *i.e.* the focal plane (x, x'') touches the locus of focal points at the point (x, x').

245. Degree and Class of the Focal Surface. We proceed to give the application of this method to the determination of the degree and class of the focal surface. The degree of the focal surface is equal to the number of times two consecutive rays meet on any line p which meets s; and to the lines which meet p correspond the points of a hyperplane Σ of S_4, tangent to Φ at some point P at which the generators of Φ are (say) σ_1 and σ_2; hence since the degree of the focal surface is equal to the number of sheaves belonging to p each of which contains two consecutive rays, and to the sheaves which belong to p correspond the ∞^1 planes through σ_2 in Σ (Art. 225), the degree of the focal surface is equal to the number of planes of Σ through σ_2 meeting F in two consecutive points.

Now let σ_1 and σ_2 meet any hyperplane $\overset{1}{\Sigma}$ in M and N respectively, the projection of the tangent hyperplane Σ from P on $\overset{1}{\Sigma}$ being the plane α, then it follows that

the degree of the focal surface

= the number of tangents from N to the section of f by α:

similarly, the class of the focal surface

= the number of tangents from M to the section of f by α.

Now f has an m-fold point in M and an n-fold point in N*; while, the section of f by α being of degree $m + n$, the number of tangents to the section from any point is in general

$$(m+n)(m+n-1) - 2h.$$

To obtain the tangents from N distinct from the tangents *at* N, $2n$ must be deducted, since for N two tangents to each branch through N coincide into a tangent at N, hence

degree of focal surface

$$= (m+n)(m+n-1) - 2h - 2n$$

$$= (m+n)(m+n-1) - 2\left(\frac{m(m-1)}{2} + \frac{n(n-1)}{2} + r\right) - 2n$$

$$= 2n(m-1) - 2r.$$

Similarly the class of the focal surface is seen to be

$$2m(n-1) - 2r.$$

246. Double and Cuspidal curves of the focal surface. It has been seen that to the pairs of consecutive rays which meet in the points of a section of the focal surface by a plane π through s, correspond the pairs of consecutive points of F which lie in ∞^1 planes through σ_2, where σ_2 is the generator of Φ associated with π (Art. 223).

These planes are projected from a point of σ_2 into the generators of the tangent cone V from N to f; corresponding to a double point of the section of the focal surface by π there is a plane through σ_2 which contains *two* pairs of consecutive points of F, which gives on projection a double edge of V: a plane which contains *three* consecutive points of F is projected into a cuspidal edge of V; to this there corresponds in the section of the focal surface a point through which three consecutive rays pass; such a point is a cuspidal point, and hence arise on the focal surface a double curve and a cuspidal curve.

* Since σ_1 contains m points of F_{m+n} and σ_2 contains n points of F_{m+n} (Art. 231).

247. Rank of the Focal Surface. The degree of the tangent cone to the focal surface, or the class of its plane section, is called the Rank of the focal surface, and is equal to the class of f; for it has been seen (Art. 244), that to a line p which touches a section of the focal surface by a plane π through s, corresponds a tangent hyperplane of Φ which contains a tangent plane to F; also the tangent hyperplanes of Φ which correspond to the lines of a pencil whose plane π contains s, *themselves form a pencil,* and are of the form

$$\sum_1^5 a_i X_i + \lambda \sum_1^5 b_i X_i = 0,$$

where
$$\sum_1^4 a_i^2 = \sum_1^4 a_i b_i = \sum_1^4 b_i^2 = 0,$$

and pass through a generator σ_2 of Φ, where σ_2 corresponds to π; they thus contain the same plane α through σ_2, viz.

$$\sum_1^5 a_i X_i = 0, \quad \sum_1^5 b_i X_i = 0.$$

These hyperplanes are projected from any point of σ_2 into a pencil of planes in Σ, whose axis is the intersection of α and Σ; and such of them as contain tangent planes to F are projected into tangent planes to f through the line (α, Σ), hence the class of a plane section of the focal surface is equal to the class of f.

The number of tangent planes through any line l to f_{m+n} is equal to the number of intersections of the first polar f'_{m+n} of any point P on l, with the curve of contact a of the tangent cone to f_{m+n} from any point P' of l, *diminished by* the number of intersections of a with the double curve d: we proceed to find the latter number. Of the intersections of a and d, some are at ordinary points of d, and for such the tangent from P' meets one sheet of f and touches the other, and therefore meets f in three consecutive points; such points thus lie on the second polar f'' of P', and consist of all the intersections of d and f'' except those arising from multiple points of f of higher degree than the second (for through such points a will not pass in general), their number is therefore

$h(m+n-2)-$(the number of intersections of d and f'' at higher multiple points).

Now at M there are m tangent planes and hence $\dfrac{m(m-1)}{2}$

branches of d and M is an $m-2$-fold point of f'', similarly for N, also d contains t triple points, hence the number of intersections of d with f'' which lie upon a, *i.e.* the total number of intersections of d and a, is

$$h(m+n-2) - \frac{m(m-1)(m-2)}{2} - \frac{n(n-1)(n-2)}{2} - 3t$$

$$= \frac{m+n-2}{2}(mn+2r) - 3t.$$

In the next place it is seen that d and a touch in $2h-2r$ points, viz. at the cuspidal points of d, for at such a point *every* plane through the tangent to d at the point is a tangent plane of f, hence the element of d is also an element of the curve of contact of the tangent cone of *any* point.

Now since the order of a is

$$\overline{m+n}\,.\,\overline{m+n-1} - 2h = 2mn - 2r,$$

the required class of f is

$$(2mn-2r)(m+n-1)$$

$$-\left\{\frac{m+n-2}{2}(mn+2r) - 3t + 2\left(m\,.\,\overline{m-1} + n\,.\,\overline{n-1}\right)\right\},$$

which is therefore the Rank of the Focal Surface.

248. Determination of r and t for the intersection of two complexes. When the congruence is the complete intersection of two complexes of degrees μ and ν respectively, *i.e.* a congruence, whose order and class are $\mu\nu$, the values of r and t can be determined.

To one complex corresponds a 'space' $S^{2\mu}$ of degree 2μ which contains Φ μ-fold; to the other a 'space' $S^{2\nu}$ of degree 2ν which contains Φ ν-fold; the complete intersection of these spaces consists of Φ counted $\mu\nu$ times and a 'surface' F whose points correspond to the rays of the congruence.

It has been seen that r is the number of apparent double points of the curve of points of F which lie in any hyperplane.

Take therefore a tangent hyperplane Σ of Φ which meets it in the generators σ_1^0, σ_2^0 whose intersection is P; Σ meets $S^{2\mu}$ in a surface $f^{2\mu}$ of which σ_1^0, σ_2^0 are μ-fold lines, and Σ meets $S^{2\nu}$ in a surface $f^{2\nu}$ of which σ_1^0, σ_2^0 are ν-fold lines; the complete intersection of $f^{2\mu}$ and $f^{2\nu}$ consists of σ_1^0 and σ_2^0 each taken $\mu\nu$ times, and a curve C of order $2\mu\nu$ which meets the lines σ_1^0, σ_2^0 in $2\mu\nu$ points.

The number r is thus the number of chords of C which pass through P, excluding σ_1^0 and σ_2^0, *i.e.* the number of generators of the cone K, or (P, C), which meet C twice. Now K meets $f^{2\mu}$ in (i) the curve C which contains $2\mu\nu$ points of σ_1^0 and σ_2^0, (ii) the lines σ_1^0 and σ_2^0 counted $2\mu\nu \,.\, \mu$ times (for each of them is μ-fold on $f^{2\mu}$), (iii) a curve C' whose degree is therefore

$$2\mu\nu \,.\, 2\mu - 2\mu\nu - 2\mu\nu \,.\, \mu = 2\mu\nu\,(\mu - 1).$$

A double edge of K occurs when, as Q describes the curve C, the generator PQ of K to the point Q contains a point of C' coinciding with a former point of C. The intersections of C and C' arise (i) from such a double edge, (ii) when PQ meets $f^{2\mu}$ in a point Q' consecutive to Q, *i.e.* where C meets the first polar of P with regard to $f^{2\mu}$. Denoting therefore by A_1 the number of intersections of C' with $f^{2\nu}$ which do not lie upon σ_1^0 or σ_2^0, and by A_2 the number of intersections of C with the first polar of P with regard to $f^{2\mu}$, also not upon σ_1^0 or σ_2^0, then $A_1 - A_2$ is *double* the required number (since *two* points of intersection of C and C' lie on each double edge), *i.e.*

$$r = \tfrac{1}{2}\,(A_1 - A_2).$$

Now $$A_1 = 2\mu\nu\,(\mu - 1)\,2\nu - 2\mu\nu\,(\mu - 1)\,\nu,$$

since C' meets σ_1^0 and σ_2^0, which are ν-fold on $f^{2\nu}$, in $2\mu\nu\,(\mu - 1)$ points; also

$$A_2 = 2\mu\nu\,(2\mu - 1) - 2\mu\nu \,.\, \mu,$$

since C meets σ_1^0 and σ_2^0 in $2\mu\nu$ points which are easily seen to be μ-fold on the first polar of P for $f^{2\mu}$, hence

$$r = \tfrac{1}{2}\,(A_1 - A_2) = \mu\nu\,(\mu - 1)\,(\nu - 1).$$

It follows that

$$h = \frac{m\,(m-1)}{2} + \frac{n\,(n-1)}{2} + r$$

$$= \mu\nu\,(\mu\nu - 1) + r = \mu\nu\,(2\mu\nu - \mu - \nu).$$

The number t can also be found. For we have seen that $S^{2\mu}$ and $S^{2\nu}$ contain Φ respectively μ-fold and ν-fold, and that their intersection consists of Φ taken $\mu\nu$ times and a locus F whose points correspond to the rays of the given congruence. If K is the cone of chords for F of any point P on Φ, its intersection with $S^{2\mu}$ consists, (i) of a curve C of which two points lie on each generator, (ii) σ_1^0 and σ_2^0 which are $\dfrac{\mu\nu\,(\mu\nu - 1)}{2}$-fold edges of K and μ-fold lines of $S^{2\mu}$, since σ_1^0 and σ_2^0 each contain $\mu\nu$ points of C, of which each is a $\overline{\mu\nu - 1}$-fold point of C (Arts. 231, 242), (iii) a curve C' whose degree is therefore

$$h \,.\, 2\mu - 2h - \mu\nu\,(\mu\nu - 1)\,\mu.$$

A point of intersection of C and C' which does not lie upon σ_1^0 or σ_2^0 gives *either* a tangent to $S^{2\mu}$ *or* a triple secant of F. Hence three times the number of triple secants through P is equal to the number B_1 of intersections of C and C' not in σ_1^0 or σ_2^0, diminished by the number B_2 of generators of K which touch $S^{2\mu}$. Now B_1 is the number of intersections of C' with $S^{2\nu}$ which do not lie upon σ_1^0 or σ_2^0, hence

$$B_1 = \{h\,(2\mu - 2) - \mu^2\nu\,(\mu\nu - 1)\}\,2\nu - \{h\,(2\mu - 2) - \mu^2\nu\,(\mu\nu - 1)\}\,\nu$$

$$= 2h\,(\mu - 1) - \mu^2\nu\,(\mu\nu - 1)\,\nu.$$

B_2 is the number of intersections, outside Φ, of C with the 'first polar' of P with regard to $S^{2\mu}$, diminished by the number $2h-2r$ of tangents from P to C (Art. 242), which touch both $S^{2\mu}$ and $S^{2\nu}$. Now the first polar of P for $S^{2\mu}$ contains Φ, $(\mu-1)$ times, and in particular the generators $\sigma_1{}^0$ and $\sigma_2{}^0$ μ times (as may be seen by taking the section of $S^{2\mu}$ by any hyperplane through $\sigma_1{}^0$ or $\sigma_2{}^0$); the total number of intersections of C with the first polar of P for $S^{2\mu}$ is $2h(2\mu-1)$; also of the intersections of C with Φ, $2\mu\nu(\mu\nu-1)$ lie upon $\sigma_1{}^0$ and $\sigma_2{}^0$ and are to be taken μ times, while $2r$ lie upon Φ at the double points of F (Art. 242), these must be taken $\mu-1$ times, hence

$$B_2=2h(2\mu-1)-\{2r(\mu-1)+2\mu\nu(\mu\nu-1)\mu\}-(2h-2r),$$
$$=2h\mu-4(h-r).$$

Thus
$$3t=B_1-B_2=h(\mu-2)(\nu-2)+r(\mu\nu-r).$$

Hence inserting the values of r, h and t in the expressions found for the Degree, Class and Rank of the focal surface we find

$$\text{Degree}=\text{Class}=2\mu\nu(\mu+\nu-2),$$
$$\text{Rank}=2\mu\nu\{(\mu+\nu-1)^2-\mu\nu+1\}.$$

CHAPTER XV.

THE CONGRUENCES OF THE SECOND ORDER WITHOUT SINGULAR CURVES.

249. AMONG congruences, those of the second order are the most interesting and fully investigated*. In the present chapter and the one which follows, we discuss the congruences $(2, n)$ which possess a finite number of singular points.

The degree of the focal surface, Φ, is four and its class $2n$; for each ray p touches the focal surface twice, and if it met the surface again, then through the latter point there would pass three rays, viz., p and the (two coincident) rays through the point; hence the degree of the focal surface is four, and (Art. 237),

$$4 = 2n - 2r, \text{ hence, } r = n - 2,$$

therefore the class of the focal surface, being $4(n-1) - 2r$, is equal to $2n$.

250. The Surfaces (P). To each point is assigned one plane by the system, viz., that of the two rays through the point; this plane will be termed the *null-plane* of the point. A surface (P) is therefore the locus of points whose null-planes pass through P; every surface (P) passes through each singular point S_h and is of degree $n-1$ (Art. 235); the point P is not a singular point of (P) since here $\dfrac{m(m-1)}{2} = 1$. Any ray l through S_h meets a surface (P) in $n-h$ points exclusive of S_h, viz., in its intersection with the $n-h$ rays of the plane (P, l) which do not pass through S_h; hence S_h is a singular point on (P) of order $n-1-(n-h) = h-1$. The line S_hP meets (P) in $n-h-1$

* See Kummer's important memoir, "Ueber die algebraischen Strahlensysteme insbesondere über die der ersten und zweiten Ordnung," *Berliner Abh.* (1866).

points exclusive of S_h and P, all of whose null-planes pass through P, and since P may be *any* point, it is seen that for a line through a singular point S_h, $r = n - h - 1$.

251. Each singular point of the congruence is a double point of Φ. If we form for any line l through S_h the correspondence of Art. 237, we obtain an involution $[n-h]$ on the pencil of planes of axis l, disregarding the lines of (S_h); and of the $2(n-h)$ coincidences $2(n-h-1)$ are due to pairs of rays in the same pencil with l, thus leaving only two coincidences due to intersections of l with Φ, the Focal Surface, apart from S_h, hence S_h must be a double point of Φ. Since by each point P one null-plane is determined, the equation of such a plane is

$$P_1 x_1' + P_2 x_2' + P_3 x_3' + P_4 x_4' = 0 \quad \text{..............} \quad \text{(i)},$$

where x' is any point on the plane, and the P_i are functions of the coordinates of P. If x' be fixed and the x_i be variable, we therefore obtain the equation of the surface (P) corresponding to x_i'; the degree of the P_i is therefore $n-1$; between them the following identity exists:

$$P_1 x_1 + P_2 x_2 + P_3 x_3 + P_4 x_4 \equiv 0 \quad \text{..............} \quad \text{(ii)}.$$

The surface (i) contains each of the rays through P, since the null-plane of any point on these rays passes through P; they will thus lie on any polar of (P) with regard to P, and are the intersections of the polar plane and polar quadric of (P) for P.

252. Double rays of the congruence. The null-plane of any point on the focal surface is determinate, except for a point S_h, where $h > 1$. If two rays coincide without any definite point of ultimate intersection we have a *double ray*; for each point of such a double ray the null-plane is indeterminate, hence such lines do not belong to the focal surface.

The points whose null-planes are indeterminate are given as the intersections of three of the surfaces P_i (through which therefore all four surfaces must pass, from (ii)); thus the double rays when they exist are common to the surfaces P_i and hence to each surface (P). These surfaces cannot intersect in a *curve*, since a curve of singular points is excluded.

The curve of intersection of two surfaces (P), *e.g.* for the points A and B, consists of the curve $|AB|$ together with the double rays; hence, the number of double rays

$$= (n-1)^2 - \{\tfrac{1}{2}n(n-1) + n - 2\} = \tfrac{1}{2}(n-2)(n-3).$$

Conversely, if a line p lies upon a surface (P) and is independent of the position of P it is a double ray.

A point of intersection of a ray with a double ray must be a singular point S_h with a cone of rays (S_h) of which the double ray is a double edge, hence the surface (l) of degree $n+2$ becomes, when l is a double ray, a certain number of cones (S_h); in fact it consists of *two* such cones, for if k is the number of singular points on a double ray which possesses cones of degree $h_1 \ldots h_k$, a plane through the double ray will cut these cones in rays distinct from the double ray whose number is $n-2$, and is also

$$h_1 - 2 + h_2 - 2 + \ldots + h_k - 2 = \Sigma h - 2k;$$

hence $\Sigma h = 2k + n - 2$, but $\Sigma h = n + 2$, *i.e.*, $k = 2$.

Thus if h_1 and h_2 are the degrees of these cones,

$$h_1 + h_2 = n + 2.$$

The degree of any cone (S_h) *cannot exceed* $n-1$, for in any plane through S_h and a ray which does not pass through S_h there are n rays, hence $h \not> n-1$; thus the degree of a cone (S_h) having its vertex on a double ray cannot be less than three. For each of such a pair of cones the double ray is a double edge, and if a cone has a double edge it is a double ray.

In all cases except when $n = 6$ *the double rays are concurrent.* For if p and p' are two non-concurrent double rays, while S_{h_1}, S_{h_2} are the vertices of ray-cones on p, and S_{h_3}, S_{h_4} on p', (S_{h_1}), being at least of degree three must have a double edge, otherwise it would meet p' in more than two points, *i.e.*, there would be *three* singular points on p'; and this double edge passes through S_{h_3} or S_{h_4} say S_{h_3}; similarly for S_{h_2}; thus there must in this case be four double rays and hence at least six (since the number of double rays is $\frac{1}{2}(n-2)(n-3)$), *i.e.* n, the class of the system, must be at least six, so that $h_1 + h_2$ is at least eight. Thus one of the cones (S_{h_1}), (S_{h_2}) is at least of the fourth degree, and it cannot be of greater degree than the fourth, otherwise there would be more than two singular points on p', hence both (S_{h_1}) and (S_{h_2}) are of the fourth degree and have two double edges, similarly for (S_{h_3}) and (S_{h_4}). The double rays form a tetrahedron. The class of the system is six.

For all other values of n the double rays are concurrent in a point which is necessarily a singular point, since the number of double rays, if such exist, is at least three, except when $n = 4$, in which case there is only one double ray. The singular points of a congruence (2, n), therefore, in general belong to one of three classes:

(i) The point S through which *all* double rays pass.

(ii) Those through which *one* double ray passes; the degree of such has been shown to be at least three, and it must be *equal* to three, for if not, its ray-cone would *either* meet any double ray in more than one point besides S, *or* would have an additional double edge, *i.e.* double ray, which latter double ray would not pass through S. It follows that the *degree of the ray-cone of S is $n-1$*; the degree of the ray-cone of the other point on each double ray is *three*.

(iii) Those through which no double ray passes; the ray-cones of such points cannot be of greater degree than the second in order to avoid meeting the double rays in more than two points.

253. The class of a congruence (2, n) cannot be greater than seven. On the generators of a cone (S_h) the vertex S_h is one focal point, the other focal points form the curve a of contact of (S_h) and Φ; the order of a is $2h$. Since each generator has, besides S_h, only one point in common with a, S_h is an h-fold point on a; the generators of (S_h) which touch at S_h the h branches of a, are rays having four coincident points in common with Φ; but through a double point of a surface there pass six lines having four-point contact with the surface, hence $h \leqslant 6$, hence the degree of (S_{n-1}) is not greater than 6, *i.e.* $n \leqslant 7$.

254. Number of singular points. It has been seen (Art. 238), that every surface (l) has each singular point S_h as an h-fold point, and that two such surfaces have $n+2$ rays in common. Every point common to three such surfaces is *either* a singular point *or* a point where a ray common to two of the three surfaces meets the third; this occurs $3(n+2)^2$ times. A singular point with a ray-cone of degree h counts as h^3 points of intersection of the three surfaces, hence if α_h is the number of points with a ray-cone of degree h, we have

$$(n+2)^3 = 3(n+2)^2 + \alpha_1 + 2^3\alpha_2 + 3^3\alpha_3 + \dots \quad \dots\dots\dots\text{(A)}.$$

A curve $|l|$, being the locus of points of intersection of rays in each plane through l, is a double curve on (l); each point of intersection of $|l|$ and any surface (l') is *either* a singular point*, *or* at the point one of the rays for $|l|$ coincides with the ray for (l'), *i.e.* it is a point in which one of the $n+2$ rays common to (l) and (l') is met by one of the remaining $n-1$ rays in its plane through l. A point S_h is an h-fold point on (l') and an $\dfrac{h(h-1)}{2}$-fold point on

* Since through it pass two rays for $|l|$ and one ray for (l').

$|l|$, and hence counts as $\frac{h^2(h-1)}{2}$ points of the intersection of $|l|$ and (l'), therefore

$$(n+2)\{\tfrac{1}{2}n(n-1)+n-2\}$$
$$=(n+2)(n-1)+2\alpha_2+9\alpha_3+\ldots+\frac{h^2(h-1)}{2}\alpha_h+\ldots \quad \ldots(\text{B}).$$

It will now be shown that the equations (A) and (B) are sufficient to determine the number of singular points. For it has been seen that, except for $n=6$, there is one point S_{n-1} through which all the double rays pass, and except for $n=4$, when there are two points S_3, there cannot be another S'_{n-1}, since in that case $S'_{n-1}S_3$ would have to be a double ray to secure that in any plane through $S'_{n-1}S_3$ there should not be more than n rays. Each point S_3 lies on a double ray, therefore

$$\alpha_{n-1}=1, \quad \alpha_3=\tfrac{1}{2}(n-2)(n-3).$$

When $n=5$ we have one point S_4; when $n=6$ we have *either* four points S_4, and therefore no point S_5, for if S_5 existed, (S_4S_5) would be of degree at least nine, *i.e.* $n+3$; *or* one point S_5 and therefore no point S_4 as before. In no other case is there a point S_4 or S_5, while $n=6$ thus gives two different congruences. The equations (A) and (B)* are thus sufficient in all cases to determine the numbers α_1 and α_2; solving them we obtain the results embodied in the following Table.

	(2, 2)	(2, 3)	(2, 4)	(2, 5)	$(2, 6)_{\text{I}}$	$(2, 6)_{\text{II}}$	(2, 7)
α_1	16	10	6	3	1	0	0
α_2		5	6	6	4	8	0
α_3			2	3	6	0	10
α_4				1	0	4	0
α_5					1		0
α_6							1
$\Sigma\alpha$	16	15	14	13	12	12	11

From the Table it is seen that the number of singular points is $18-n$ which is the number of double points of Φ required to reduce its class to $2n$. The double points of Φ are therefore identical with the singular points of the congruence.

* A third equation

$$\Sigma\alpha_h \,.\, h=4(n+2),$$

due to U. Masoni, exists between the numbers α_i. See *Rendiconti dell' Accademia di Napoli*, vol. XXII. p. 145.

255. Distribution of the singular points*. The points of contact of the rays give rise to an involutory (1, 1) correspondence of points on Φ. But in this correspondence, to each point S_h there will correspond all the points of the curve a whose order is $2h$, in which (S_h) touches Φ, (Art. 253). If Q, Q' are the points of contact of a ray with Φ, since the null-plane of Q touches Φ at Q', the points of the curve of contact σ' of Φ and the tangent cone to Φ from any point P will correspond to the points of the curve σ which is the intersection of Φ and the surface (P). Since any point S_h is of order $h-1$ on (P) and 2 on Φ, it is of order $2(h-1)$ on σ, also σ' is the intersection of Φ and the first polar of P with regard to Φ. Now σ passes $2(h-1)$ times through S_h, hence $2(h-1)$ of the intersections of σ' and a correspond to S_h; but the first polar of P for Φ meets a in $6h$ points, of which h coincide with S_h, since S_h is an h-fold point on a, (Art. 253); deducting the previous $2(h-1)$ points there remain $3h+2$ of the $6h$ points on a and σ' other than S_h which have corresponding points on σ other than S_h; moreover these $3h+2$ points as being on (S_h) have S_h also as corresponding point, *i.e.* each has *more than one* corresponding point and is therefore a singular point on (S_h); and these $3h+2$ points are the *only* singular points on (S_h) other than (S_h), since (S_h) and Φ have only the curve a in common.

In the case of a cone (S_h) with a double ray, the curve a has the other singular point on the double ray as a double point, so that this point counts as two intersections of the first polar and a; hence, since the number of double rays through S_h is easily seen from the Table to be $\frac{1}{2}(h-1)(h-2)$, the number of singular points on (S_h), including S_h, is

$$3(h+1)-\tfrac{1}{2}(h-1)(h-2).$$

Thus, for instance, each singular plane contains 6 singular points,

each cone (S_2) „ 9 „ „

each cone (S_3) „ 11 „ „

i.e. all except $7-n$.

In (2, 7) each (S_3) passes through all singular points; in $(2, 6)_{\text{II}}$ each (S_4) passes through all singular points.

256. Conjugate singular points. Two singular points are said to be *conjugate* if the line joining them is a ray. Each singular plane, since it meets each double ray, must do so either in

* For a detailed investigation of the singular points see Sturm, *Liniengeometrie*, Bd. II. S. 43—60.

a point S_3 or in the point S_{n-1}, hence in each such plane σ there lies *one* such point, but not both, since then at each point P of the double ray S_3S_{n-1} there would be an additional ray PS_1, where S_1 is the centre of the pencil of rays in σ, which is impossible.

Each S_3 or S_{n-1} is conjugate to each S_2 and S_3, for if not, the planes $S_{n-1}S_3S_3'$, $S_{n-1}S_3S_2$ would contain at least $n+1$ rays, which is impossible since these planes are, by the foregoing, not singular. A plane σ, the centre of whose pencil of rays is S_1, which contains a point S_{n-1} will contain $n-2$ *points* S_2 *and* $6-n$ *other points* S_1; for if l be any line of intersection of (S_{n-1}) and σ, except S_1S_{n-1}, (l) consists of (S_{n-1}), the pencil (S_1, σ), and a quadric surface, but at each point P of l the rays are l and S_1P, hence this quadric can only arise as a cone (S_2) whose vertex is on l. No S_2 in σ can lie outside such a line l since S_{n-1} is conjugate to each S_2; hence there are $n-2$ points S_2 in σ, and therefore $6-n$ points S_1 other than that for which σ is the null-plane.

In the congruence (2, 3) we notice that all the points S_2 are conjugate; to each of the 10 pairs of points S_2 there is *one* S_1 conjugate as being required to make up the order five of (S_2S_2'); thus each of the 10 singular planes includes two points S_2.

257. Equation of a surface (P). If the point S_{n-1} of a congruence $(2, n)$ be taken as the vertex A_1 of the tetrahedron of reference, since it is a point of order $n-2$ upon the surface (P) of any point, the equation of such a surface must be of the form

$$x_1\phi + \psi = 0,$$

where ϕ and ψ are cones having their vertices at A_1 and of degrees $n-2$ and $n-1$ respectively.

If the vertex A_2 of the tetrahedron of reference be taken as the point P, and the rays through P as the edges A_2A_3, A_2A_4, it follows that ψ must have x_2 and x_3 as factors, and (P) will have as its equation

$$x_1\phi + x_3x_4\psi' = 0.$$

It is then clear that the surface contains the $2(n-2)$ lines

$$x_3 = 0,\ \phi = 0;\ x_4 = 0,\ \phi = 0;$$

(P) therefore contains at least

$$2 + \tfrac{1}{2}(n-2)(n-3) + 2(n-2)$$

lines. The existence of the latter lines is also shown by the fact that the plane section of (P) through a ray of P and S_{n-1} meets it in this ray and a curve of degree $n-2$ having an $\overline{n-2}$-fold point (at S_{n-1}), the curve must therefore break up into $n-2$ lines through S_{n-1}*. The rays which meet any line p through S_{n-1} form a ruled cubic surface of which p is the double directrix, they will therefore all meet a single directrix. This result also follows from the form of the null-plane of any point which is

$$x_1' P_1 + x_2' (x_1\phi_2 + \psi_2) + x_3' (x_1\phi_3 + \psi_3) + x_4' (x_1\phi_4 + \psi_4) = 0,$$

the null-planes of the points of any line through S_{n-1} are obtained by keeping x_2, x_3, x_4 constant and varying x_1, and are therefore seen to form a pencil.

258. Tetrahedral complexes of the congruences (2, n). If in the congruence $(2, 6)_{\text{II}}$ the four points S_4 be taken as the vertices of the tetrahedron of reference, since the surface (P) of each vertex includes the ray-cone Q of that vertex, the surface (P) of any point x_i' is therefore represented by

$$x_1'\alpha_1 Q_1 + x_2'\alpha_2 Q_2 + x_3'\alpha_3 Q_3 + x_4'\alpha_4 Q_4 = 0;$$

where the α_i are linear in the coordinates.

This equation also represents the null-plane of any point x_i; but since the edges of the tetrahedron of reference are double rays, $\alpha_1 Q_1$, $\alpha_2 Q_2$, $\alpha_3 Q_3$, $\alpha_4 Q_4$ vanish identically for any point on such an edge, this requires that

$$\alpha_1 \equiv x_1, \; \alpha_2 \equiv x_2, \; \alpha_3 \equiv x_3, \; \alpha_4 \equiv x_4,$$

and the surface (P) has as its equation

$$x_1' x_1 Q_1 + x_2' x_2 Q_2 + x_3' x_3 Q_3 + x_4' x_4 Q_4 = 0.$$

The cone $Q_1 = 0$ has A_1A_2, A_1A_3, A_1A_4 as double edges, hence

$$Q_1 \equiv a_{23}x_2^2x_3^2 + a_{34}x_3^2x_4^2 + a_{42}x_4^2x_2^2 + x_2x_3x_4\beta,$$

where $\beta = 0$ is a plane through A_1.

Similarly

$$Q_2 \equiv b_{13}x_1^2x_3^2 + \ldots, \; Q_3 \equiv c_{12}x_1^2x_2^2 + \ldots, \; Q_4 \equiv d_{12}x_1^2x_2^2 + \ldots.$$

The null-plane of any point P in the plane $x_4 = 0$ is therefore

$$a_{23}x_2x_3x_1' + b_{13}x_1x_3x_2' + c_{12}x_1x_2x_3' = 0,$$

and this plane meets $x_1' = 0$ in the line

$$b_{13}x_3x_2' + c_{12}x_2x_3' = 0.$$

Now the coordinates of P being $(x_1, x_2, x_3, 0)$ and the coordinates of the point in which one ray through P meets α_1 being $(0, x_2', x_3', x_4')$, we have

$$p_{12} = x_1x_2', \; p_{34} = x_3x_4', \; p_{13} = x_1x_3', \; p_{42} = -x_2x_4',$$

* See Sturm, *Lin. Geom.* Bd. II. S. 48.

hence each ray of the congruence belongs to the complex

$$b_{13}p_{12}p_{34} - c_{12}p_{13}p_{42} = 0.$$

It follows that *the congruence belongs to a tetrahedral complex for which the four points S_4 form the fundamental tetrahedron.*

The congruence (2, 5). Take as vertices of the tetrahedron of reference the point S_4 and the three points S_3, it follows as in the last case that the null-plane of x_i has as its equation

$$x_1'Q_1 + x_2'x_2Q_2 + x_3'x_3Q_3 + x_4'x_4Q_4 = 0;$$

where $Q_1 = 0$ is the ray-cone of A_1, etc.; since in this case A_1 is a fourfold point on each surface (P), the cones $Q_2 = 0$, $Q_3 = 0$, $Q_4 = 0$ each contain x_1 in the first degree only (Art. 257); moreover since Q_2 passes through A_3 it cannot contain x_3^3, thus the result of putting $x_4 = 0$ in Q_2 gives merely a term $a_2x_1x_3^2$; similarly from Q_3 arises $a_3x_1x_2^2$; hence the null-plane of a point P in $x_4 = 0$ is

$$x_1'Q_1 + x_1x_2x_3(a_2x_3x_2' + a_3x_2x_3') = 0.$$

The trace of this plane on $x_1' = 0$ is

$$a_2x_3x_2' + a_3x_2x_3' = 0,$$

i.e. as in the previous case, the null-planes of the points of a line of the pencil (A_1, α_4) pass through a line of the pencil (A_4, α_1) and hence *the rays of the congruence belong to a tetrahedral complex of which the point S_4 and the three points S_3 form the fundamental tetrahedron.*

The congruence (2, 4). Take as vertices A_1 and A_3 of reference the two points S_3, and two non-conjugate points S_2 as A_2 and A_4; the equation of the null-plane of any point x is then

$$x_1'Q_1 + x_2'\beta Q_2 + x_3'Q_3 + x_4'\delta Q_4 = 0.$$

Since A_2 and A_4 are non-conjugate points, Q_2 contains a term x_4^2, and Q_4 a term x_2^2; as before β and δ pass through A_1A_3, moreover $\beta \equiv x_2$, $\delta \equiv x_4$, for if β contained a term x_4, then in the identity

$$x_1Q_1 + x_2\beta Q_2 + x_3Q_3 + x_4\delta Q_4 \equiv 0,$$

a term $x_2x_4^3$ would arise which could not be cancelled, similarly for δ.

Again the term which does not involve x_4 in Q_2 is bx_1x_3, in Q_3 it is $ax_1x_2^2$, therefore the null-plane of any point in $x_4 = 0$ is

$$Q_1x_1' + bx_1x_2x_3x_2' + ax_1x_2^2x_3' = 0.$$

The trace of this plane on $x_1' = 0$ is $bx_3x_2' + ax_2x_3' = 0$, *i.e.* the system is contained in a tetrahedral complex which has the tetrahedron of reference as its fundamental tetrahedron.

A pair of non-conjugate singular points S_2 can be chosen in three ways; for through either point S_3 and a point S_2 there passes one singular plane σ, viz. that of the pencil required to complete the degree of (S_2S_3), let S_1 be the centre of this pencil; then in σ there is one other point of the second order $\overline{S}_2$ and two points of the first order (Art. 256); each of these latter three points is non-conjugate to S_2; also the plane σ' through S_2, $\overline{S}_2$ and the other point of the third order S_3 is singular since it contains at least five rays; let the centre of the pencil of rays in it be S_1', then in σ' there are two additional points of the first order each non-conjugate to S_2, therefore the points non-conjugate to S_2 are $\overline{S}_2$ and four points S_1; thus S_2 being conjugate to eight points (Art. 256), must have four points S_2 conjugate to it, *i.e. to each point S_2 there is one other point S_2 non-conjugate to it*; this gives three pairs of non-conjugate points S_2, hence the congruence (2, 4) is contained in three tetrahedral complexes.

The congruence (2, 3). Since all points S_2 are here conjugate, and each cone (S_2) contains eight singular points exclusive of the vertex, such a cone must contain four points S_1. Any two points S_2, S_2' have one point S_1 conjugate to each of them, viz., the centre of the pencil required to complete the degree of (S_2S_2'); it follows that three cones S_2 together contain nine points S_1, and hence that there is one point S_1 non-conjugate to any three points S_2. Since these three points may be chosen in 10 ways, there are 10 tetrahedra whose vertices are three points S_2 and a point S_1 non-conjugate to them.

The null-plane of any point x for such a tetrahedron is

$$x_1'Q_1 + x_2'Q_2 + x_3'Q_3 + x_4'\alpha\beta = 0,$$

where $\alpha = 0$ is the null-plane of S_1, and has the form

$$Ax_1 + Bx_2 + Cx_3 = 0.$$

Since $\alpha\beta$ can only contain x_1, x_2, x_3 in the first degree (Art. 257), it follows that $\beta \equiv x_4$: the part of Q_2 which does not involve x_4 is ax_1x_3 and that of Q_3 is bx_1x_2; hence the trace of the null-plane of any point in $x_4 = 0$, upon $x_1' = 0$, is $ax_3x_2' + bx_2x_3' = 0$, which shows that the congruence is contained in a tetrahedral complex whose fundamental tetrahedron is that of reference; it follows from the foregoing that *the congruence* (2, 3) *is contained in* 10 *tetrahedral* *complexes.* It will be seen in the following chapter that the congruence (2, 2) is contained in 40 tetrahedral complexes*.

259. Non-conjugate singular points. If S_h, $S_{h'}$ are non-conjugate singular points, the curve of intersection of (S_h) and $(S_{h'})$ meets the focal surface in singular points only, since through such a point of the focal surface there pass two non-consecutive rays; the number of such points is $2hh'$, since the order of the curve of intersection of (S_h) and Φ is $2h$.

Taking the two singular points as being each of the first degree, we observe that on the line of intersection of the null-planes of two non-conjugate points S_1 and S_1' there are two singular points S_h, $S_{h'}$ (say), and the rays through any point P of $S_hS_{h'}$ being PS_1, PS_1' it follows that $(S_hS_{h'})$ consists of (S_h), $(S_{h'})$ and the pencils whose centres are S_1 and S_1'; therefore $h+h'=n$. The points S_h, $S_{h'}$ are non-conjugate, since, if $S_hS_{h'}$ were a ray, three rays would pass through each of its points.

If S_h and $S_{h'}$ are any two non-conjugate singular points the surface $(S_hS_{h'})$ breaks up into two surfaces; for the surface (P) for S_h consists of (S_h) together with a surface Q of degree $n-h-1$, and Q contains $\frac{1}{2}(n-2)(n-3)-\frac{1}{2}(h-1)(h-2)$ double rays. Similarly the surface (P) for $S_{h'}$ consists of $(S_{h'})$ together with a surface Q' of degree $n-h'-1$. In the next place we observe that the curve $|l|$ for $S_hS_{h'}$ consists of the intersection of Q and Q' apart from the double rays,

$$\tfrac{1}{2}(n-2)(n-3)-\tfrac{1}{2}(h-1)(h-2)-\tfrac{1}{2}(h'-1)(h'-2)$$

in number, which Q and Q' have in common; thus $|l|$ is of the degree

$$(n-h-1)(n-h'-1)-\tfrac{1}{2}(n-2)(n-3)$$
$$+\tfrac{1}{2}(h-1)(h-2)+\tfrac{1}{2}(h'-1)(h'-2),$$

which is equal to

$$\tfrac{1}{2}(n-h-h'+2-1)(n-h-h'+2-2).$$

Now $|l|$ is a double curve on $(S_hS_{h'})$, which latter surface is of the degree $n-h-h'+2$, after subtraction of (S_h) and $(S_{h'})$, but l is also a part of the double curve of $(S_hS_{h'})$, *i.e.* this surface possesses a double curve whose order is greater at least by unity than that possible for the double curve of a surface whose degree is

$$n-h-h'+2.$$

Therefore $(S_hS_{h'})$ must break up into two surfaces.

* See also Arts. 123, 146.

It follows from this result that for two non-conjugate points S_h, $S_{h'}$ such that $h+h'=n$, there are two points S_1 conjugate to S_h and to $S_{h'}$; since $(S_h S_{h'})$ which is here of degree 2 must split up into two plane pencils. We notice that if two points S_1 are both conjugate to the same S_h they must be non-conjugate to each other.

260. Reguli of the congruences (2, n). A regulus of rays is formed by such as intersect a line l which passes through a point S_{n-1} and lies in a singular plane σ; the various lines of the pencil (S_{n-1}, σ) give rise to ∞^1 reguli of the system of lines.

In $(2, 6)_I$ the cone (S_5) contains twelve, *i.e.* *all* the singular points, and hence passes through the single point S_1; thus through S_5 there passes one singular plane, giving ∞^1 reguli formed by rays of the congruence.

In (2, 5) the cone (S_4) passes through two of the three points S_1.

In (2, 4), (2, 3), (2, 2) through each point S_{n-1} there pass several singular planes.

In the case of each system of reguli, two of these reguli pass through a given point P, viz. those determined by the lines l, l' in which the two rays through P meet the pencil (S_{n-1}, σ); any plane touches n of these reguli, viz. those determined by the n rays which lie in the plane. Thus the reguli determined by the pencil (S_{n-1}, σ) are contained in a "net," *i.e.* consist of the ∞^1 quadrics $\lambda^2 u+\lambda v+w=0$, where u, v, w are given quadrics. In fact we will now show that each regulus corresponding to the pencil (S_{n-1}, σ) passes through eight points, which are the following:—*the point S_{n-1}, the pole S_1 of σ, all points S_3, all points S_2 not in σ, all points S_1 not in σ nor upon (S_{n-1})*; for each regulus clearly passes through S_{n-1}, and at the point P in which l meets the conic of contact of σ and Φ the rays coincide, and one of them must be S_1P, hence the regulus passes through S_1; again each cone (S_3) passes through S_{n-1} and therefore meets l in one other point, *i.e.* the regulus passes through each point S_3; the same remark applies to each S_2 not in σ, and lastly the null-plane of each S_1 not in σ nor conjugate to S_{n-1} meets l in one point which is distinct from S_{n-1}; thus each regulus determined by a line l of the pencil (S_{n-1}, σ) passes through the points which have been stated. That the number of these points is eight may be seen as follows: the number of points S_3 is $\frac{1}{2}(n-2)(n-3)$, and it is easily seen from the Table of Art. 254 that the total number of points S_2 is $(n-2)(7-n)$*, and therefore the number of points S_2 not in σ is $(n-2)(7-n)-(n-2)$, *i.e.* $(n-2)(6-n)$; none of the $6-n$ points S_1 in σ besides its null-point lie upon (S_{n-1}); again (S_{n-1}) contains $3n-\frac{1}{2}(n-2)(n-3)-1$ singular points exclusive of the vertex (Art. 255), deducting the points S_3 and S_2 we obtain the number of points S_1 upon (S_{n-1}) as being

$$3n-\tfrac{1}{2}(n-2)(n-3)-1-\tfrac{1}{2}(n-2)(n-3)-(n-2)(7-n)=7-n;$$

it is seen from the Table that the number of points S_1 is

$$\tfrac{1}{2}(7-n)(8-n)\dagger,$$

and hence the number of points S_1 not upon (S_{n-1}) nor in σ is

$$\tfrac{1}{2}(7-n)(8-n)-(6-n)-(7-n)=\tfrac{1}{2}(n-5)(n-6);$$

thus the total number of the specified points through which each regulus passes is

$$1+1+\tfrac{1}{2}(n-2)(n-3)+(n-2)(6-n)+\tfrac{1}{2}(n-5)(n-6)=8.$$

When $n=4$ it is seen from the foregoing that there is one point S_1' which is non-conjugate both to the given S_1 and to the S_{n-1} (S_3 in this case); its null-plane σ' is therefore a tangent plane of each regulus ρ, hence the trace of ρ upon σ' is a line l' and l' must therefore contain the point S_3'; now each regulus ρ determines one line l of (S_3, σ) and also one line of (S_3', σ'), therefore these two plane pencils are in (1, 1) correspondence; whence the theorem follows that (2, 4) is contained in a tetrahedral complex; this has been already proved in Art. 258. The four vertices of the fundamental tetrahedron are S_3, S_3' and the two singular points on the line (σ, σ'), which must be two non-conjugate points S_2.

Again when $n=3$ there are three points S_1 which are non-conjugate to either the given S_1 or to the given S_2 (S_{n-1} is here S_2); the null-planes of each of these three points S_1 touch each of the ∞^1 reguli ρ, let S_1' be one of them and σ' its null-plane, then ρ meets σ' in a line containing a singular point S_2', also on the line (σ, σ') there are two singular points $\overline{S}_2$ and S_1^0 (Art. 259), then we know there are only two points of the second degree in σ', viz. $\overline{S}_2$ and S_2', and S_1^0 is non-conjugate to S_2, S_2' and $\overline{S}_2$: hence as before, the congruence (2, 3) is contained in a tetrahedral complex whose tetrahedron has for vertices three points S_2 and a point S_1 non-conjugate to them; and the pencils (S_2, σ), (S_2', σ') are made

* Except in $(2, 6)_{II}$. † Except for $n=2$, or $(2, 6)_{II}$.

projective by the reguli ρ. Similarly for the two other points S_1'', S_1''' non-conjugate to S_2 or S_1. Thus the ∞^1 reguli determine on four planes, pencils which are mutually projective; again as before on the line $(\sigma\sigma'')$ there is a singular point of the second degree conjugate to S_1 and S_1'' which must be $\bar{S}_2$, since σ only contains two points of the second degree; hence the four planes $\sigma, \sigma', \sigma'', \sigma'''$ meet in $\bar{S}_2$. Taking any three of the four mutually projective pencils, which have been seen to have three corresponding lines meeting in $\bar{S}_2$, the congruence (2, 3) may be defined as *the ∞^2 lines which meet corresponding lines of three projective pencils having three corresponding lines concurrent.*

If this latter condition be not fulfilled a congruence (3, 3) is obtained; for let the sections by any plane a of the planes of three given projective pencils be a, b, c, then upon a, b, c are determined three projective rows of points $P\ldots$, $Q\ldots$, $R\ldots$; and having given any point R of c, the join of the corresponding points P, Q of a and b meets c in a fourth point R', *i.e.* we have upon c a correspondence (1, 2); for having given R then R' is uniquely determined, but having given R' there are *two* points R, viz. those corresponding to the two pairs P, Q; P', Q' where $R'PQ$, $R'P'Q'$ are the two tangents through R' to the conic enveloped by lines joining corresponding points on a and b. This correspondence (1, 2) has three *united points*, *i.e.* in a there lie three lines of the system.

Similarly if any point P be joined to the centres A, B, C of the three pencils we have determined three projective pencils of planes, and as before on the axis PA we have a correspondence (1, 2) of planes, hence through P there pass three lines of the system.

261. In $(2, 6)_\text{I}$, (2, 5), (2, 4), (2, 3) we have therefore the following sets of ∞^1 reguli:—

one set of ∞^1 reguli in $(2, 6)_\text{I}$;

two sets of ∞^1 reguli in (2, 5), viz. one for each of the two points S_1 conjugate to S_4;

three sets of ∞^1 reguli in (2, 4); for on a cone (S_3) lie three points S_1, as is seen by deducting from the eleven points on (S_3) the two points S_3 and the six points S_2, and on the other cone (S_3') lie the three other points S_1'; thus S_3 and the three points S_1 determine three systems of ∞^1 reguli; each of these systems passes through S_3' and *one* of the points S_1', so that there is no other system of reguli than these three;

five sets of ∞^1 reguli in (2, 3); for it has been seen, (Art. 260), that the group of eight points associated with a set of ∞^1 reguli are the singular points which lie on a cone $(\bar{S}_2)$; thus we obtain five groups of eight associated points.

262. *Reguli of* $(2, 6)_{II}$. Since the system $(2, 6)_{II}$ has no singular plane its reguli are determined by entirely different considerations. In the first place it is to be noticed that the singular points of the second degree fall into the two groups S_2', S_2'' such that to each point S_2' the four points S_2'' are conjugate, and to each point S_2'' the four points S_2' are conjugate. To see this, we observe that to a point S_2' there are four conjugate points of the second order S_2'', since on (S_2') there lie eight singular points excluding S_2' and four of these are the points S_4, while the singular points common to any two cones (S_2') are eight in number, (Art. 259), *i.e.* are the *same* points, viz. the four points S_4 and the points S_2''.

This proves the required result. Thus through the eight points S_4 and S_2'' pass four quadric cones, the (S_2'); hence through them there will pass ∞^2 quadrics, similarly for the eight points S_4 and S_2'.

The ∞^2 quadrics through eight points, which have an equation of the form $\lambda u + \mu v + \nu w = 0$, where λ, μ, ν are variable and $u = 0$, $v = 0$, $w = 0$ are any three given quadrics through the given points, are said to form a 'net.'

It will now be shown that *the generators of the quadrics of a net form a cubic complex which includes the eight sheaves whose centres are the eight fundamental points of the net.* For, ∞^1 quadrics of the net pass through any point P and form a 'pencil' of quadrics of the form $f + \rho\phi = 0$; each quadric of the pencil has a generator through P, which meets the curve $f = 0$, $\phi = 0$ in one other point besides P, *i.e.* these ∞^1 generators are those of the cone which projects this curve from P, the cone is therefore *cubic*; hence in any pencil there are three lines of the complex, which is therefore cubic. Since there are ∞^2 quadrics through each of the eight fundamental points, every line through these points belongs to the cubic complex.

The congruence $(2, 6)_{II}$ is contained in each of the cubic complexes $\{S_4, S_2'\}$, $\{S_4, S_2''\}$; for if not, it will have in common with each of them a ruled surface of degree $3(2+6) = 24$, (Art. 234); but the congruence $(2, 6)_{II}$ has in common with the cubic complex $\{S_4, S_2'\}$ ruled surfaces whose degrees together amount to more than 24, *e.g.*, the four cones (S_2'') and the 12 reguli which are the loci of rays intersecting the join of two points S_2'', (Art. 259): similarly, the congruence is contained in the cubic complex $\{S_4, S_2''\}$.

Now it was seen that the congruence $(2, 6)_{II}$ is contained in a tetrahedral complex T^2, (Art. 258); hence with the four sheaves whose centres are the points S_4, it is the complete intersection of T^2 and either $\{S_4, S_2'\}$ or $\{S_4, S_2''\}$.

Now T^2 has ∞^4 reguli which pass through the points S_4 (Art. 94), and of these ∞^1 will pass through three points S_2', but a quadric through seven fundamental points of a net will also pass through the eighth point*; hence these ∞^1 reguli of T also pass through the fourth point S_2', and therefore belong to the cubic complex (S_4, S_2'). We thus obtain ∞^1 reguli of $(2, 6)_{II}$. In the same way there are ∞^1 reguli of $(2, 6)_{II}$ which pass through the eight points S_4 and S_2''. Thus there are two sets of ∞^1 reguli of $(2, 6)_{II}$.

263. Confocal congruences. The class of any plane section of Φ or the degree of the enveloping cone of Φ is 12, since Φ possesses no double curve.

This cone has 24 cuspidal edges†, and since the class of the surface and also of the cone is $2n$, if δ is the number of double edges of the cone, we have

$$12 \times 11 - 3 \times 24 - 2\delta = 2n,$$

hence

$$\delta = 30 - n.$$

Of these double edges $18 - n$ pass through the double points of Φ, since the curve of contact of the enveloping cone, being the intersection of Φ and the first polar of Φ for the vertex of the cone, will have two branches through each double point. Deducting these double edges there remain twelve, which is therefore the number of double tangents of Φ which pass through any point.

Any plane section of Φ has 28 double tangents, and if N' be the number of singular tangent planes of Φ, it is clear that $28 - N'$ is the number of double tangents of Φ which lie in any plane, excluding the lines of the singular tangent planes which are not proper double tangents of Φ. Thus the complete system of double tangents of Φ forms a congruence $(12, 28 - N')$; of this congruence the given $(2, n)$ forms part, leaving after its removal a congruence of double tangents $(10, 28 - N' - n)$.

Now since for

	$(2, 3)$,	$(2, 4)$,	$(2, 5)$,	$(2, 6)_I$,	$(2, 6)_{II}$,	$(2, 7)$
$N' =$	10,	6,	3,	1,	0,	0,

* Salmon, *Geom. of Three Dimensions*, Art. 131. † Salmon, Art. 279.

we obtain in these respective cases, *residual congruences* of double tangents which are

(10, 15), (10, 18), (10, 20), (10, 21), (10, 22), (10, 21):

these will now be investigated.

The origin of these additional systems of double tangents is explained in part by the following theorem: *the complementary reguli* ρ' *of the reguli* ρ *of* (2, n) *determine a congruence which has* Φ *for its focal surface.* The truth of this appears from the fact that since each generator of a regulus ρ touches Φ twice, ρ touches Φ along a curve k which must be of the fourth order; thus the plane through any generator of ρ and any generator of ρ' meets k in four points of which two points lie on the generator of ρ, and therefore two points on the generator of ρ'; hence each generator of ρ' meets k twice; at each of these latter points the generator of ρ' lies in the common tangent plane of ρ and Φ, *i.e. each generator of* ρ' *touches* Φ *twice.* The congruence formed by the generators of the reguli ρ' is of the second order and nth class, since as many reguli ρ' pass through a given point as reguli ρ, (Art. 260); similarly the congruences formed by generators of the reguli ρ and of the reguli ρ' have the same class. In this way the systems of reguli ρ possessed by the system (2, n) give rise to ∞^2 double tangents (generators of the ρ'), arranged as follows:

for	$(2, 6)_{\text{I}}$	$(2, 6)_{\text{II}}$	(2, 5)	(2, 4)	(2, 3)
a congruence	(2, 6)	(4, 12)	(4, 10)	(6, 12)	(10, 15).

The congruence confocal with (2, 3) is thus accounted for; in the other cases there remain ∞^2 double tangents of Φ which do not belong to the given (2, n), nor are generators of ρ', and which form respectively,

in	(2, 7)	$(2, 6)_{\text{I}}$	$(2, 6)_{\text{II}}$	(2, 5)	(2, 4)
the congruence	(10, 21)	(8, 15)	(6, 10)	(6, 10)	(4, 6).

It will now be shown that these systems are formed, in all cases except $(2, 6)_{\text{II}}$, *by the single directrices* l' *of the ruled cubics* (l), *where* l *is a line of the sheaf whose centre is* S_{n-1}. For, each generator of such a cubic surface ρ^3 touches Φ twice, hence ρ^3 and Φ touch along a curve, so that at a point of intersection of l' and Φ, l' must lie in the tangent plane to Φ at the point, *i.e.* l' touches Φ twice.

Now the two rays of (2, n) through any point P determine a surface ρ^3 whose double directrix is $S_{n-1}P$; hence the $\frac{1}{2}n(n-1)$

pairs of rays of (2, n) which lie in any plane determine as many surfaces ρ^3, whose single directrices l' lie in the plane, *i.e.*, *the class of the congruence of lines* l' *is* $\frac{1}{2}n(n-1)$.

Again it was seen, (Art. 257), that each surface (P) contains $2(n-2)$ lines through S_{n-1}, and the null-plane of every point on such a line passes through P, hence the single directrix of the surface ρ^3 for such a line passes through P, *i.e.*, *through* P *there pass* $2(n-2)$ *lines of the congruence* l'. The ∞^2 double tangents l' thus form a congruence $\{2(n-2), \frac{1}{2}n(n-1)\}$, *i.e.* they form the residual congruence of double tangents after deduction of the given (2, n) and the generators of the reguli ρ'.

The case of $(2, 6)_{II}$, in which there is a residual congruence (6, 10), remains to be discussed. The congruence $(2, 6)_{II}$ is contained in a tetrahedral complex T^2, the vertices of whose fundamental tetrahedron are the points S_4, (Art. 258); now there are ∞^3 twisted cubics r passing through the points S_4 all the chords of which belong to T^2, (Art. 95); these chords for any cubic r form a system (1, 3), which will have in common with the cubic complex (S_4, S_2') a ruled surface of degree $3(1+3)=12$, (Art. 234); but this surface is in part composed of the four cones of the second degree (S_4, r); there remains after their removal a ruled quartic; hence each of the ∞^3 chord-congruences of T^2 contains one ruled quartic of $(2, 6)_{II}$. Each generator of such a quartic meets r twice, and through each point P of r proceed two such generators, viz., the intersections of the cone (P, r) and the cubic complex (S_4, S_2'), excluding the four lines joining P to the points S_4; hence r is a double curve of the quartic which is thus the general quartic of class III. Now since any three lines p, p', p'' of T^2 determine a twisted cubic* r, we see that any three rays of $(2, 6)_{II}$ determine such a ruled quartic.

If moreover these three rays are coplanar, the ruled quartic possesses also a single directrix l', (Art. xvi), and is of class IV; one plane of the pencil whose axis is l' passes through any given point A, *i.e.* such a quartic of class IV has three generators in the plane (l', A), and since there are ∞^2 planes through A, there are ∞^2 ruled quartics of $(2, 6)_{II}$ which have also a single directrix l'. As before this quartic touches Φ and l' touches Φ twice, so that the

* For three of the points S_4 make the three pencils of planes whose axes are p, p', p'' projective to each other, and the locus of intersection of three corresponding planes is a twisted cubic, (Art. xii), which passes through the fourth point S_4 and whose chords belong to T^2.

required system of ∞^2 double tangents is formed by the single directrices of the ∞^2 ruled quartics of class IV which belong to $(2, 6)_{II}$.

If l' is the single directrix determined by three coplanar rays of $(2, 6)_{II}$, the surface (l') breaks up into the aforesaid quartic of class IV and another ruled quartic which must also be of class IV. Now sets of three rays can be made in twenty ways out of the six rays in any plane; hence there are determined twenty such ruled quartics but only ten simple directrices l', since each l' belongs to two quartics; hence *the class of the system of lines l' is ten.*

And its order must be six, for if it were less than six, one or more sheaves of double tangents would exist, which is not the case. Hence for $(2, 6)_{II}$ the double tangents of Φ consist of $(2, 6)_{II}$, the generators of the two sets of reguli ρ', and the single directrices of the ruled quartics of class IV which belong to $(2, 6)_{II}$.

CHAPTER XVI.

THE CONGRUENCE OF THE SECOND ORDER AND SECOND CLASS.

264. THE congruence (2, 2), of the second order and class, is the one of the series $(2, n)$ which has been most fully investigated: an account of this congruence is given in the present chapter. From Arts. 249, 254 it follows that the congruence (2, 2) has a focal surface of the fourth degree and fourth class, which has 16 double points and 16 singular tangent planes, *i.e.* is a Kummer's Surface. It will be shown, (Art. 265), that any congruence (2, 2) is the complete intersection of a linear and a quadratic complex, and is therefore identical with the congruence (2, 2) already discussed in Chapter VIII.

To each point of space one plane is assigned by the system (2, 2), viz. that of the two rays through the point, while to each plane one point is assigned, viz. the intersection of the rays in the plane, so that by the system (2, 2) an involutory reciprocity is established in which corresponding elements are united, *i.e.* a linear complex is determined, (Art. 37), to each pencil of which two rays belong; therefore each system (2, 2) is contained in a linear complex C_1; and only one such linear complex is thus related to any given congruence (2, 2).

This result may also be seen as follows:—let S and S' be two conjugate singular points, (Art. 256), σ, σ' their null-planes, and l any line of the pencil (S', σ); then the surface (l) consists of the pencils (S, σ), (S', σ'), together with a regulus ρ which has one generator belonging to the pencil (S, σ) and one to (S', σ'). If any generator of ρ meets σ' in P, the line SP, since it meets three lines of ρ, is a directrix of ρ, and therefore σ' is a tangent plane to ρ, *i.e.* the trace of ρ on σ' is the line SP. Since all the rays of the system can be grouped into such reguli, together with the pencils

(S, σ), (S', σ'), it is clear that they establish a (1, 1) correspondence on the lines of the pencils (S', σ), (S, σ'), viz. that of pairs similar to l and SP, and in this correspondence SS' corresponds to itself; hence the rays are included in a linear complex, (Art. 39).

This may again be seen from the fact that the surfaces (P) are of the form

$$x_1'P_1 + x_2'P_2 + x_3'P_3 + x_4'P_4 = 0,$$

with the identity

$$x_1P_1 + x_2P_2 + x_3P_3 + x_4P_4 \equiv 0;$$

but since the P_i are linear in x_i

$$P_i = a_{i1}x_1 + a_{i2}x_2 + a_{i3}x_3 + a_{i4}x_4,$$

i.e. we must have $a_{ii} = 0$, $a_{ik} + a_{ki} = 0$, hence (P) has the form

$$\Sigma a_{ik}(x_i x_k' - x_i' x_k) = 0,$$

which is the bilinear equation connecting two points of a line of a linear complex.

265. Confocal congruences (2, 2). The equation of any Kummer surface has been seen, (Art. 85), to be reducible to the form

$$A\Sigma y_1^4 + 2B(y_1^2y_2^2 + y_3^2y_4^2) + 2C(y_1^2y_3^2 + y_2^2y_4^2) + 2D(y_1^2y_4^2 + y_2^2y_3^2) + 4E\, y_1y_2y_3y_4 = 0;$$

while its double points and singular tangent planes form a system described in Art. 29, such that each point is the pole for six complexes mutually in involution of the six singular planes through it, and the six points in each singular plane are the poles of the plane for these six complexes; hence the singular points and singular planes of the congruence (2, 2), being the double points and singular tangent planes of a Kummer surface, form such a system. In each singular plane there is one pencil of rays which therefore belongs to the linear complex C_1 of the congruence, and thus C_1 is *one* of the above six linear complexes in involution.

Moreover the double tangents which belong to C_1 belong also to a quadratic complex, (Art. 83); hence, *any congruence* (2, 2) *is the complete intersection of a linear with a quadratic complex.*

Now it was seen that the double tangents of a Kummer surface form six congruences (2, 2), (Art. 83), hence, as we have already seen, (Art. 126), associated with any congruence (2, 2) there are five others having the same focal surface as the given congruence (2, 2).

Certain properties of the singular points and planes of a congruence (2, 2) will now be investigated, taking as starting point the fact that any congruence (2, 2) is contained in a linear complex C_1; and that its singular points and planes form a system described in Art. 29 for six linear complexes in mutual involution, of which C_1 is one.

266. Distribution of the Singular Points. Any two singular points S_1 and S_2 are conjugate in *two* of the six fundamental linear complexes associated with the focal surface; for S_2 must be contained in one of the six singular planes π through S_1; let S_1 be the pole for C_i of this plane π and S_2 for C_j, then from the involution of C_i and C_j there is one plane π' through S_1S_2 for which S_1 is the pole for C_j and S_2 the pole for C_i; thus S_1S_2 belongs to each of the complexes C_i and C_j, and it is seen that through the join of any two singular points there pass *two* singular planes.

Any three singular points S_1, S_2, S_3, which are non-conjugate to each other in any particular complex C_i, must lie in a singular plane; for let S_1 and S_2 be conjugate in C_j and C_k, S_2 and S_3 in C_l and C_m, S_3 and S_1 in C_n and C_r, then all these six complexes cannot be different from C_i unless at least two of them are the same, *e.g.* $C_j \equiv C_n$, here S_1S_2 and S_1S_3 belong to C_j, *i.e.* S_1 is the pole of $S_1S_2S_3$ for C_j, hence the plane $S_1S_2S_3$, or π, is a singular plane. Moreover in this case S_2 is the pole of π for C_k and S_3 the pole of π for C_r, *i.e.* S_2 and S_3 are conjugate in C_k and C_r. It follows that there is a plane π' through S_2S_3 for which S_2 is the pole for C_r and S_3 the pole for C_k, and S_1, S_2, S_3 form a system described in Art. 26, in which the three singular planes through S_1S_2, S_2S_3, S_3S_1 respectively, distinct from π, meet in a singular point S_4, which is the pole of these respective planes for C_r, C_j, C_k. Hence the four points S_1, S_2, S_3, S_4 are mutually non-conjugate for C_i.

There is no point outside π, except S_4, which is non-conjugate to S_1, S_2, S_3 for C_i; for if S were such a point then the plane SS_1S_2 would be a singular plane, *i.e.* S must lie in the second singular plane through S_1S_2, similarly it must lie in the other singular planes through S_2S_3, S_3S_1, *i.e.* it must coincide with S_4.

The number of tetrahedra whose vertices are non-conjugate to each other for C_i is 40; for in any singular plane three points of the system non-conjugate in C_i may be selected in 10 ways, and each selection determines a fourth point outside the plane non-conjugate to them in C_i; thus the number of such tetrahedra is $10 \times 16 = 160$, but in this process each tetrahedron occurs four times, hence the number of the required tetrahedra is 40.

267. Every (2, 2) is included in 40 tetrahedral complexes*. Let the points S_1, S_2, S_3, S_4 be non-conjugate in C_1, and denote by σ the singular plane $S_1S_3S_4$ having S as its pole in C_1, and by σ' the singular plane $S_2S_3S_4$ having S' as its pole in C_1; take also any line S_1P of the pencil (S_1, σ); then of the two rays from any point of S_1P one belongs to (S, σ) and the other to a regulus ρ. The line $S'P$ belongs to ρ, hence σ' is a tangent plane of ρ; also since S_1 and S_2 are non-conjugate, the null-plane of S_2 will meet S_1P in a point different from S_1, so that ρ passes through S_2, and the trace of ρ on σ' is $S'P$ and a line through S_2; hence *the reguli ρ make the pencils (S_1, σ), (S_2, σ') projective, these reguli therefore belong to a tetrahedral complex.*

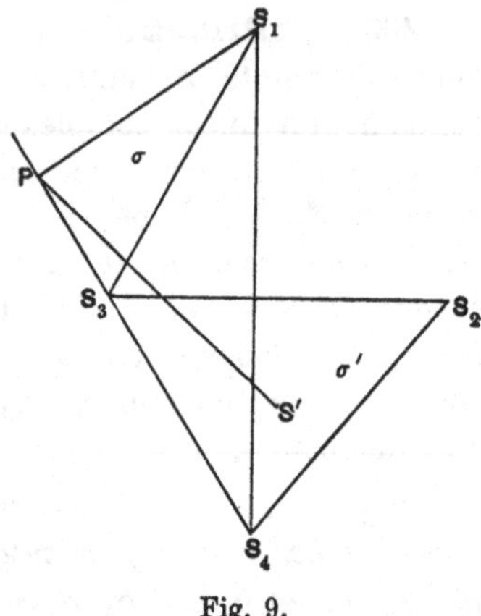

Fig. 9.

The other two vertices of the fundamental tetrahedron must be S_3 and S_4, since when P coincides with one of them, the regulus (S_1P) must break up into two pencils of which one has P for centre, *i.e.* these vertices must be S_3 and S_4.

Each set of four points mutually non-conjugate in C_1 gives rise to a tetrahedral complex to which the given system (2, 2) belongs, hence any (2, 2) is contained in 40 tetrahedral complexes.

268. The Kummer Configuration. The closed system of 16 points and planes determined by six complexes mutually in involution has been already investigated, (Art. 14); a table showing the configuration of the system can now be constructed. The following notation is due to Weber†; denote by (1) any singular plane and let its poles in $C_1, \ldots C_6$ be denoted respectively by 0, 12, 13, 14, 15, 16; let also the null-plane of 0 in C_2 be (2), then, from the involution of C_1 and C_2, 12 will be the pole of (2) in C_1;

* See Arts. 123, 146. † *Crelle's Journal*, Bd. LXXXIV. S. 349.

similarly for the planes (3), (4), (5), (6) the null-planes of 0 in C_3, C_4, C_5, C_6 respectively. Again denote by 23 the pole of (2) in C_3, then since 0 is the pole of (2) in C_2 and of (3) in C_3 it follows that 23 is the pole of (3) in C_2; similarly for the points 24, 25, 26, the poles of (2) in C_4, C_5, C_6 respectively. By a process identical with the preceding, 34, 35, 36, 45, 46, 56 will denote similar points, and the first six columns of the table are completed. All the singular points have now been accounted for. To complete the table, denote by (123) the plane whose pole in C_1 is 23, etc.; the singular planes are now all designated and the top row completed.

Again from consideration of the columns beneath (3) and (123), it is clear, from the involution of the complexes, that the second place in the latter column must be filled up by 13, and from comparing (2) and (123) that the third place is occupied by 12; in this way to the points 12, 13, 14, 15, 16 are assigned their places in each of the remaining columns, each of which has now three places occupied. Lastly, the plane (123) which contains 12, 13, 23 can contain no other point of which one member is 1, 2 or 3, *e.g.* if it contained 14 it would have three points in common with (1), if it contained 24 it would have three points in common with (2) and so on, hence it can only contain 45, 46, 56 whose places are at once determined from consideration of the involution of the complexes and the arrangement of the first six columns; thus the column below (123) is filled up; similarly for each of the other singular planes.

	(1)	(2)	(3)	(4)	(5)	(6)	(123)	(124)	(125)	(126)	(134)	(135)	(136)	(145)	(146)	(156)
I	0	12	13	14	15	16	23	24	25	26	34	35	36	45	46	56
II	12	0	23	24	25	26	13	14	15	16	56	46	45	36	35	34
III	13	23	0	34	35	36	12	56	46	45	14	15	16	26	25	24
IV	14	24	34	0	45	46	56	12	36	35	13	26	25	15	16	23
V	15	25	35	45	0	56	46	36	12	34	26	13	24	14	23	16
VI	16	26	36	46	56	0	45	35	34	12	25	24	13	23	14	15

269. The Weber groups. It may be shown that if six points selected in a certain manner from a Kummer configuration be given, the remaining points and planes of the system are determined. The following considerations will make this clear. If P, Q, R be any three points of the system, P and Q are conjugate in two complexes, say C_α and C_β, while P and R

are also conjugate in two complexes, of which either *one* complex or *neither* complex is the same as C_α or C_β. In the first case let C_α and C_γ be the complexes in which P and R are conjugate, then P, Q, R lie in a plane of the system, viz., in that for which P is the pole for C_α. In the second case let P and R be conjugate in C_γ and C_δ; here P, Q, R do not lie in a plane of the system, for if so, P being then the pole of the plane in one complex, PQ, PR would both have to belong to one complex of the system, which is here not the case; it follows that Q and R are conjugate in the two remaining complexes C_ϵ, C_η.

It should also be observed that if upon the line of intersection of two planes σ, σ' of the system there are two points which are conjugate in two complexes C_β and C_γ, then the poles of σ and σ' for any third complex C_α are conjugate in C_β and C_γ; for if S, S' are the latter points and S_2 the pole of σ for C_β and of σ' for C_γ, then S, S_2 are conjugate in C_α and C_β and S', S_2 conjugate in C_α and C_γ, therefore SS' are conjugate in C_β and C_γ.

Six points S_1, S_2, S_3, S_4, S_5, S which will be seen to determine the rest of the system are now to be chosen as follows:—take *any* point S_1 and let its null-plane in C_1 be σ_1, in σ_1 let S_2, S_5 be the points whose joins to S_1 are conjugate in C_1, C_α and C_1, C_ϵ respectively; the null-plane of S_2 in C_1 being σ_2, in σ_2 take S_3 so that S_2, S_3 are conjugate in C_1 and C_β, and in σ_3, the null-plane of S_3 for C_1, take S_4 so that S_3, S_4 are conjugate in C_1 and C_γ.

We shall express the fact that any two points SS' of the system are conjugate in two complexes C_α, C_β by saying that SS' is $\overline{\alpha\beta}$; with this notation it is seen that

$$S_1S_2 \text{ is } \overline{1\alpha},\quad S_2S_3 \text{ is } \overline{1\beta},\quad S_3S_4 \text{ is } \overline{1\gamma},\quad S_5S_1 \text{ is } \overline{1\epsilon}.$$

It follows that S_2S_5 is $\overline{\alpha\epsilon}$ and therefore that S_3S_5 is $\overline{\gamma\delta}$, hence S_4S_5 must be $\overline{1\delta}$.

Thus the points $S_1 \ldots S_5$ are such that the null-plane of each for C_1 is that of the lines joining it to the two adjacent points S. Again, take the lines in σ_1 which are $\overline{1\beta}$ and $\overline{1\delta}$, and let S_β, S_δ be the other points of the system on these lines respectively; then through $S_\beta S_\delta$ pass two planes of the system of which one is σ_1 and the other is a plane which will be denoted by σ; let the pole for C_1 of σ be S. Then $S_\beta S_\delta$ being $\overline{\beta\delta}$, so also is SS_1 the join

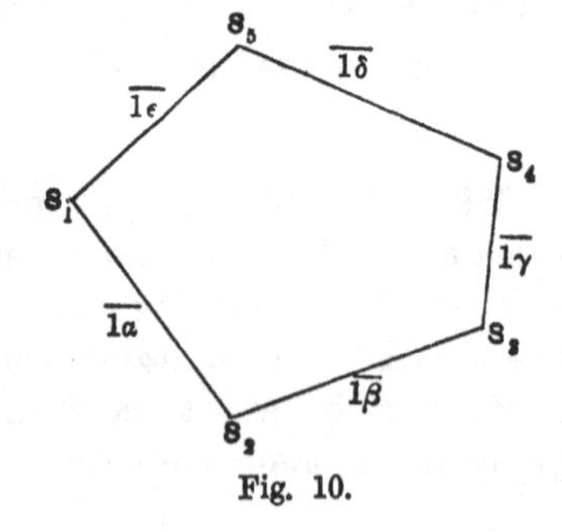

Fig. 10.

of the poles of σ and σ_1 for C_1; therefore S_1S_2 being $\overline{1\alpha}$, SS_2 must be $\overline{\gamma\epsilon}$, similarly, SS_3 is $\overline{\alpha\delta}$, SS_4 is $\overline{\beta\epsilon}$, SS_5 is $\overline{\alpha\gamma}$; thus all the points $S_1 \ldots S_5$ lie outside σ.

Again since S and S_2 are the poles of σ and σ_2 for C_1, and SS_2 is $\overline{\gamma\epsilon}$, the line of intersection of σ and σ_2 is $\overline{\gamma\epsilon}$, so that the connexion between σ_2 and σ is similar to that between σ_1 and σ, and so for all the planes $\sigma_3 \ldots \sigma_5$; hence all the points $S_1 \ldots S_5$ are related in the same manner to σ.

It will now be shown that having given the points $S, S_1, S_2, S_3, S_4, S_5$ the remaining points and planes of the system may be linearly determined; for the planes $\sigma_1, \ldots \sigma_5$ belong to the system, and so also does each plane such as SS_2S_5, for the lines SS_2, S_2S_5, S_5S being $\overline{\gamma\epsilon}$, $\overline{\alpha\epsilon}$, $\overline{\alpha\gamma}$, the points SS_2S_5 are non-conjugate in C_1 and hence the plane SS_2S_5 is a plane of the system; and so for each plane through S and the five diagonals of the pentagon $S_1 \ldots S_5$. Again SS_2 being $\overline{\gamma\epsilon}$ and SS_5 being $\overline{\alpha\gamma}$, the plane SS_2S_5 will meet σ in a line $\overline{1\gamma}$, *i.e.* in the line SS_1', if S_1' is the pole of σ for C_γ; hence $S_1'S_2$ is $\overline{1\epsilon}$ and $S_1'S_5$ is $\overline{1\alpha}$, *i.e.* S_1' lies in σ_2 and σ_5 and is therefore the point $(\sigma_2\sigma_5\sigma_1')$, if σ_1' is (SS_2S_5); similarly the other points of the system in σ are determined. Lastly, if the singular point in σ_1, not yet referred to, be denoted by $\bar{S}_1$, then $S_1\bar{S}_1$ is $\overline{1\gamma}$ and hence $S_2\bar{S}_1$ is $\overline{\alpha\gamma}$, but for σ_1 the point S_5 is the pole for C_ϵ and for σ_3' the point S is the pole for C_ϵ, hence the line of intersection of σ_1 and σ_3', (which passes through S_2), must be $\overline{\alpha\gamma}$, *i.e.* this line is $S_2\bar{S}_1$; similarly $\bar{S}_1$ lies in the intersection of σ_4' and σ_1, therefore $\bar{S}_1$ is the point $(\sigma_1\sigma_3'\sigma_4')$. The four points $\bar{S}_2$, $\bar{S}_3$, $\bar{S}_4$, $\bar{S}_5$ are similarly determined, and the positions of the 16 singular points are now known. To determine the five remaining planes, we observe that since $S_1\bar{S}_1$ is $\overline{1\gamma}$ and S_1S_3 is $\overline{\alpha\beta}$, therefore $\bar{S}_1S_3$ is $\overline{\delta\epsilon}$; also $S_3\bar{S}_3$ is $\overline{1\epsilon}$, hence $\bar{S}_1\bar{S}_3$ is $\overline{1\delta}$; similarly $\bar{S}_1\bar{S}_4$ is $\overline{1\beta}$, thus $\bar{S}_1\bar{S}_3\bar{S}_4$ is the null-plane of $\bar{S}_1$ for C_1; this plane with the four other similar planes completes the system. Observe that the relationship of the pentagon $\bar{S}_1\bar{S}_3\bar{S}_5\bar{S}_2\bar{S}_4$ to σ is similar to that of $S_1S_2S_3S_4S_5$ to σ.

Each pentagon $S_1S_2S_3S_4S_5$ determines one plane σ; if σ be given there are 12 such pentagons, for S_1 may be taken in 10 ways, and then S_2, S_5 in three ways, then S_3, S_4 in two ways giving $\frac{1}{5} \times 10 \times 3 \times 2 = 12$ ways; thus the total number of pentagons is $16 \times 12 = 192$.

270. Reguli of the congruence. It has been seen, (Art. 264), that if l is any line of the pencil (S', σ), where S' is any singular point in a singular plane σ whose pole for C_1 is S, the surface (l) is a regulus ρ, and that, if σ' is the null-plane of S' for C_1, the trace of ρ on σ' is a line l' of the pencil (S, σ'). This regulus ρ will pass through the six singular points not in σ or σ', and will have their null-planes for C_1 as tangent planes; on each of these six planes therefore, part of the trace of ρ is a line which must pass through a singular point in the plane; hence the rays of the congruence (2, 2), distributed in reguli, make eight pencils (S_i', σ_i) projective to each other.

Taking the original pencils (S, σ'), (S', σ) and any other of the remaining six pencils (S_i', σ_i); the lines which meet corresponding lines of these three pencils form the congruence: a congruence (2, 2) may therefore be regarded as *the locus of lines which meet corresponding lines of three projective pencils, of which two pencils have a common self-corresponding line.*

Another method of formation of this congruence is the following:—take any two points S_1 and S_2 which are $\overline{\alpha\beta}$, take also points S_3, S_4 such that S_1S_3 is $\overline{\beta\gamma}$ and S_2S_4 is $\overline{\beta\gamma}$, then S_1S_4 is $\overline{\alpha\gamma}$, S_2S_3 is $\overline{\alpha\gamma}$ &c. and each pair of opposite edges of the tetrahedron $S_1S_2S_3S_4$ are conjugate for the same two complexes, (Art. 26). Then, as in Art. 260, the regulus ρ, which is the surface (S_1P), passes through S_1 and the pole $\bar{S}$ of $\bar{\sigma}$ (or $S_1S_2S_4$) for C_1, and also through six other singular points of which S_3 and the pole S for C_1 of the plane σ (or $S_2S_3S_4$), are two; hence, since σ contains a line of ρ, the trace of ρ on σ is a line through S_3. These reguli ρ make the pencils $(S_1, \bar{\sigma})$, (S_3, σ) projective, and determine on S_2S_4 a (1, 1) correspondence of points of which S_2 and S_4 are the united points.

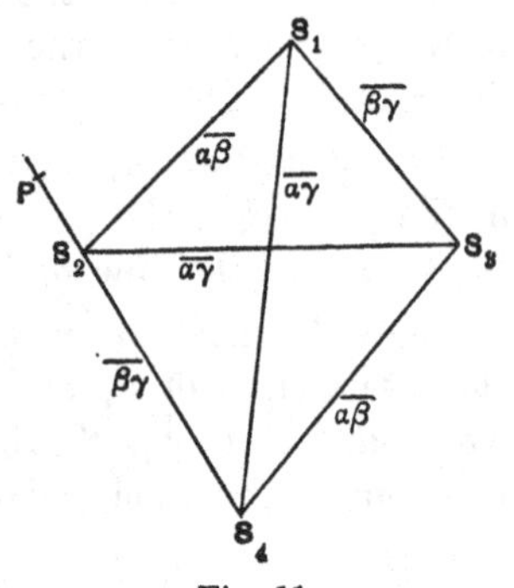

Fig. 11.

In the same manner if S_1S_3' be $\overline{\beta\delta}$ and S_2S_4' be $\overline{\beta\delta}$, then S_4' lies in $\bar{\sigma}$, and the surfaces (SP) or ρ determine on S_2S_4' a (1, 1) correspondence of which S_2 and S_4' are united points; lastly if S_1S_3'' and S_2S_4'' be $\overline{\beta\epsilon}$, S_4'' lies in $\bar{\sigma}$ and a (1, 1) correspondence is established on S_2S_4'' of which S_2 and S_4'' are united points.

Thus it is clear that the eight singular points through which the reguli ρ pass are S_1, S_3, S_3', S_3'' and $\bar{S}$, S, S', S''; the reguli make the four pencils $(S_1, \bar{\sigma})$, (S_3, σ), (S_3', σ'), (S_3'', σ'') projective to each other.

Moreover since S_1, S_2, S_3, S_3', S_3'' lie in the same plane, viz. the null-plane of S_1 for C_β, the four projective pencils have the corresponding lines S_1S_2, S_3S_2, $S_3'S_2$, $S_3''S_2$ concurrent and coplanar; hence, since three of these pencils are sufficient to determine the congruence, *a congruence* (2, 2) *is the locus of the* ∞^2 *lines which meet corresponding lines of three projective pencils which have three corresponding lines concurrent and coplanar**.

271. A congruence (2, 2) includes ten sets of ∞^1 reguli†. Taking one of the tetrahedral complexes T^2, in which the congruence is contained, as having the equation $p_{12}p_{34} - kp_{13}p_{42} = 0$, the equations

$$p_{12} = k\lambda p_{13}, \quad \lambda p_{34} = p_{42}, \quad p_{23} = \rho p_{13} + \sigma p_{14} + \tau p_{42},$$

give ∞^4 reguli which belong to T^2, the first two complexes being the special ones whose directrices are two lines of the respective pencils (A_4, α_1), (A_1, α_4). If $C \equiv \Sigma c_{ik} p_{ik}$ is the linear complex to which the congruence belongs, any of the preceding reguli of T^2 which belong to C must identically satisfy the equation $\Sigma c_{ik} p_{ik} = 0$; hence substituting in the latter equation we obtain

$$c_{13} + k\lambda c_{12} + \rho c_{23} = 0, \quad c_{14} + \sigma c_{23} = 0, \quad c_{42} + \frac{c_{34}}{\lambda} + \tau c_{23} = 0.$$

These equations show that σ is a constant, and determine ρ and τ in terms of λ, and hence give the ∞^1 reguli of the system corresponding to the two pencils (A_1, α_4), (A_4, α_1). In a similar manner we obtain five other sets of reguli, each of which passes through two vertices of the tetrahedron and touches two faces of it: six varieties of ∞^1 reguli thus arise.

Again taking as the equation of T^2

$$A(x_1^2 + x_2^2) + B(x_3^2 + x_4^2) + C(x_5^2 + x_6^2) = 0,$$

the substitution

$$x_1 = \frac{y_1}{\sqrt{A+\mu}}, \quad x_3 = \frac{y_3}{\sqrt{B+\mu}}, \quad x_5 = \frac{y_5}{\sqrt{C+\mu}},$$

$$x_2 = \frac{y_2}{\sqrt{A+\mu}}, \quad x_4 = \frac{y_4}{\sqrt{B+\mu}}, \quad x_6 = \frac{y_6}{\sqrt{C+\mu}},$$

* Compare with Art. 260.

† This has been already shown in Art. 118.

gives as the locus of y a tetrahedral complex T_μ^2 having the same fundamental tetrahedron as T^2, and whose complex cones and complex conics are "images" of reguli of T^2, (Art. 116).

Since each complex cone of T_μ^2 has a generator through each vertex of the tetrahedron, and hence contains the four lines

$$y_1 = \pm iy_2, \quad y_3 = \pm iy_4, \quad y_5 = \pm iy_6,$$

(where the signs are to be taken *all positive* or *two negative*), the corresponding regulus of T^2 will also contain four lines of this description, *i.e.* will pass through the vertices of the tetrahedron. Similarly the regulus which corresponds to a complex conic of T_μ^2 will have a generator in each of the four faces of the fundamental tetrahedron; we thus obtain ∞^4 reguli of T^2 through each vertex, and ∞^4 reguli touching each face of the fundamental tetrahedron.

To a regulus of T^2 which belongs to the complex C, whose equation may be taken as $\Sigma c_i x_i = 0$, will correspond a cone or complex conic of T_μ^2 which belongs to the complex

$$\frac{c_1}{\sqrt{A+\mu}}x_1 + \frac{c_2}{\sqrt{A+\mu}}x_2 + \frac{c_3}{\sqrt{B+\mu}}x_3 + \frac{c_4}{\sqrt{B+\mu}}x_4 + \frac{c_5}{\sqrt{C+\mu}}x_5 + \frac{c_6}{\sqrt{C+\mu}}x_6 = 0;$$

while the latter complex must be special, since it contains a cone or the tangents of a conic, hence

$$\frac{c_1^2 + c_2^2}{A+\mu} + \frac{c_3^2 + c_4^2}{B+\mu} + \frac{c_5^2 + c_6^2}{C+\mu} = 0.$$

The last equation gives two values of μ, having roots μ_1 and μ_2; thus the cones or conics of $T_{\mu_1}^2$, $T_{\mu_2}^2$ whose vertices or planes are *united* to the respective lines $\left(\frac{c_1}{\sqrt{A+\mu_1}}, \ldots\right)$, $\left(\frac{c_1}{\sqrt{A+\mu_2}}, \ldots\right)$, give rise to four sets of reguli of T^2 which belong to C. Thus there are in all ten varieties of ∞^1 reguli which belong to the congruence.

272. Focal surface of the intersection of any two complexes. The focal surface of the intersection of any two complexes may be determined analytically as follows: let $f=0$, $\phi=0$ be any two complexes, then if x is a ray of the system determined by them, all rays consecutive to x satisfy the equations

$$\left(y\frac{\partial f}{\partial x}\right) = 0, \quad \left(y\frac{\partial \phi}{\partial x}\right) = 0.$$

The directrices of this linear congruence being z, z', it is clear that

$$z_i = \frac{\partial f}{\partial x_i} + \lambda_1 \frac{\partial \phi}{\partial x_i}, \quad z_i' = \frac{\partial f}{\partial x_i} + \lambda_2 \frac{\partial \phi}{\partial x_i},$$

where λ_1, λ_2 are the roots of the equation

$$\Sigma \left(\frac{\partial f}{\partial x_i}\right)^2 + 2\lambda \Sigma \frac{\partial f}{\partial x_i} \cdot \frac{\partial \phi}{\partial x_i} + \lambda^2 \Sigma \left(\frac{\partial \phi}{\partial x_i}\right)^2 = 0.$$

The intersection of x with z and z' gives the points of contact, P, P' of x with the focal surface, and the planes (xz), (xz') are its tangent planes at P and P'; hence if y is any tangent to the focal surface

$$\rho \cdot y_i = \mu x_i + \frac{\partial f}{\partial x_i} + \lambda \frac{\partial \phi}{\partial x_i},$$

where λ is either λ_1 or λ_2.

Applying to the congruence (2, 2), we have

$$f \equiv A(x_1^2 + x_2^2) + B(x_3^2 + x_4^2) + C(x_5^2 + x_6^2),$$
$$\phi \equiv \Sigma c_i x_i,$$

and the equations for the determination of y are

$$\rho \cdot y_1 = (\mu + A) x_1 + \lambda c_1, \quad \rho \cdot y_3 = (\mu + B) x_3 + \lambda c_3, \quad \rho \cdot y_5 = (\mu + C) x_5 + \lambda c_5,$$
$$\rho \cdot y_2 = (\mu + A) x_2 + \lambda c_2, \quad \rho \cdot y_4 = (\mu + B) x_4 + \lambda c_4, \quad \rho \cdot y_6 = (\mu + C) x_6 + \lambda c_6.$$

The elimination between these equations and the equations $f = 0$, $\phi = 0$, of the quantities x_i, λ, μ, leads to the equation of the focal surface in line coordinates.

273. Double rays of special congruences (2, 2)*. If the complex $(cx) = 0$ contains an edge of the fundamental tetrahedron of the tetrahedral complex, a particular case of the congruence (2, 2) arises; *e.g.* let C contain the edge $A_1 A_2$, then we have $c_1 - ic_2 = 0$; in this case $A_1 A_2$ is a *double ray* of the congruence, since at each point of it there is only one ray, viz. the line $A_1 A_2$. For those tangents y of the focal surface which meet $A_1 A_2$ we have $y_1 - iy_2 = 0$, hence for such tangents $\mu + A = 0$ in the above set of equations, substituting for μ and eliminating x_3, x_4, x_5, x_6 and λ by aid of the last four of the equations of the last Article, we find that y belongs to the quadratic complex

$$\frac{(c_1 y_3 - c_3 y_1)^2 + (c_1 y_4 - c_4 y_1)^2}{B - A} + \frac{(c_1 y_5 - c_5 y_1)^2 + (c_1 y_6 - c_6 y_1)^2}{C - A} = 0,$$

and also to the linear complex $y_1 - iy_2 = 0$.

Hence the tangents to the focal surface in any plane through

* The following classification of congruences (2, 2) which have a double ray is due to W. Stahl.

A_1A_2 envelope a conic, therefore the focal surface is a Plücker's surface of which A_1A_2 is the double line.

(ii) If the complex $(cx)=0$ also passes through the edge A_1A_3, then $c_3-ic_4=0$, and the focal surface is a Plücker surface in which A_1A_3 is also a double line: the congruence has two intersecting double rays.

(iii) If the complex $(cx)=0$ passes through A_1A_2 and A_3A_4, then $c_1=c_2=0$, and tangents y of the focal surface which meet A_1A_2 also meet A_3A_4, and are therefore *generators* of the focal surface, since they meet it in the point of contact of x and the double lines A_1A_2, A_3A_4; such lines y also belong to the quadratic complex $Y_3^2+Y_4^2+Y_5^2+Y_6^2=0$, where Y_3, Y_4, Y_5, Y_6 are linear functions of y_3, y_4, y_5, y_6; as is easily seen by eliminating x_3, x_4, x_5, x_6, λ, ρ from the last four equations of the last Article. Thus these lines y are generators of a ruled quartic with two double directrices, *i.e.* of the class I; this surface is here the focal surface: the congruence has two non-intersecting double rays.

(iv) If $(cx)=0$ passes through A_1A_2, A_2A_3, A_3A_4, the focal surface is a ruled quartic which has A_1A_2, A_3A_4 as double directrices and A_2A_3 as double generator, *i.e.* belongs to class VII: the congruence has three double rays.

(v) If $(cx)=0$ passes through A_1A_2, A_2A_3, A_3A_4, A_4A_1, the focal surface has four double lines and hence must consist of two quadrics which have two generators of each system in common. Hence the congruence (4, 4) of the double tangents of two quadrics becomes, when the quadrics have two generators of each system in common, *two congruences* (2, 2).

For the species (iii) the following theorem holds: *any regulus through A_1A_2, A_3A_4 and one other ray of the congruence belongs entirely to the congruence*; for if X is the additional ray, such a regulus is given by the equations

$$\rho . x_1=\alpha+\beta+X_1, \quad \rho . x_3=X_3, \quad \rho . x_5=X_5,$$
$$\rho . x_2=(\alpha-\beta)i+X_2, \quad \rho . x_4=X_4, \quad \rho . x_6=X_6.$$

Now since it is given that

$$c_3X_3+c_4X_4+c_5X_5+c_6X_6=0,$$
$$(B-A)(X_3^2+X_4^2)+(C-A)(X_5^2+X_6^2)=0;$$

it follows that two equations of the same form as the last are also satisfied by x, *i.e.* the regulus of lines x belongs entirely to the congruence.

Thus from the two given double rays d_1 and d_2 and any regulus ρ of the system not containing d_1 and d_2 the whole system can be constructed by forming the ∞^1 reguli determined by d_1, d_2 and any generator of ρ.

CHAPTER XVII.

THE GENERAL COMPLEX.

274. MANY of the leading characteristics of the quadratic complex are seen to belong also to a complex $f(x)=0$, of any degree n. For instance, the lines of this general complex through any point form a cone whose degree is n, and those in any plane envelope a curve whose class is n. Since, x and y being any two intersecting lines, the equation $f(x+\lambda y)=0$, when expanded becomes

$$f(x)+\lambda\Delta f+\frac{\lambda^2}{2!}\Delta^2 f+\ldots\ldots=0,$$

where $\Delta=\Sigma y_i\frac{\partial}{\partial x_i}$, this gives n values for λ, *i.e. there are n lines of the complex in any plane pencil.*

The equation $\Delta f\equiv\Sigma y_i\frac{\partial f}{\partial x_i}=0$, in which x is a given line, is said to be a linear polar complex of $f=0$. If x belongs to $f=0$, the equation $\Delta f=0$ is one member of the singly infinite set of linear complexes

$$\Sigma y_i\left(\frac{\partial f}{\partial x_i}+\mu\frac{\partial\omega}{\partial x_i}\right)=0;$$

they are called the tangent linear complexes of $f(x)=0$*.

Each of these complexes contains x and every line $x+dx$ which is consecutive to x in the given complex $f(x)=0$; since for such consecutive lines we have

$$\Sigma\frac{\partial f}{\partial x_i}dx_i=0,\quad \Sigma\frac{\partial\omega}{\partial x_i}dx_i=0.$$

On any line x of $f=0$, a correlation is established between its points and their polar planes in a tangent linear complex for x: the *same correlation* is determined by each of the tangent linear

* See Art. 74.

complexes of x, the polar plane of any point P of x being the tangent plane through x to the complex cone of P.

275. The Singular Surface. Again, as in the case of the quadratic complex, we consider such tangent linear complexes as are *special*; and as in Arts. 76, 157 we find as the necessary condition

$$\Omega\left(\frac{\partial f}{\partial x}\right)=0.$$

This becomes $\Sigma\left(\frac{\partial f}{\partial x_i}\right)^2=0$, if we take $\omega(x)=0$ as being $\Sigma x_i^2=0$. Hence the lines of $f(x)=0$ whose tangent linear complexes are special satisfy the equations

$$f(x)=0, \quad \Sigma\left(\frac{\partial f}{\partial x_i}\right)^2=0.$$

These ∞^2 lines are called, as before, *singular lines* of $f(x)=0$.

If x be a singular line, the pencil $\left(x, \frac{\partial f}{\partial x}\right)$ consists of directrices of *special* tangent linear complexes. These directrices therefore form a complex, which consists of ∞^2 plane pencils.

If all the lines of such a pencil intersect any given line a, we have

$$(ax)=0, \quad \left(a\frac{\partial f}{\partial x}\right)=0, \quad f(x)=0, \quad \Sigma\left(\frac{\partial f}{\partial x_i}\right)^2=0.$$

These equations determine, if a is given, $4n(n-1)^2$ singular lines x, for each of which the pencil $\left(x, \frac{\partial f}{\partial x}\right)$ meets a. This may occur in two ways; *either* on account of a passing through the centre of the pencil, *or* on account of a lying in the plane of the pencil; from the duality of the subject, there will be as many solutions of one kind as of the other. Hence the locus of the points $\left(x, \frac{\partial f}{\partial x}\right)$ is a surface of degree $2n(n-1)^2$, and the envelope of the planes $\left(x, \frac{\partial f}{\partial x}\right)$ is a surface whose class is $2n(n-1)^2$.

*These surfaces are identical**; for, denoting $\frac{\partial f}{\partial x}$ by ξ, if P, P' are any two consecutive points of the first locus, let P be the point (x, ξ) and P' the point (x', ξ'), where

$$x'=x+dx, \quad \xi'=\xi+d\xi.$$

Then $$\Sigma x_i\xi_i=0, \quad \Sigma(x_i+dx_i)(\xi_i+d\xi_i)=0,$$

* This was shown by Pasch; see reference on page 92.

and neglecting small quantities of the second order, since we have $\Sigma\xi_i dx_i = 0$, therefore $\Sigma x_i d\xi_i = 0$, hence

$$\Sigma x_i' \xi_i = 0, \quad \Sigma x_i \xi_i' = 0;$$

so that the four lines x, ξ, x', ξ' form a twisted quadrilateral, and the point P' lies in the plane (x, ξ), if small quantities of the second order are neglected.

We therefore obtain one surface, the Singular Surface of the complex, which is both the locus of the *singular points* (x, ξ) and the envelope of the *singular planes* (x, ξ).

In the tangent linear complexes of a singular line x, the plane (x, ξ) is the polar plane of each point of x; therefore the complex cones of $f = 0$, whose vertices lie on x, touch (x, ξ) along x.

276. If y is any line through the singular point, we have $\Sigma y_i x_i = 0$, $\Sigma y_i \xi_i = 0$, hence

$$f(x + \lambda y) = \frac{\lambda^2}{2} \Sigma y_i y_k f_{ik} + \ldots\ldots + \lambda^n f(y);$$

therefore in the pencil (x, y) there are only $n-2$ lines of f distinct from x, hence the complex cone of f for a singular point has the singular line for double edge.

If y also lies on the cone of the complex $\Sigma y_i y_k f_{ik} = 0$ for the singular point, the pencil (x, y) contains only $n-3$ lines of f distinct from x, hence the complex cone of $\Sigma y_i y_k f_{ik} = 0$ for the singular point must split up into two planes; they are the pair of tangent planes through the singular line to the complex cone of f for the singular point.

Reciprocally, the complex curves in the planes through any singular line x have the singular point as point of contact with x, except in the case of the singular plane. Let y lie in the singular plane and also satisfy the equation $\Sigma y_i y_k f_{ik} = 0$, then, as before, the curve of the latter complex in the singular plane must split up into a pair of points whose centres lie on the singular line; the complex curve of f in the singular plane touches the singular line in these two points, *i.e.* has the singular line as a bitangent.

277. If ϵ_1, ϵ_2 are the two planes into which the complex cone of $\Sigma y_i y_k f_{ik} = 0$ breaks up for a singular point P, and a the polar plane for P of a tangent linear complex for x of the complex $F \equiv \Sigma \left(\frac{\partial f(y)}{\partial y_i}\right)^2 = 0$, then ϵ_1, ϵ_2, a, *and the singular plane form a harmonic pencil*; for let v be any line of the pencil (P, a), then $\Sigma v_i \frac{\partial F}{\partial x_i} = 0$; also the lines of $\Sigma y_i y_k f_{ik} = 0$ which are contained in the pencil (ξ, v) are obtained by substituting $\xi+\lambda v$ for y in the last equation, giving

$$\Sigma\xi_i\xi_k f_{ik}+2\lambda\Sigma v_i\frac{\partial F}{\partial x_i}+\lambda^2\Sigma v_i v_k f_{ik}=0,$$

which, since $\Sigma v_i\frac{\partial F}{\partial x_i}=0$, reduces to

$$\Sigma\xi_i\xi_k f_{ik}+\lambda^2\Sigma v_i v_k f_{ik}=0,$$

whence the result follows as stated above.

Reciprocally if E_1, E_2 are the points into which the complex conic of $\Sigma y_i y_k f_{ik}=0$ breaks up for the singular plane, and A the vertex of the cone of F upon x for which the singular plane is the tangent plane, the points E_1, E_2, A and the singular point are four harmonic points.

278. The Principal Surfaces. The significance of the Principal Surfaces, as being the analogues of the lines of curvature of a hypersurface, has already been noticed (Art. 228). In relation to them a result may be given here which is the extension of a theorem shown in connexion with the linear complex (Art. 41). On each line x of any ruled surface of the complex $f(x)=0$, a (1, 1) correspondence exists between the point of contact with the surface of any plane π through x and the point of contact of x with the complex curve in π; the latter point being the pole of π in the tangent linear complexes of x. There are, therefore, two planes π for each of which the point of contact of π with the surface coincides with the pole of π in these tangent complexes. The locus of such points, of which there are thus *two* Q, Q' *on each generator*, is a curve k. If now the given surface is a principal surface, the complex $f(x)=0$ is 'touched' by a tangent linear complex along x and also along a generator consecutive to x (Art. 132), so that this tangent complex contains three consecutive generators of the surface; hence, by Art. 41, since two consecutive tangents of k at Q belong to the same linear complex, the osculating plane of k at Q is the tangent plane of the surface at Q, *i.e., the curve k is a principal tangent curve of the surface.*

That the (1, 1) correspondence of points upon x just noticed is *an involution* may be shown as follows:—the coordinates of x are functions of one variable θ, and the complex $\Sigma y_i\left(x_i+\sigma\frac{\partial x_i}{\partial\theta}\right)=0$ has at each point P of x the tangent plane π as its polar plane; for the lines of this complex through P intersect both x, the generator through P, and the consecutive generator $x+\frac{dx}{d\theta}d\theta$.

Also this complex is in involution with any tangent linear complex of x; for

$$\Sigma\left(x_i+\sigma\frac{dx_i}{d\theta}\right)\left(\frac{\partial f}{\partial x_i}+\mu x_i\right)=nf(x)+\mu(x^2)+\tfrac{1}{2}\sigma\mu\frac{d(\Sigma x_i^2)}{d\theta}+\sigma\Sigma\frac{dx_i}{d\theta}\frac{\partial f}{\partial x_i}$$

$$=0, \text{ since } f\left(x+\frac{dx}{d\theta}d\theta\right)=0.$$

Hence the (1, 1) correspondence of points is an involution, a pair of corresponding points being, as stated, the point of contact of π with the surface and the pole of π in any tangent linear complex of x; the double points of the involution are the two points Q, Q' of k which lie upon x.

279. Independent constants of the complex. A homogeneous equation $f(x)=0$ of the nth degree in six variables contains

$$\frac{(n+1)(n+2)(n+3)(n+4)(n+5)}{5\,!}$$

terms; but the complex represented by $f(x)=0$, is also represented by $\psi(x)\equiv f(x)+\omega(x).\phi(x)=0$, where ϕ is any expression of degree $(n-2)$ in the variables, and $\omega(x)=0$ is the fundamental relation. So that the complex contains

$$\frac{(n-1)n(n+1)(n+2)(n+3)}{5\,!}$$

arbitrary constants, viz. the coefficients of ϕ.

The expression $\Sigma A_{ik}\dfrac{\partial^2\psi(x)}{\partial x_i\partial x_k}$ is a covariant of ψ (n being supposed greater than two), where the A_{ik} are the first minors of Δ the discriminant of $\omega(x)$. Hence if we assume $\Sigma A_{ik}\dfrac{\partial^2\psi(x)}{\partial x_i\partial x_k}\equiv 0$, we have as many linear equations between the coefficients of f and ϕ as there are coefficients of ϕ. The indeterminateness of the equation of the complex is now removed. The equation of the complex $\psi=0$, under these conditions, is said to be in its *normal* form. When $\omega(x)\equiv\Sigma x_i^2$, this condition becomes $\psi_{ii}=0$ (see Art. 87).

280. The Special Complex. Any pencil of lines defines a surface element. The condition for united position of two consecutive pencils will now be investigated. Let

$$(a_i,\ b_i),\quad (a_i+da_i,\ b_i+db_i)$$

be two consecutive pencils; if the plane of the first pencil passes through the centre of the second pencil, there is one line of the second pencil which meets all the lines of the first. Now since

$$a_i + da_i, \; b_i + db_i$$

are two intersecting lines we see that

$$\Sigma a_i da_i, \; \Sigma b_i db_i, \; \Sigma (a_i db_i + b_i da_i)$$

are each small quantities of the second order; also for some value of the ratio ρ/σ and for all values of λ

$$\Sigma (a_i + \lambda b_i)(\rho . \overline{a_i + da_i} + \sigma . \overline{b_i + db_i}) = 0,$$

hence

$$\rho \Sigma a_i da_i + \sigma \Sigma a_i db_i = 0,$$

$$\rho \Sigma b_i da_i + \sigma \Sigma b_i db_i = 0;$$

therefore, ignoring small quantities of the second order, the required condition is seen to be that either of the following equivalent equations should hold, viz.

$$\Sigma a_i db_i = 0, \; \Sigma b_i da_i = 0.$$

If the condition is satisfied we have in the most general case ∞^2 pencils whose centres lie on a surface which is touched by their planes; a more special case is that of the ∞^1 pencils of planes through the tangents of a curve, or the tangent planes of a developable and their points of contact.

Consider a line complex $\phi = 0$ for each of whose lines the condition $\Sigma \left(\frac{\partial \phi}{\partial x_i}\right)^2 = 0$ is satisfied either identically, or, as a consequence of $\phi = 0$, $\Sigma x_i^2 = 0$; the condition expresses that $\frac{\partial \phi}{\partial x_i}$ is a line; and considering the pencil $\left(x_i, \frac{\partial \phi}{\partial x_i}\right)$, since for all lines of the complex consecutive to x_i we have $\Sigma \frac{\partial \phi}{\partial x_i} dx_i = 0$, this pencil satisfies the above condition; hence in general all the lines x_i are tangents to one surface; the complex consists therefore of the ∞^3 tangents to this surface. Such a complex is said to be *special*.

281. Congruences and their Focal Surfaces*. The lines common to two complexes $f = 0$, $\phi = 0$ of degrees m and n respectively, form a doubly infinite set of lines or a congruence†. Through any point there pass mn lines of the congruence, viz.

* On the subject of this Article, see Voss, "Ueber Complexe und Congruenzen," *Math. Ann.* IX.

† The chief properties of a line-congruence have been already discussed in Chapter XIV. The congruences here considered are those which consist of the *complete* intersection of two complexes.

the intersections of the complex cones of f and ϕ; in any plane there lie mn lines of the congruence, viz. the common tangents of the complex curves of f and ϕ.

On any line x of the congruence there are two sets of points in (1, 1) correspondence, for any plane π through x has two points P and P' as its poles in the tangent linear complexes of x for f and ϕ respectively; hence there are two united points F and F' at which these tangent complexes have a common polar plane. Analytically, these points are obtained as follows:—if (F, π) is a pencil common to the tangent complexes and y the line of the pencil which meets any line α, we have

$$\Sigma y_i x_i = \Sigma y_i \frac{\partial f}{\partial x_i} = \Sigma y_i \frac{\partial \phi}{\partial x_i} = \Sigma y_i \alpha_i = \Sigma y_i^2 = 0;$$

hence F and F' are determined.

All the lines of the congruence consecutive to x satisfy the equations

$$\Sigma y_i \frac{\partial f}{\partial x_i} = 0, \quad \Sigma y_i \frac{\partial \phi}{\partial x_i} = 0^*;$$

the directrices of this linear congruence are z and z', where

$$z_i = \frac{\partial f}{\partial x_i} + \lambda_1 \frac{\partial \phi}{\partial x_i}, \quad z_i' = \frac{\partial f}{\partial x_i} + \lambda_2 \frac{\partial \phi}{\partial x_i};$$

λ_1, λ_2 being the roots of

$$\Sigma \left(\frac{\partial f}{\partial x_i}\right)^2 + 2\lambda \Sigma \frac{\partial f}{\partial x_i} \cdot \frac{\partial \phi}{\partial x_i} + \lambda^2 \Sigma \left(\frac{\partial \phi}{\partial x_i}\right)^2 = 0.$$

The points F, F' are called 'focal points' and the corresponding planes 'focal planes' of the congruence, hence for F the plane (x, z) is the focal plane and for F' the plane (x, z').

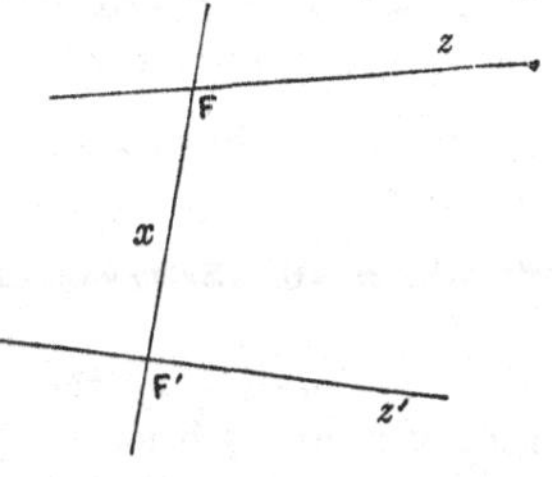

Fig. 12.

The locus of focal points and the envelope of focal planes give the same surface†; for consider a consecutive line $x' \equiv x + dx$ of the congruence, then

$$\Sigma x_i dx_i = \Sigma \frac{\partial \phi}{\partial x_i} dx_i = \Sigma \frac{\partial f}{\partial x_i} dx_i = 0;$$

to x' will correspond as directrices ζ, ζ', where

$$\zeta_i = \frac{\partial f}{\partial x_i} + \lambda_1 \frac{\partial \phi}{\partial x_i} + d\frac{\partial f}{\partial x_i} + \lambda_1 d\frac{\partial \phi}{\partial x_i} + \frac{\partial \phi}{\partial x_i} d\lambda,$$

$$\zeta_i' = \frac{\partial f}{\partial x_i} + \lambda_2 \frac{\partial \phi}{\partial x_i} + d\frac{\partial f}{\partial x_i} + \lambda_2 d\frac{\partial \phi}{\partial x_i} + \frac{\partial \phi}{\partial x_i} d\lambda.$$

* See Art. 272. † See Art. 236.

Hence $\Sigma x_i \zeta_i = \Sigma x_i z_i = \Sigma x_i' \zeta_i = \Sigma x_i' z_i = 0;$

$$\Sigma x_i \zeta_i' = \Sigma x_i z_i' = \Sigma x_i' \zeta_i' = \Sigma x_i' z_i' = 0.$$

Therefore the lines x, x', ζ, z form a twisted quadrilateral and the focal point (x', ζ) lies in the focal plane (x, z), neglecting throughout small quantities of the second order. A similar result holds for the focal plane (x, z'). Hence the focal planes (x, z), (x, z') envelope the surface which is the locus of the focal points, but the focal plane of F is the tangent plane at F' and *vice versâ*; *i.e., the lines of the congruence are double tangents of the focal surface, the focal planes of F and F' being the tangent planes of this surface at F' and F respectively.*

If y is any tangent of the focal surface we have

$$\rho \cdot y_i = \mu x_i + \frac{\partial f}{\partial x_i} + \lambda \frac{\partial \phi}{\partial x_i};$$

the equation of the focal surface in line-coordinates is obtained by eliminating x, λ, μ, ρ from the last set of equations and $f = 0$, $\phi = 0$, $\Sigma x_i^2 = 0$, and

$$\Sigma \left(\frac{\partial f}{\partial x_i}\right)^2 + 2\lambda \Sigma \frac{\partial f}{\partial x_i}\frac{\partial \phi}{\partial x_i} + \lambda^2 \Sigma \left(\frac{\partial \phi}{\partial x_i}\right)^2 = 0.$$

282. Any line x of the congruence touches the complex curves of f and ϕ in any plane π through x in two points P and Q, hence there are two complex cones of f and ϕ respectively, having their vertices on x, for which π is a tangent plane.

These points P and Q have with the focal points F, F' upon x a definite double ratio independent of the plane π.

For draw through F and F' respectively two lines r and s in π, then we have

$$\Sigma r_i x_i = 0, \quad \Sigma s_i x_i = 0, \quad \Sigma r_i \frac{\partial f}{\partial x_i} + \lambda_1 \Sigma r_i \frac{\partial \phi}{\partial x_i} = 0, \quad \Sigma s_i \frac{\partial f}{\partial x_i} + \lambda_2 \Sigma s_i \frac{\partial \phi}{\partial x_i} = 0;$$

also the lines joining the point (r, s) to P and Q being $r + \mu_1 s$, $r + \mu_2 s$, since they respectively belong to the tangent linear complexes

$$\Sigma y_i \frac{\partial f}{\partial x_i} = 0, \quad \Sigma y_i \frac{\partial \phi}{\partial x_i} = 0,$$

we have $$\Sigma r_i \frac{\partial f}{\partial x_i} + \mu_1 \Sigma s_i \frac{\partial f}{\partial x_i} = 0, \quad \Sigma r_i \frac{\partial \phi}{\partial x_i} + \mu_2 \Sigma s_i \frac{\partial \phi}{\partial x_i} = 0.$$

But the double ratio of the lines r, s, $r + \mu_1 s$, $r + \mu_2 s$ being $\frac{\mu_1}{\mu_2}$,

$$(FPF'Q) = \frac{\mu_1}{\mu_2} = \frac{\Sigma r_i \frac{\partial f}{\partial x_i}}{\Sigma r_i \frac{\partial \phi}{\partial x_i}} \div \frac{\Sigma s_i \frac{\partial f}{\partial x_i}}{\Sigma s_i \frac{\partial \phi}{\partial x_i}} = \frac{\lambda_1}{\lambda_2}.$$

283. Degree and class of the Focal Surface. The focal surface being the locus of points whose complex cones in f and ϕ touch, and also the envelope of planes whose complex curves touch, it follows that the *degree* and *class* of the focal surface are equal. If a line a_i meets two consecutive and intersecting lines x, $x+dx$ of the congruence, then $\Sigma a_i x_i = 0$, $\Sigma a_i dx_i = 0$, also we have

$$\Sigma \frac{\partial f}{\partial x_i} dx_i = \Sigma \frac{\partial \phi}{\partial x_i} dx_i = \Sigma x_i dx_i = \Sigma dx_i^2 = 0.$$

Hence the dx_i are proportional to the coordinates of a line which belongs to the regulus

$$\Sigma a_i y_i = 0, \quad \Sigma \frac{\partial f}{\partial x_i} y_i = 0, \quad \Sigma \frac{\partial \phi}{\partial x_i} y_i = 0;$$

moreover because x belongs to this regulus and since $\Sigma x_i dx_i = 0$, the regulus must consist of two pencils, hence (Art. 59),

$$\begin{vmatrix} 0 & \Sigma a_i \frac{\partial f}{\partial x_i} & \Sigma a_i \frac{\partial \phi}{\partial x_i} \\ \Sigma a_i \frac{\partial f}{\partial x_i} & \Sigma \left(\frac{\partial f}{\partial x_i}\right)^2 & \Sigma \frac{\partial f}{\partial x_i} \cdot \frac{\partial \phi}{\partial x_i} \\ \Sigma a_i \frac{\partial \phi}{\partial x_i} & \Sigma \frac{\partial f}{\partial x_i} \cdot \frac{\partial \phi}{\partial x_i} & \Sigma \left(\frac{\partial \phi}{\partial x_i}\right)^2 \end{vmatrix} = 0.$$

On account of this equation and the equations

$$\Sigma a_i x_i = f = \phi = 0,$$

we obtain

$$4mn(m+n-2)$$

lines x such that a meets x and a line of the congruence consecutive to x; so that a *either* passes through the intersection of x and $x+dx$ *or* lies in their plane, therefore

$$\text{degree of focal surface} = \text{class of focal surface} = 2mn(m+n-2).$$

Since the order and class of the congruence are each equal to mn, if r is the rank of the congruence, we see from Art. 237

$$2mn(m+n-2) = 2mn(mn-1) - 2r,$$

hence

$$r = mn(m-1)(n-1)^*.$$

The focal surface may split up into two surfaces; this will occur if

$$\Sigma \left(\frac{\partial f}{\partial x_i}\right)^2 . \Sigma \left(\frac{\partial \phi}{\partial x_i}\right)^2 - \left(\Sigma \frac{\partial f}{\partial x_i} \cdot \frac{\partial \phi}{\partial x_i}\right)^2$$

* See Art. 248.

is a perfect square, the two focal points being then given rationally. For instance in the case of the congruence of the singular lines

$$f=0, \quad \Sigma\left(\frac{\partial f}{\partial x_i}\right)^2 \equiv F=0,$$

the determinant of the last article becomes

$$\left(\Sigma a_i \frac{\partial f}{\partial x_i}\right)\left\{2\Sigma a_i \frac{\partial F}{\partial x_i}\,\Sigma \frac{\partial f}{\partial x_i}\cdot\frac{\partial F}{\partial x_i} - \Sigma\left(\frac{\partial F}{\partial x_i}\right)^2 \Sigma a_i \frac{\partial f}{\partial x_i}\right\}.$$

The focal surface breaks up into two surfaces, *of which one is the singular surface of* $f=0$; since, if the point lies on the singular surface, two of the singular lines through it coincide; similarly for each tangent plane of the singular surface; so that the equations

$$f=F=\Sigma a_i x_i = \Sigma a_i \frac{\partial f}{\partial x_i} \equiv 0$$

relate to the points and planes of the singular surface which are united to a; *thus the degree and class of the singular surface are seen to be* $2n(n-1)^2$ *.

The solutions of

$$f=F=\Sigma a_i x_i = \Sigma a_i \frac{\partial f}{\partial x_i}\,\Sigma\left(\frac{\partial F}{\partial x_i}\right)^2 - 2\Sigma a_i \frac{\partial F}{\partial x_i}\,\Sigma \frac{\partial f}{\partial x_i}\frac{\partial F}{\partial x_i} = 0$$

relate to the other portion of the focal surface, which is therefore of degree and class $2n(n-1)(5n-7)$ †.

The lines which satisfy the equations

$$f=0, \quad \phi=0, \quad \Sigma\left(\frac{\partial f}{\partial x_i}\right)^2 . \,\Sigma\left(\frac{\partial \phi}{\partial x_i}\right)^2 - \Sigma \frac{\partial f}{\partial x_i}\cdot\frac{\partial \phi}{\partial x_i} = 0,$$

form a ruled surface; for each of them the focal points *coincide.*

If, by virtue of the form of $f=0$, $\phi=0$, we have

$$\Sigma\left(\frac{\partial f}{\partial x_i}\right)^2 \Sigma\left(\frac{\partial \phi}{\partial x_i}\right)^2 - \Sigma \frac{\partial f}{\partial x_i}\cdot\frac{\partial \phi}{\partial x_i} \equiv 0$$

the congruence consists of the tangents to one set of principal tangent curves of a surface.

284. The ruled surface common to three complexes‡. If $f=0$, $\phi=0$, $\psi=0$ be three complexes of degrees m, n and p

* See Art. 275.

† This surface has been termed the Accessory Surface, Voss, *Math. Ann.* IX. The Rank of the focal surface has been found by Voss to be

$$2mn\{(m+n-1)^2 - mn+1\};$$

his investigation is too long for insertion here; see Art. 248.

‡ See Voss, "Zur Theorie der windschiefen Flächen," *Math. Ann.* VIII.

respectively, the ruled surface which is their intersection is of degree and class $2mnp$, since this is the number of lines whose coordinates satisfy the equations

$$f = \phi = \psi = \Sigma x_i^2 = \Sigma a_i x_i = 0,$$

where a_i is any line.

We may take $\Sigma k_i x_i = 1$ as an equation connecting the variables, and, for purposes of symmetry, assume $\Sigma \alpha_i x_i = t$, where the α_i are arbitrary constants and t a new variable. The coordinates of the generator $x + dx$ consecutive to x are then given by the equations

$$\Sigma \frac{\partial f}{\partial x_i} dx_i = \Sigma \frac{\partial \phi}{\partial x_i} dx_i = \Sigma \frac{\partial \psi}{\partial x_i} dx_i = \Sigma k_i dx_i = \Sigma x_i dx_i = 0;$$

$$\Sigma \alpha_i dx_i = dt;$$

whence

$$dx_i = \frac{\partial \Pi}{\partial \alpha_i} \cdot \frac{dt}{\Pi},$$

where

$$\Pi = \left| \frac{\partial f}{\partial x_i}, \quad \frac{\partial \phi}{\partial x_i}, \quad \frac{\partial \psi}{\partial x_i}, \quad x_i, \quad k_i, \quad \alpha_i \right|.$$

If two consecutive generators intersect, we have $\Sigma dx_i^2 = 0$, and expressing that the five complexes

$$\Sigma y_i \frac{\partial f}{\partial x_i} = 0, \quad \Sigma y_i \frac{\partial \phi}{\partial x_i} = 0, \quad \Sigma y_i \frac{\partial \psi}{\partial x_i} = 0, \quad \Sigma y_i k_i = 0, \quad \Sigma y_i x_i = 0$$

have a common line, we obtain

$$\begin{vmatrix} \Sigma \left(\frac{\partial f}{\partial x_i}\right)^2 & \Sigma \frac{\partial f}{\partial x_i}\frac{\partial \phi}{\partial x_i} & \Sigma \frac{\partial f}{\partial x_i}\frac{\partial \psi}{\partial x_i} & \Sigma \frac{\partial f}{\partial x_i} k_i & \Sigma \frac{\partial f}{\partial x_i} x_i \\ \Sigma \frac{\partial f}{\partial x_i}\frac{\partial \phi}{\partial x_i} & \Sigma \left(\frac{\partial \phi}{\partial x_i}\right)^2 & \Sigma \frac{\partial \phi}{\partial x_i}\frac{\partial \psi}{\partial x_i} & \Sigma \frac{\partial \phi}{\partial x_i} k_i & \Sigma \frac{\partial \phi}{\partial x_i} x_i \\ \Sigma \frac{\partial f}{\partial x_i}\frac{\partial \psi}{\partial x_i} & \Sigma \frac{\partial \phi}{\partial x_i}\frac{\partial \psi}{\partial x_i} & \Sigma \left(\frac{\partial \psi}{\partial x_i}\right)^2 & \Sigma \frac{\partial \psi}{\partial x_i} k_i & \Sigma \frac{\partial \psi}{\partial x_i} x_i \\ \Sigma \frac{\partial f}{\partial x_i} k_i & \Sigma \frac{\partial \phi}{\partial x_i} k_i & \Sigma \frac{\partial \psi}{\partial x_i} k_i & \Sigma k_i^2 & \Sigma k_i x_i \\ \Sigma \frac{\partial f}{\partial x_i} x_i & \Sigma \frac{\partial \phi}{\partial x_i} x_i & \Sigma \frac{\partial \psi}{\partial x_i} x_i & \Sigma k_i x_i & \Sigma x_i^2 \end{vmatrix} = 0;$$

or, since $\Sigma x_i \frac{\partial f}{\partial x_i} = \Sigma x_i \frac{\partial \phi}{\partial x_i} = \Sigma x_i \frac{\partial \psi}{\partial x_i} = \Sigma x_i^2 = 0$, this becomes

$$(\Sigma k_i x_i)^2 \begin{vmatrix} \Sigma \left(\frac{\partial f}{\partial x_i}\right)^2 & \Sigma \frac{\partial f}{\partial x_i}\frac{\partial \phi}{\partial x_i} & \Sigma \frac{\partial f}{\partial x_i}\frac{\partial \psi}{\partial x_i} \\ \Sigma \frac{\partial f}{\partial x_i}\frac{\partial \phi}{\partial x_i} & \Sigma \left(\frac{\partial \phi}{\partial x_i}\right)^2 & \Sigma \frac{\partial \phi}{\partial x_i}\frac{\partial \psi}{\partial x_i} \\ \Sigma \frac{\partial f}{\partial x_i}\frac{\partial \psi}{\partial x_i} & \Sigma \frac{\partial \phi}{\partial x_i}\frac{\partial \psi}{\partial x_i} & \Sigma \left(\frac{\partial \psi}{\partial x_i}\right)^2 \end{vmatrix} = 0.$$

This determinant is of degree $2(m+n+p-3)$; hence there are $4mnp(m+n+p-3)$ generators which are intersected by consecutive generators*. Such a generator is said to be 'singular.'

285. Rank of the surface. The Rank of a surface, being the degree of its tangent cone or the class of its plane section, is *the degree of the surface in line coordinates*; *i.e.* the degree of the complex formed by the tangents of the surface. To find the equation of the surface in line coordinates, we have, if y is any tangent, $\Sigma y_i x_i = 0$, $\Sigma y_i dx_i = 0$, hence eliminating the differentials between the equations

$$\Sigma \frac{\partial f}{\partial x_i} dx_i = \Sigma \frac{\partial \phi}{\partial x_i} dx_i = \Sigma \frac{\partial \psi}{\partial x_i} dx_i = \Sigma k_i dx_i = \Sigma x_i dx_i = \Sigma y_i dx_i = 0,$$

we obtain $\left| \frac{\partial f}{\partial x_i}, \frac{\partial \phi}{\partial x_i}, \frac{\partial \psi}{\partial x_i}, k_i, x_i, y_i \right| = 0$, and squaring,

$$(\Sigma k_i x_i)^2 \begin{vmatrix} \Sigma \left(\frac{\partial f}{\partial x_i}\right)^2 & \Sigma \frac{\partial f}{\partial x_i}\frac{\partial \phi}{\partial x_i} & \Sigma \frac{\partial f}{\partial x_i}\frac{\partial \psi}{\partial x_i} & \Sigma y_i \frac{\partial f}{\partial x_i} \\ \Sigma \frac{\partial f}{\partial x_i}\frac{\partial \phi}{\partial x_i} & \Sigma \left(\frac{\partial \phi}{\partial x_i}\right)^2 & \Sigma \frac{\partial \phi}{\partial x_i}\frac{\partial \psi}{\partial x_i} & \Sigma y_i \frac{\partial \phi}{\partial x_i} \\ \Sigma \frac{\partial f}{\partial x_i}\frac{\partial \psi}{\partial x_i} & \Sigma \frac{\partial \phi}{\partial x_i}\frac{\partial \psi}{\partial x_i} & \Sigma \left(\frac{\partial \psi}{\partial x_i}\right)^2 & \Sigma y_i \frac{\partial \psi}{\partial x_i} \\ \Sigma y_i \frac{\partial f}{\partial x_i} & \Sigma y_i \frac{\partial \phi}{\partial x_i} & \Sigma y_i \frac{\partial \psi}{\partial x_i} & 0 \end{vmatrix} = 0.$$

It follows that $\Sigma k_i x_i$ divides out of the previous equation leaving an equation of degree $m+n+p-3$ in x and unity in y. Eliminating the x_i from this equation and

$$f=0, \quad \phi=0, \quad \psi=0, \quad \Sigma x_i y_i = 0, \quad \Sigma x_i^2 = 0,$$

the equation of the surface in line coordinates is seen to be of degree $2mnp(m+n+p-3)$.

If for any generator the equations which give the consecutive generator are not linearly independent, we take as an additional equation $\Sigma f_{ik} dx_i dx_k = 0$, thus obtaining *two* sets of values for the dx_i; in this case a double generator exists, for each such double generator the rank is diminished by two, since the class of any plane section is diminished by two for each double point.

286. Clifford's Theorem. Investigations into the general ruled surface are outside the scope of the present treatise, but on

* Klein, *Math. Ann.* v.

account of their interest we add a sketch of some ideas contained in the memoir on *Classification of Loci* due to Prof. Clifford*

He denotes by a *curve* a continuous one-dimensional aggregate of any sort of elements, which includes not merely a curve in the ordinary sense (an aggregate of points), but also a ruled surface, or indeed a singly infinite system of curves, surfaces, complexes, &c. such that *one condition* is sufficient to determine a finite number of the elements. "The elements may be regarded as determined by k coordinates; and then if these be connected by $k-1$ equations of any order, the curve is either the whole aggregate of common solutions of these equations, or, when this breaks up into algebraically distinct parts, the curve is one of these parts. It is thus convenient to employ still farther the language of geometry, and to speak of such a curve as the complete or partial intersection of $k-1$ loci in flat space† of k dimensions." "If a certain number, say h, of the equations are linear, it is evidently possible by a linear transformation to make these equations equate h of the coordinates to zero; it is then convenient to leave these coordinates out of consideration altogether, and to regard only the remaining $k-h-1$ equations between $k-h$ coordinates. In this case the curve will, therefore, be regarded as a curve in flat space of $k-h$ dimensions. And, in general, when we speak of a curve as in flat space of k dimensions, we mean that it cannot exist in flat space of $k-1$ dimensions."

The whole aggregate of linear complexes may be regarded as constituting a space of five dimensions, in which *straight lines* constitute a quadric locus‡.

A ruled surface is a 'curve' lying in a quadric locus in five dimensions. If, however, the generators of the ruled surface belong to the same linear complex, the ruled surface is a 'curve' in a quadric locus in four dimensions. If the ruled surface has two linear directrices, it is a 'curve' on an ordinary quadric surface in three dimensions. *So that the theory of ruled surfaces which have two directrices is identical with that of curves on a quadric.* Hence, such ruled *quartic* surfaces correspond either to quadri-quadric curves of deficiency unity (*elliptic* curves whose coordinates are expressible as elliptic functions of one variable), or to the

* See *Collected Works*, p. 305.

† By a *flat* space is meant one which is intersected by a line which does not belong to it in *one point only*.

‡ See Art. 229.

curves of deficiency zero which are the partial intersections of a quadric and a cubic surface.

"Similar considerations apply to surfaces. By a *surface* we shall mean, in general, a continuous two-dimensional aggregate (which may also be called a *two-spread* or *two-way* locus) of any elements whatever."

Theorem. A curve of order n in flat space of k dimensions (and no less) may be represented, point for point, on a curve of order $n-k+2$ in a plane.

"The proposition is obvious when $k=3$. The cone standing on a curve of order n (in ordinary space of three dimensions), and having its vertex at a point of the curve, is of order $n-1$; if then we cut this cone by a plane, we have the tortuous curve represented, point for point, on a plane curve of order $n-1$. Now this process is applicable in general. Starting with an arbitrary point P of a curve in any number of dimensions, let us join this point to all the other points of the curve; we shall thus get a *cone* of order $n-1$. For, any flat locus of $k-1$ dimensions drawn through the point P, must meet the curve in n points of which P is one; and therefore it must meet the cone in $n-1$ lines. Hence, if we cut this cone by such a flat $(k-1)$-way locus *not* passing through P, we shall get a curve of order $n-1$ in flat space of $k-1$ dimensions, which is a point for point representation of the original curve. By continuing this process we may go on diminishing the order of the curve and the number of dimensions by equal quantities, until we have subtracted $k-2$ from each; when we are left with a curve of order $n-k+2$ in a plane."

It follows, by taking $k=n$, that *a curve of order n in flat space of n dimensions is unicursal*, since it has (1, 1) correspondence with a *conic*. By taking $k=n-1$, we obtain the result that *every curve of order n in flat space of $n-1$ dimensions is either unicursal or elliptic*; for it has (1, 1) correspondence with a plane *cubic*.

Two applications of these last results will now be made: since a ruled *quintic* surface which does not lie in a linear complex, corresponds to a curve of order five in flat space of five dimensions such a surface is *unicursal*: similarly, a ruled *sextic* which does not lie in a linear complex is by the second result seen to be either unicursal or elliptic.

287. Symbolic form of the equation of the complex*. The complex

$$f(p) \equiv \Sigma a_{ih,kl,} \ldots p_{ih} p_{kl} \ldots = 0$$

is always uniquely capable of symbolical representation. For the coefficient $a_{ih,kl} \ldots$ has the property that if two pairs of suffixes are interchanged the coefficient is not altered in value, hence the sum of all the terms arising from such interchanges effected upon a given term is equal to that term multiplied by the number of such arrangements of its pairs of suffixes, hence we may write symbolically

$$a_{ih,kl,} \ldots = a_{ih} \cdot a_{kl} \ldots\ldots$$

and thus
$$f(p) \equiv (\Sigma a_{ih} p_{ih})^n.$$

It should be noticed, however, that if in $a_{ih,kl,} \ldots$ two suffixes of the same pair are interchanged the coefficient is altered in sign, hence this must also be true of the symbolic quantities a_{ih}, etc. Hence *the complex is represented symbolically by the result of equating to zero the nth power of a linear expression in the line-coordinates.*

Moreover in the equation of the complex, which is generally of the form $f + \phi \cdot \omega = 0$, where $\phi = 0$ is any complex of degree $n-2$, ϕ may be so chosen as to make this symbolic form arise from a *special* linear complex; *i.e.* to make the six symbolic quantities a_{ih} satisfy the equation

$$a_{12} \cdot a_{34} + a_{13} \cdot a_{42} + a_{14} \cdot a_{23} = 0 \ldots\ldots\ldots\ldots\ldots \text{(i)}.$$

This involves that quantities a_1, a_2, a_3, a_4; b_1, b_2, b_3, b_4 can be found such that $a_{ih} = a_i \cdot b_h - a_h \cdot b_i$, and hence that

$$a_{ih,kl,} \ldots = (a_i b_h - a_h b_i)(a_k b_l - a_l b_k) \ldots\ldots\ldots\ldots \text{(ii)}.$$

The last equation is seen to satisfy the condition that if in $a_{ih,kl,} \ldots$ two suffixes of the same pair are interchanged the sign of the coefficient is changed.

For, in order that the substitution (ii) may hold, on multiplying the left side of (i) by $a_{mn,} \ldots$ to $n-2$ factors it must follow that

$$a_{12,34,mn,} \ldots + a_{13,42,mn,} \ldots + a_{14,23,mn,} \ldots = 0 \ldots\ldots\ldots \text{(iii)}.$$

The number of equations of the form (iii) is that of the possible combinations of $n-2$ pairs of indices, *i.e.* the number of coefficients of a complex ϕ of degree $n-2$; hence if in (iii) we suppose the quantities to be the coefficients of $f + \phi \cdot \omega$, we obtain a system of

* The developments which follow were given by Clebsch, "Ueber die Complexflächen und die Singularitätenflächen der Complexe," *Math. Ann.* v.

linear equations with the coefficients of ϕ as the quantities to be determined, the number of the equations being that of the coefficients of ϕ. Hence the coefficients of ϕ are thus uniquely found so as to satisfy (iii), and hence to make

$$f+\phi\,.\,\omega \equiv \{\Sigma\,(a_i b_h - a_h b_i)\,p_{ih}\}^n.$$

This form of the equation of the complex, *i.e.* that for which ϕ is thus determined, is the Normal Form (Art. 279); for

$$\Sigma A_{ik}\frac{\partial^2}{\partial x_i \partial x_k}(f+\phi\,.\,\omega)\equiv 0$$

becomes in Plücker coordinates

$$\left(\frac{\partial^2}{\partial p_{12}\partial p_{34}}+\frac{\partial^2}{\partial p_{13}\partial p_{42}}+\frac{\partial^2}{\partial p_{14}\partial p_{23}}\right)\{\Sigma\,(a_i b_h - a_h b_i)\,p_{ih}\}^n \equiv 0.$$

That such a form can be determined, and only in one way, may also be seen as follows: denote by Δ the operator

$$\frac{\partial^2}{\partial p_{12}\partial p_{34}}+\frac{\partial^2}{\partial p_{13}\partial p_{42}}+\frac{\partial^2}{\partial p_{14}\partial p_{23}},$$

and apply it and its successive powers to each side of the equation

$$f+\phi\,.\,\omega \equiv \{\Sigma\,(a_i b_h - a_h b_i)\,p_{ih}\}^n;$$

the operators Δ, $\Delta^2\ldots$ reduce the right side of this equation to zero on account of the identity

$$(a_1 b_2 - a_2 b_1)(a_3 b_4 - a_4 b_3) + \ldots + \ldots = 0,$$

hence ϕ has to be so determined that

$$\Delta(f+\phi\,.\,\omega)=0,\ \ \Delta^2(f+\phi\,.\,\omega)=0,\ \text{etc.};$$

but we have

$$\Delta(\phi\,.\,\omega)=\omega\Delta\phi+\phi\Delta\omega+\frac{\partial\phi}{\partial p_{12}}\cdot\frac{\partial\omega}{\partial p_{34}}+\frac{\partial\phi}{\partial p_{34}}\cdot\frac{\partial\omega}{\partial p_{12}}+\ldots$$

$$=\omega\Delta\phi+3\phi+p_{12}\frac{\partial\phi}{\partial p_{12}}+\ldots,$$

i.e. $$\Delta(\phi\,.\,\omega)=\omega\Delta\phi+(n+1)\,\phi.$$

Similarly $$\Delta(\omega\Delta\phi)=\omega\Delta^2\phi+(n-1)\,\Delta\phi,$$

$$\Delta(\omega\Delta^2\phi)=\omega\Delta^3\phi+(n-3)\,\Delta^2\phi,\ \text{etc.}$$

To determine $\Delta^{k+1}(\phi\omega)$, operate on the first k of these equations respectively with Δ^k, $\Delta^{k-1},\ldots\Delta$ and form their sum with the $(k+1)$th; we then obtain

$$\Delta(\phi\omega)=\omega\Delta\phi+(n+1)\,\phi,$$
$$\Delta^2(\phi\omega)=\omega\Delta^2\phi+2n\Delta\phi,$$
$$\Delta^3(\phi\omega)=\omega\Delta^3\phi+3\,(n-1)\,\Delta^2\phi,$$
$$\Delta^4(\phi\omega)=\omega\Delta^4\phi+4\,(n-2)\,\Delta^3\phi,\ \text{etc.}$$

The equations $\Delta(f+\phi\omega)=0$ etc. thus become

$$\Delta f+\omega\Delta\phi+(n+1)\phi=0,$$
$$\Delta^2 f+\omega\Delta^2\phi+2n\Delta\phi=0,$$
$$\Delta^3 f+\omega\Delta^3\phi+3(n-1)\Delta^2\phi=0,$$
$$\Delta^4 f+\omega\Delta^4\phi+4(n-2)\Delta^3\phi=0, \text{ etc.}$$

In the last of these equations the middle term is zero, hence beginning with the last, the successive $\Delta^k\phi$ are determined, and hence, from the first equation, ϕ itself.

Finally if to these equations we join $f+\phi.\omega=f_1$, where $f_1=0$ is the required normal form of the complex, multiply the first, second, etc. of them by

$$-\frac{\omega}{1.(n+1)},\quad \frac{\omega^2}{1.2(n+1)n},\quad -\frac{\omega^3}{1.2.3(n+1)n(n-1)}, \text{ etc.}$$

and add them all together, we obtain

$$f_1=f-\frac{\omega}{1.(n+1)}\Delta f+\frac{\omega^2}{1.2.(n+1)n}\Delta^2 f$$
$$-\frac{\omega^3}{1.2.3(n+1)n(n-1)}\Delta^3 f+\ldots.$$

The right side of this equation when put equal to zero gives the normal form of equation of the given complex.

288. Symbolic forms for the Complex surface and Singular surface. As in Art. 87 the expression

$$\Sigma(a_ib_h-a_hb_i)p_{ih}$$

may be written in either of the forms $a_xb_y-a_yb_x$, (α, β, x, y); hence the symbolic form of the normal equation of the complex is $(\alpha, \beta, x, y)^n=0$, or $(a_xb_y-a_yb_x)^n=0$. If we write

$$p_{12}=\pi_{34}=u_3v_4-u_4v_3, \text{ etc.,}$$

the symbolic form of equation becomes

$$(\alpha_u\beta_v-\alpha_v\beta_u)^n=0, \text{ or } (a, b, u, v)^n=0.$$

If in either of the first two forms we consider the x_i as constant, we obtain the symbolic form of equation of the complex cone of the point x; if in either of the second the u_i be taken as constant, we obtain the complex curve of the plane u in plane coordinates.

The Complex surface of a line a is the locus of intersection of consecutive complex cones whose vertices lie on a. If $y+\lambda z$ is any point of the line (y, z), the equation of its complex cone is, by the foregoing,

$$(a, b, x, y+\lambda z)^n=0, \text{ or, } \{(a, b, x, y)+\lambda(a, b, x, z)\}^n=0.$$

The locus of intersection of consecutive complex cones is therefore obtained by forming the discriminant of this equation for λ. Now the discriminant of the expression $(p_y + \lambda p_z)^n$ is equal to $\Sigma C\Pi (pp')$*; replacing (pp') by $p_y p_z' - p_z p_y'$, and p_y by (a, b, x, y) etc., we obtain as the equation of the complex surface of the line (y, z)

$$\Sigma C\Pi \{(a, b, x, y)(a', b', x, z) - (a, b, x, z)(a', b', x, y)\} = 0.$$

The number of the symbolic pairs ab, $a'b'$, ... is equal to the degree of the discriminant of a binary form of the nth degree, *i.e.* $2(n-1)$, while each pair enters n times into each term of the sum Σ, hence the last equation is of degree $2n(n-1)$ in x, or, *the Complex surface of a complex of the nth degree is of degree* $2n(n-1)$.

We may take as a definition of the singular surface, either that it is the locus of points whose complex cone has a double edge, or that it is the envelope of planes whose complex curve has a double tangent. The symbolic form of equation of the singular surface may be obtained as follows:—denoting symbolically a curve of the nth degree by

$$\gamma_x^n = \gamma_x'^n = \ldots\ldots = 0,$$

its discriminant Δ (whose vanishing is the condition for a double point) may be written symbolically in the form

$$\Delta = \Sigma C\Pi (\gamma\gamma'\gamma'').$$

The degree of Δ is $3(n-1)^2$ in the coefficients of the curve, hence this must be the number of the sets of symbols γ which appear; each γ must enter to the power n in each term of Δ, hence each such term must contain $n(n-1)^2$ determinant factors. Thus the condition that the section by the plane u of the surface of the nth degree

$$f = \gamma_x^n = \gamma_x'^n = \ldots\ldots = 0,$$

should have a double point, is

$$F \equiv \Sigma C\Pi (\gamma, \gamma', \gamma'', u) = 0.$$

The last equation represents, in general, the equation of the surface in plane coordinates. If f is a cone, this equation becomes

$$M \, . \, u_x^{n(n-1)^2} = 0,$$

where x is the vertex of the cone, and M involves x but not u.

* See Clebsch, *Vorlesungen über Geometrie*, Bd. I.

To apply to a complex cone, we observe that the equation of the latter being

$$(a_x b_y - a_y b_x)^n = 0,$$

where the y_i are current coordinates, we have, writing

$$\gamma_i = a_x b_i - b_x a_i,$$

and noticing that $\gamma_x = 0$,

$$\begin{aligned}(\gamma, \gamma', \gamma'', u) &= (a_x b - b_x a, \gamma', \gamma'', u)\\ &= a_x (b, \gamma', \gamma'', u) - b_x (a, \gamma', \gamma'', u)\end{aligned}$$

or, by a known theorem (Art. 91),

$$\begin{aligned}&= \gamma_x'' (a, b, \gamma', u) - \gamma_x' (a, b, \gamma'', u) - u_x (a, b, \gamma', \gamma'')\\ &= -(a, b, \gamma', \gamma'')\, u_x,\end{aligned}$$

since $$\gamma_x' = \gamma_x'' = 0.$$

Thus for the complex cone

$$F = u_x^{n(n-1)^2} . M, \text{ where } M = \Sigma C \Pi (b, a, \gamma', \gamma'').$$

Now if a plane not passing through the vertex cut a cone in a curve having a double point the cone must have a double edge, the condition for a double edge is therefore $M = 0$; each symbolic determinant factor is of the second degree in x, hence M is of degree $2n(n-1)^2$ in x, *i.e. the singular surface is of degree* $2n(n-1)^2$.

CHAPTER XVIII.

DIFFERENTIAL EQUATIONS CONNECTED WITH THE LINE COMPLEX.

289. LIE* has shown that with the line complex there is associated a partial differential equation of the first order, whose characteristic curves are principal tangent curves on Integral surfaces; and has also investigated certain types of partial differential equations of the second order by aid of conceptions drawn from both line- and sphere-geometry. The present chapter consists of a sketch of his researches.

In the partial differential equation of the first order $F(x, y, z, p, q)=0$, we may consider x, y, z to be the Cartesian coordinates of a point and $p, q, -1$ to be proportional to the direction cosines of the normal to a surface element (Art. 218) of the point; then the given differential equation selects ∞^4 surface elements, and the problem of integration is to determine surfaces whose surface elements shall satisfy the equation $F=0$. Through every point there pass ∞^1 surface elements which satisfy the given condition; these elements envelope a cone and if l, m, n are proportional to the direction cosines of any generator of this cone, since this generator is the intersection of the planes of two consecutive surface elements, we have

$$lp+mq-n=0, \quad ldp+mdq=0, \quad \frac{\partial F}{\partial p}dp+\frac{\partial F}{\partial q}dq=0;$$

hence

$$\frac{l}{F_p}=\frac{m}{F_q}=\frac{n}{pF_p+qF_q}.$$

Eliminating p and q from these equations and the equation $F=0$, we obtain an equation of the form

$$f(l, m, n, x, y, z)=0,$$

* See the memoir quoted on p. 233.

which is homogeneous in the quantities l, m, n, and hence represents the cone connected with any point (x, y, z).

Conversely, if the last equation be given, we obtain the corresponding equation $F = 0$, by eliminating l, m, n from the equations

$$p = -\frac{\dfrac{\partial f}{\partial l}}{\dfrac{\partial f}{\partial n}}, \quad q = -\frac{\dfrac{\partial f}{\partial m}}{\dfrac{\partial f}{\partial n}}, \text{ and } f = 0.$$

A differential equation of the form $f(x, y, z, dx, dy, dz) = 0$, which is homogeneous in dx, dy, dz, is usually termed a Monge equation. It assigns to each point of space a cone of directions, and hence leads, by the foregoing process, to a partial differential equation of the first order. A curve, such that the direction cosines of the tangents at all its points satisfy this Monge equation, is called an Integral curve.

290. The characteristic curves of a partial differential equation. It will be convenient to recall for reference some well-known results in the theory of partial differential equations of the first and second orders.

An equation of the form $F(x, y, z, p, q) = 0$ has three types of solutions:—

(i) A *complete* solution, viz. a solution of the form

$$z = f(x, y, a, b),$$

where a and b are arbitrary constants.

(ii) A *general* solution, obtained by making b a function of a and eliminating a from the equations

$$z = f(x, y, a, \phi(a)), \quad \frac{\partial f}{\partial a} + \frac{\partial f}{\partial \phi}\phi'(a) = 0.$$

These two equations represent, for any given value of a, a curve called a *characteristic* along which the surface $z = f$ is touched by its envelope, the general solution.

(iii) A *singular* solution, obtained by eliminating a and b from the equations

$$z = f(x, y, a, b), \quad \frac{\partial f}{\partial a} = 0, \quad \frac{\partial f}{\partial b} = 0.$$

There are ∞^3 characteristics of any given non-linear equation $F(x, y, z, p, q) = 0$. Through any point there pass ∞^1 characteristics whose tangents at the point are the generators for the point of the Monge equation connected with $F=0$. *At any point of an Integral surface of $F=0$, it is touched by the cone for the point of this Monge equation in the direction of a characteristic.* On each Integral surface there are ∞^1 characteristics, and through each characteristic there pass an unlimited number of integral surfaces: this is the distinguishing property of a characteristic, from it the equations determining the characteristics are seen to be

$$\frac{dx}{F_p}=\frac{dy}{F_q}=\frac{dz}{pF_p+qF_q}=\frac{dp}{-F_x-F_z p}=\frac{dq}{-F_y-F_z q}.$$

If two Integral surfaces touch along a curve, the curve is necessarily a characteristic, and an infinite number of Integral surfaces touch along a given characteristic*.

A *linear* equation $Pp+Qq-R=0$ has only ∞^2 characteristics, which are the curves $u=a$, $v=b$, the integrals of the equations

$$\frac{dx}{P}=\frac{dy}{Q}=\frac{dz}{R}.$$

Through any point there passes *one* characteristic; the surface elements determined at the point by the differential equation all contain the tangent to this characteristic.

The characteristics of the differential equation of the second order

$$Rr+Ss+Tt+U(rt-s^2)=V$$

have the same special property as those of the equation

$$F(x, y, z, p, q)=0;$$

on each Integral surface there is an infinite number of characteristics, whose differential equation is

$$U(dp\,dx+dq\,dy)+Rdy^2+Tdx^2-Sdx\,dy=0;$$

this is an equation of the second degree, showing that through each point of the surface there pass in general *two* characteristics, which however coincide if $S^2=4(RT+UV)$. If

$$F(x, y, z, p, q)=0$$

is a first integral of the equation, each of its characteristics is a characteristic of the equation of the second order.

291. The Monge equation of a line complex. The equation

$$f(ydz-zdy, zdx-xdz, xdy-ydx, dx, dy, dz)=0\ \ldots(\text{A}),$$

homogeneous in these six quantities, is known to give a line

* A discussion of the above results will be found in Lie and Scheffers' *Berührungstransformationen*, S. 498–511.

complex. It is easy to see that if the cones of a Monge equation for each point are those of a line complex, its equation must be of this form; for, regarding the ratios $dx : dy : dz$ as *given* and equal to $h : k : l$, the Monge equation gives the locus of points whose cones include the direction

$$\frac{dx}{h} = \frac{dy}{k} = \frac{dz}{l};$$

but if the cones are those of a line complex, this locus must be a *cylinder* whose axis has the given direction; and the equation of all cylinders whose axes are in this given direction is of the form

$$x - \frac{h}{l} z = \phi\left(y - \frac{k}{l} z\right)^*,$$

or, generally, $F(ly - kz, hz - lx, kx - hy) = 0$.

Hence, a Monge equation which represents a line complex must be of the form (A). Such an equation has among its integral curves the ∞^3 lines of the complex.

The equations which enable us to pass from a Monge equation to the corresponding partial differential equation $F = 0$, apply, of course, also in this case. The characteristics of the latter equation through any point have as tangents at the point the lines of the complex. The complex cone of any point touches at that point any Integral surface of F through it; continuing along one such surface in the direction thus indicated at each point, we obtain a characteristic c on this Integral surface; there are ∞^1 such curves c on the surface.

292. The characteristics on an Integral surface are principal tangent curves. Taking $f(x, y, z, x', y', z') = 0$ as the equation of a line complex, where $dx : dy : dz = x' : y' : z'$, it was seen that, corresponding to any given values of x', y', z', we obtain a complex cylinder the direction cosines of whose axis are proportional to x', y', z'; the direction cosines of the normal at any point (xyz) of this cylinder are proportional to f_x, f_y, f_z; hence

$$x'f_x + y'f_y + z'f_z \equiv 0.$$

Again taking the point (x, y, z) of any complex curve, its coordinates are functions of one parameter t, hence $dx = x'dt$, etc., and $\frac{df}{dt} = 0$, therefore

$$f_x x' + f_y y' + f_z z' + f_{x'} x'' + f_{y'} y'' + f_{z'} z'' = 0,$$

hence

$$x''f_{x'} + y''f_{y'} + z''f_{z'} = 0.$$

* Salmon, *Geom. of Three Dimensions*, p. 375.

Moreover, since f is homogeneous in x', y', z' we have

$$x'f_{x'} + y'f_{y'} + z'f_{z'} = 0,$$

so that
$$\frac{y'z'' - z''y'}{f_{x'}} = \frac{z'x'' - z''x'}{f_{y'}} = \frac{x'y'' - x''y'}{f_{z'}}.$$

Now these numerators are proportional to the direction cosines of the osculating plane of the given complex curve at the point (xyz), hence *all complex curves which touch at the same point have the same osculating plane at that point*: this plane touches the complex cone of the point.*

Considering any Integral surface S of $F=0$, at any point P of S the tangent plane, common to S and the complex cone of P, thus osculates the characteristic c on S which passes through P, hence *c is a principal tangent curve on S.* On any Integral surface, therefore, the characteristics of $F=0$ form one set of principal tangent curves.

Conversely, if in a Monge equation the osculating plane at each point of every Integral curve touches the cone assigned to the point by the Monge equation, the latter equation is that of a line complex; for if $f(x, y, z, x', y', z')=0$ is the given Monge equation, by the given condition we have

$$\frac{y'z'' - y''z'}{f_{x'}} = \frac{z'x'' - z''x'}{f_{y'}} = \frac{x'y'' - x''y'}{f_{z'}};$$

hence
$$x''f_{x'} + y''f_{y'} + z''f_{z'} = 0.$$

But the points of the Integral curve, being functions of one parameter t, satisfy the condition $\frac{df}{dt} = 0$, hence

$$x'f_x + y'f_y + z'f_z + x''f_{x'} + y''f_{y'} + z''f_{z'} = 0,$$

i.e. the equation $x'f_x + y'f_y + z'f_z = 0$ is satisfied by each line element of the Monge equation; this shows that the locus of points, which have a line element of the Monge equation in a given direction, is a cylinder, and therefore, by the foregoing, the Monge equation is seen to be that of a line complex.

293. Form of the partial differential equation corresponding to a line complex†. The differential equation of the principal tangents on any Integral surface of $F=0$

is
$$dp\,dx + dq\,dy = 0;$$

* Compare with the result obtained for a linear complex (Art. 41). These curves have also the same *torsion*, see L. and S. *Berühr.* S. 309.

† See L. and S. *Berühr.* S. 640.

expressing that this equation is satisfied by a characteristic of $F=0$ we obtain

$$F_x F_p + F_y F_q + F_z(pF_p + qF_q) = 0$$

as an equation which is identically satisfied by any partial differential equation $F=0$ thus derived from a line complex.

Conversely it may be shown that the Monge equation of every non-linear equation $F=0$, which satisfies the above identity, is that of a line complex. For consider the points at which a generator of the cone of the Monge equation has a given direction l, m, n; then we have by Art. 289

$$\frac{l}{F_p} = \frac{m}{F_q} = \frac{n}{pF_p + qF_q} \quad \ldots\ldots\ldots\ldots\ldots\ldots \text{(I)};$$

the points in question form a surface which is obtained by eliminating p and q from these equations and $F=0$.

Now the condition

$$F_x F_p + F_y F_q + F_z(pF_p + qF_q) = 0$$

becomes

$$lF_x + mF_y + nF_z = 0 \quad \ldots\ldots\ldots\ldots\ldots \text{(II)};$$

and since

$$lp + mq - n \equiv 0,$$

where p and q have values determined by (I) for each point of the surface considered, we have

$$l\frac{\partial p}{\partial x} + m\frac{\partial q}{\partial x} = 0, \quad l\frac{\partial p}{\partial y} + m\frac{\partial q}{\partial y} = 0, \quad l\frac{\partial p}{\partial z} + m\frac{\partial q}{\partial z} = 0,$$

over the surface, *i.e.*

$$F_p\frac{\partial p}{\partial x} + F_q\frac{\partial q}{\partial x} = 0, \text{ etc.} \quad \ldots\ldots\ldots\ldots\ldots \text{(III)}.$$

But we may regard $F(x, y, z, p, q) = 0$ as the equation of this surface, provided p and q have the values assigned to them from (I); the direction cosines of its normal at any point are proportional to

$$F_x + F_p\frac{\partial p}{\partial x} + F_q\frac{\partial q}{\partial x}, \quad F_y + F_p\frac{\partial p}{\partial y} + F_q\frac{\partial q}{\partial y}, \quad F_z + F_p\frac{\partial p}{\partial z} + F_q\frac{\partial q}{\partial z};$$

thus the conditions (II) and (III) show that the normal at each point is perpendicular to the given direction l, m, n; *i.e.* the surface is a *cylinder*; hence, as before, the Monge equation is that of a line complex.

Hence, if a non-linear partial differential equation $F=0$ has the property that on every Integral surface the ∞^1 characteristics are principal tangent curves, its Monge equation is that of a line complex.

294. If the differential equation is linear, it has ∞^2 characteristics (Art. 290), *which must in the present case be straight lines*; for the planes of the surface elements of each point form a pencil whose axis is that of the tangent to the *one* characteristic through the point; and if each plane of the pencil is to be the osculating plane of the characteristic at the point and this is to occur for *each point* of the characteristic, the latter must be a straight line. The Integral surfaces are therefore in this case ruled surfaces of the complex.

295. Contact transformations of space*. If we adopt a different notation in the equations (vi) of Art. 217, viz. by writing X, $-Y$, Z for α, β, γ respectively, and $-z$, y, x for x, y and z respectively, these equations assume the form

$$\left.\begin{aligned} X + iY + xZ + z &= 0, \\ x(X - iY) - Z - y &= 0\dagger \end{aligned}\right\} \quad \text{(i).}$$

The linear complex corresponding to the points (XYZ) of Σ is then

$$xdy - ydx + dz = 0.$$

Proceeding as in Art. 219 we find that the surface element $(X, Y, Z; P, Q, -1)$ of Σ corresponding to the surface element $(x, y, z; p, q, -1)$ of Λ is determined by the equations (i) together with the equations

$$Z = -\frac{px + qy}{q + x}, \quad P = \frac{xq + 1}{q - x}, \quad Q = -i\frac{xq - 1}{q - x}.$$

By differentiation of the equations (i) we find that

$$\begin{aligned} dz - pdx - qdy &= -(dX + idY + xdZ + Zdx) \\ &\quad - pdx - q\{x.\overline{dX - idY} - dZ + \overline{X - iY}.dx\} \\ &= (q - x)\left\{dZ - \frac{xq + 1}{q - x}dX + i\frac{qx - 1}{q - x}dY\right\}, \end{aligned}$$

since the coefficient of dx is

$$\begin{aligned} -Z - p - q(X - iY) &= q\left(\frac{y - p}{q + x} - \overline{X - iY}\right) \\ &= q\left(\frac{Z + y}{x} - \overline{X - iY}\right) = 0. \end{aligned}$$

Hence $\quad dz - pdx - qdy = (q - x)\{dZ - PdX - QdY\}.$

If $dz - pdx - qdy$ is zero, the two consecutive surface elements

$$(x, y, z, p, q), \; (x + dx, \; y + dy, \; z + dz, \; p + dp, \; q + dq)$$

* See L. and S. *Berühr.* S. 522.

† This form of the equations is used in L. and S. *Berühr.* S. 463.

are in a 'united position,' *i.e.* the point of the second surface element lies in the plane of the first, and we therefore conclude that if two consecutive surface elements of Λ are in a united position, so are their corresponding surface elements in Σ; hence, *to the* ∞^2 *surface elements of any surface of* Λ *correspond* ∞^2 *surface elements of* Σ *which also belong to a surface.*

By the transformation considered, the equation $F=0$ gives rise to a new partial differential equation $F_1(X, Y, Z, P, Q)=0$; each surface element which satisfies $F=0$ leads to a surface element which satisfies $F_1=0$. Any two of the Integral surfaces of F which touch along a characteristic give rise to two Integral surfaces of F_1 which touch along a curve which is therefore a characteristic of F_1.

Thus to the characteristics of F there correspond characteristics of F_1; if therefore the characteristics of F are principal tangent curves, those of F_1 are lines of curvature (Art. 220); *to the line complex of principal tangents corresponds a sphere complex of principal spheres.*

296. The trajectory circle. In the correspondence of Art. 215 between the spaces Λ and Σ, the two lines which correspond to a sphere of centre (XYZ) and radius H are given by the equations

$$X+iY=s, \qquad \pm H+Z=r,$$
$$X-iY=\rho, \qquad \pm H-Z=\sigma.$$

Any equation $H=F(XYZ)$ may therefore be taken to represent *either* a line-complex *or* a sphere-complex. The tangent linear complex

$$H-H_0=\frac{\partial H_0}{\partial X_0}(X-X_0)+\frac{\partial H_0}{\partial Y_0}(Y-Y_0)+\frac{\partial H_0}{\partial Z_0}(Z-Z_0)$$

contains the sphere (X_0, Y_0, Z_0, H_0) and also every consecutive sphere of the complex. It is easy to see that this complex is formed by spheres which cut the plane

$$-H_0=\frac{\partial H_0}{\partial X_0}(X-X_0)+\frac{\partial H_0}{\partial Y_0}(Y-Y_0)+\frac{\partial H_0}{\partial Z_0}(Z-Z_0)\ldots\text{(i)}$$

at a constant angle; hence the points of contact of the sphere $(X_0Y_0Z_0H_0)$ with every consecutive sphere of the complex which touches it, lie on the circle which is the intersection of (i) with the sphere

$$(X-X_0)^2+(Y-Y_0)^2+(Z-Z_0)^2=H_0^2\ldots\ldots\ldots\text{(ii)},$$

where in equations (i) and (ii), the X, Y, Z are current Cartesian coordinates.

This circle is called the Trajectory Circle of the given sphere, and has important bearings on Lie's theory. The surface elements of the sphere at the points of this circle belong to different Integral surfaces of the differential equation $F_1 = 0$ which corresponds to the given sphere complex. The ∞^4 surface elements of the complex thus consist of ∞^1 surface elements on each of the ∞^3 spheres of the complex.

If two consecutive spheres $(X_0 Y_0 Z_0 H_0)$ and $(X_0 + dX, \ldots)$ touch each other, we have as the condition

$$dX^2 + dY^2 + dZ^2 = dH^2 = \left(\frac{\partial H_0}{\partial X_0} dX + \frac{\partial H_0}{\partial Y_0} dY + \frac{\partial H_0}{\partial Z_0} dZ\right)^2.$$

This is a Monge equation whose cone for the point $(X_0 Y_0 Z_0)$ is

$$(X - X_0)^2 + (Y - Y_0)^2 + (Z - Z_0)^2$$
$$= \left\{\frac{\partial H_0}{\partial X_0}(X - X_0) + \frac{\partial H_0}{\partial Y_0}(Y - Y_0) + \frac{\partial H_0}{\partial Z_0}(Z - Z_0)\right\}^2.$$

This cone is seen to be one of revolution and to contain the trajectory circle of the sphere $(X_0 Y_0 Z_0 H_0)$. It is termed by Lie the *elementary complex cone* of the sphere complex and gives the direction of those points consecutive to $(X_0 Y_0 Z_0)$ whose complex spheres touch the given sphere. We observe that if of the family of surfaces H = constant, the one be taken which passes through $(X_0 Y_0 Z_0)$, the normal to this surface at this point, which has direction cosines proportional to

$$\frac{\partial H_0}{\partial X_0}, \quad \frac{\partial H_0}{\partial Y_0}, \quad \frac{\partial H_0}{\partial Z_0},$$

is clearly *the axis of the elementary complex cone of* $(X_0 Y_0 Z_0)$.

297. Partial differential equations whose characteristics are geodesics. If we consider any point P, or $(X_0 Y_0 Z_0)$, which lies upon a surface $H = C$, and take a point Q, or

$$(X_0 + dX, \quad Y_0 + dY, \quad Z_0 + dZ),$$

in which the elementary complex cone of P meets a consecutive surface $H = C + dC$, then since the latter surface passes through Q, we have

$$dC = \frac{\partial H_0}{\partial X_0} dX + \frac{\partial H_0}{\partial Y_0} dY + \frac{\partial H_0}{\partial Z_0} dZ$$
$$= \sqrt{dX^2 + dY^2 + dZ^2}$$
$$= PQ.$$

The surface $H = C + dC$, therefore, cuts off from *each* generator of this cone the same length dC.

From the Monge equation of the elementary complex cones we derive a partial differential equation, which will shortly be given, whose surface elements touch these cones along characteristics, (Art. 290).

Fig. 13.

Consider a surface element of this differential equation which touches the cone of its point along PQ and meets the surface $H = C$ along PP'; then it is clear that PQ and PP' are at right angles, since the cone is one of revolution having its axis normal at P to the surface $H = C$.

On each surface element, therefore, of any Integral surface of this differential equation there are thus determined two orthogonal directions, giving two families of orthogonal curves on the surface PP', QQ', ... and PQ, $P'Q'$, ...; the former being the intersections of the Integral surface with the surfaces $H = C$, the latter the characteristics of the partial differential equation. But since it has been seen that $dC = PQ = P'Q'$, etc., the curves PP', QQ', ... are *parallel* curves, and therefore their orthogonal trajectories are *geodesics*.

Hence *the characteristics of the partial differential equation derived from the elementary complex cones are geodesics on its Integral surfaces.*

The differential equation in question is found by eliminating l, m, n from the equations

$$\Phi \equiv l^2 + m^2 + n^2 - \left(\frac{\partial H}{\partial X} l + \frac{\partial H}{\partial Y} m + \frac{\partial H}{\partial Z} n\right)^2 = 0,$$

$$P = -\frac{\partial \Phi}{\partial l} \div \frac{\partial \Phi}{\partial n}, \quad Q = -\frac{\partial \Phi}{\partial m} \div \frac{\partial \Phi}{\partial n};$$

and is therefore easily found to be

$$(PH_X + QH_Y - H_Z)^2 - (H_X^2 + H_Y^2 + H_Z^2 - 1)(P^2 + Q^2 + 1) = 0.$$

298. We shall now show that the characteristic of the surface element at any point P of the trajectory circle of any sphere of a sphere-complex, is perpendicular to the trajectory circle.

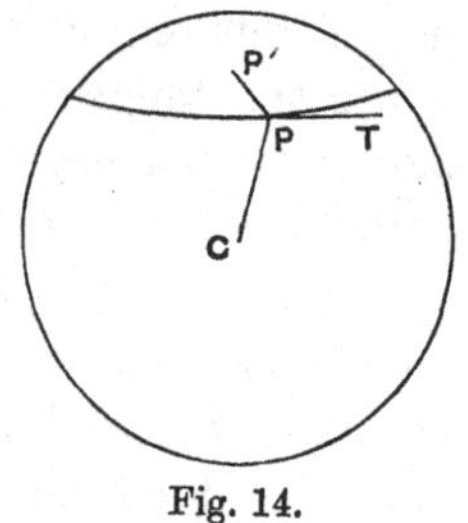

Fig. 14.

For since P is a point on each of the ∞^1 complex spheres $(X_0 Y_0 Z_0 H_0)$ which are principal spheres at the point P, or (XYZ), these spheres are determined by the equations

$$(X - X_0)^2 + (Y - Y_0)^2 + (Z - Z_0)^2 - H_0^2 = 0,$$

$$H_0 + (X - X_0)\frac{\partial H_0}{\partial X_0} + (Y - Y_0)\frac{\partial H_0}{\partial Y_0} + (Z - Z_0)\frac{\partial H_0}{\partial Z_0} = 0,$$

$$H_0 = F(X_0 Y_0 Z_0).$$

The line of intersection of two consecutive surface elements at P gives the direction of the characteristic; now this line is clearly perpendicular to the line joining the centres of the two corresponding principal spheres, *i.e.* to the line whose direction cosines are proportional to dX_0, dY_0, dZ_0. Hence, if l, m, n are the direction cosines of the required characteristic, we have

$$l dX_0 + m dY_0 + n dZ_0 = 0,$$

$$l(X - X_0) + m(Y - Y_0) + n(Z - Z_0) = 0.$$

But since the consecutive principal sphere passes through (XYZ) we have

$$(X - X_0)\, dX_0 + (Y - Y_0)\, dY_0 + (Z - Z_0)\, dZ_0 + H_0 dH_0 = 0,$$

i.e.

$$\left(X - X_0 + H_0\frac{\partial H_0}{\partial X_0}\right) dX_0 + \left(Y - Y_0 + H_0\frac{\partial H_0}{\partial Y_0}\right) dY_0 + \left(Z - Z_0 + H_0\frac{\partial H_0}{\partial Z_0}\right) dZ_0 = 0.$$

Hence the line whose direction cosines are proportional to

$$X - X_0 + H_0\frac{\partial H_0}{\partial X_0}, \quad Y - Y_0 + H_0\frac{\partial H_0}{\partial Y_0}, \quad Z - Z_0 + H_0\frac{\partial H_0}{\partial Z_0},$$

satisfies the conditions which determine the characteristic.

Hence *the tangent of the characteristic at P is perpendicular to the tangent plane of the elementary complex cone of C along CP, i.e. is perpendicular to the trajectory circle.*

The trajectory circle derives its name from this property.

It follows that the tangent of the trajectory circle at P touches at that point a line of curvature of an integral surface of the partial differential equation connected with the sphere-complex; hence the tangent plane of the elementary complex cone of C along CP touches at C the locus of centres of curvature of an Integral surface of this differential equation.

Now the differential equation whose characteristics are geodesics on Integral surfaces was derived from the Monge equation of the elementary complex cones of $H = F(XYZ)$; if $F_1 = 0$ be the partial differential equation which corresponds to this sphere-complex, it follows that each Integral surface of F_1 has as its surface of centres of curvature an Integral surface of the differential equation whose characteristics are geodesics.

299. The preceding investigations have shown the existence of three classes of partial differential equations of the first order whose solutions are mutually dependent; they are

(i) those whose characteristics are principal tangent curves on Integral surfaces, they are termed by Lie the equations D_{11};

(ii) those whose characteristics are lines of curvature on Integral surfaces, or D_{12};

(iii) those whose characteristics are geodesics on Integral surfaces, or D_{13}.

300. The complex of normals. Before leaving partial differential equations of the first order, Lie's solution of the following problem will be given, viz. the problem *to determine all surfaces whose normals belong to a given line-complex.* Let

$$f(ydz - zdy,\ zdx - xdz,\ xdy - ydx,\ dx,\ dy,\ dz) = 0$$

be the given line-complex; since the lines of each complex cone are to be perpendicular to the surface elements of the required differential equation at the vertex of the cone, we must have

$$dx : dy : dz = p : q : -1,$$

hence the differential equation is

$$f(-y - zq,\ zp + x,\ xq - yp,\ p,\ q,\ -1) = 0,$$

or, since $$xq - yp = q(x + zp) - p(y + zq),$$

the differential equation of the required surfaces has the form

$$F(y + zq,\ x + zp,\ p,\ q) = 0.$$

301. Partial differential equations of the second order connected with line- and sphere-complexes. Lie has shown that the solutions of certain classes of partial differential equations of the second order are intimately associated with complexes of lines and spheres. He considers in the first place the problem *to determine all surfaces of which one set of principal tangent curves belongs to a given line-complex.*

Let

$$f(ydz - zdy,\ zdx - xdz,\ xdy - ydx,\ dx,\ dy,\ dz) = 0$$

be the given complex, and

$$z = \phi(x, y)$$

a surface which satisfies the given condition; its principal tangents are given by the equation

$$rdx^2 + 2sdxdy + tdy^2 = 0 \quad \text{.................(i)};$$

to express that a complex curve is a principal tangent curve of this surface, we substitute in $f=0$

$$dz = pdx + qdy,$$

and thence determine one or more values of $\frac{dy}{dx}$ in the form

$$\frac{dy}{dx} = N(x, y, z, p, q).$$

Substituting this value of $\frac{dy}{dx}$ in (i) we obtain

$$r + 2Ns + N^2t = 0 \quad \text{...................(ii)}$$

as the differential equation whose Integral surfaces satisfy the given condition.

Of this equation we know already two types of solutions:

(a) ruled surfaces of the complex,

(b) surfaces which at each point are touched by the complex cone of the point, *i.e.* solutions of the D_{11} connected with the given line-complex.

Moreover there are no other solutions than these; for if an Integral surface contains ∞^1 *straight* principal tangent lines, *i.e.* generators, which belong to the complex, it is a ruled surface of the complex; if it contains ∞^1 complex curves as principal tangent curves, the osculating plane of such a curve at any point touches the surface, and we have seen that it also touches the complex cone of the point (Art. 292), so that the surface is touched at each point by the complex cone of the point, *i.e.* the surface is a solution of the corresponding D_{11}. This differential equation of the second order is one whose two characteristics through each point of an Integral surface *coincide*; the characteristics are therefore identical with one set of principal tangent curves on each Integral surface.

The symbol D_{21}' is employed by Lie to denote this class of differential equations.

302. The consideration of a sphere-complex leads to the problem *to determine all surfaces whose principal spheres belong to a given sphere-complex.*

The differential equation of the second order defining such surfaces is similarly found to be

$$(rt-s^2)R^2-[(1+p^2)t-2pqs$$
$$+(1+q^2)r]\sqrt{1+p^2+q^2}\,.\,R+(1+p^2+q^2)^2=0\,;$$

wherein R is a function of x, y, z, p and q. The symbol D_{22}' is introduced to denote such equations.

303. Partial differential equations of the second order on whose Integral surfaces both sets of characteristics are principal tangent curves or lines of curvature. We now consider two types of partial differential equations of the second order; the first type is such that on each Integral surface the two sets of characteristics are the principal tangent curves; the second type is such that the two sets of characteristics are the lines of curvature. They are denoted respectively by D_{21}'' and D_{22}''.

Since the differential equation of the characteristics of

$$rt-s^2+Ar+2Bs+Ct=F(x,y,z,p,q)$$

is
$$(A+t)\,dy^2+2(s-B)\,dxdy+(C+r)\,dx^2=0,$$

if the characteristics coincide with the principal tangent curves, whose equation is

$$rdx^2+2sdxdy+tdy^2=0,$$

we must have
$$A=B=C=0$$

and the equation of a D_{21}'' is

$$rt-s^2=F(x,y,z,p,q).$$

It has been shown by Du Bois-Reymond* that the equation of a D_{22}'' is

$$r-\frac{1+p^2}{pq}\,.\,s+\left(\frac{1+q^2}{pq}s-t\right)F=0,$$

where F is any function of x, y, z, p, q.

It is easy to see that this equation is equivalent to the following:

$$[pqt-(1+q^2)s]f^2+[(1+p^2)t-(1+q^2)r]f+[(1+p^2)s-pqr]=0,$$

where the functions f and F are connected by the equation

$$F\{(1+q^2)f+pq\}-\{f^2pq+f(1+p^2)\}=0.$$

304. The curves s and σ of a D_{22}''. If in the equation last given for a D_{22}'', $\dfrac{dx}{dy}$ be substituted for f, we obtain the

* See *Partial Differential Equations*, p. 130.

differential equation of the lines of curvature of any Integral surface; it follows that a *given* D_{22}'' assigns to any surface element of space two definite orthogonal directions; on any surface we thus obtain two sets of curves s and σ which are orthogonal.

A D_{22}'' is thus an analytical expression of the problem *to determine the most general surface whose directions of principal curvature depend by a given law on the position of the corresponding surface element.*

The characteristics of any particular integral of a D_{22}'' of the form $f(xyzpq)=0$ will be also characteristics of the D_{22}'', and therefore lines of curvature of the Integral surfaces on which they lie; hence, such a particular integral must be a D_{12}*. If a D_{22}'' has a first integral which is a D_{12}, to the latter will correspond a sphere-complex, and considering any sphere of this complex it is clear that *its trajectory circle belongs to one of the two sets of curves s and σ on this sphere.*

If we consider a singly infinite number of integrals of this kind, and therefore ∞^1 sphere-complexes, all spheres of space are thereby included. Hence for every sphere it follows that among its curves s, σ there is included one or several trajectory circles. If finally a *general* first integral exists of the form $u=f(v)$, where u and v are each functions of x, y, z, p, q, there are two conceivable cases, viz. among the curves s, σ of any sphere, there is *one or a finite number* of trajectory circles, *or* there is an *infinite* number. We shall see that the first case is impossible.

For if it were possible, let

$$H=F(XYZ)$$

be a sphere-complex whose D_{12} is a first integral of the given D_{22}''; then the trajectory circle of a sphere $(X_0Y_0Z_0H_0)$ of this complex lies on the plane

$$\frac{\partial H_0}{\partial X_0}(X-X_0)+\frac{\partial H_0}{\partial Y_0}(Y-Y_0)+\frac{\partial H_0}{\partial Z_0}(Z-Z_0)+H_0=0.$$

But this trajectory circle being assumed to be one of a *finite* number of such circles on the given sphere must be determined by the D_{22}'', *i.e.* the coefficients

$$\frac{\partial H_0}{\partial X_0},\ \frac{\partial H_0}{\partial Y_0},\ \frac{\partial H_0}{\partial Z_0}$$

* It is to be noticed, that as in the case of a D_{11}, there are two types of solutions of a D_{12}.

are determinate functions F_1, F_2, F_3 of X_0, Y_0, Z_0 and H_0, hence

$$\frac{\partial H_0}{\partial X_0} = F_1, \quad \frac{\partial H_0}{\partial Y_0} = F_2, \quad \frac{\partial H_0}{\partial Z_0} = F_3;$$

these equations give an integral involving *arbitrary constants* and not arbitrary functions; therefore the assumed complex

$$H = F(XYZ)$$

cannot relate to a *general* first integral D_{12} of the D_{22}''.

Hence *if a D_{22}'' has a general first integral, on any sphere one set of the curves s, σ consists of circles and the other of a set of orthogonal curves. If a D_{22}'' has two general first integrals the curves s, σ form two sets of orthogonal circles, and hence each member of one set passes through two fixed points, and each member of the other set through two other fixed points.*

305. To determine when a D_{21}'' or a D_{22}'' has a general first integral. The equation $F(X, Y, Z, H, \lambda) = 0$, where λ is a parameter, represents ∞^1 line- or sphere-complexes C, which we take to be linear; if we form in the usual way the envelope complex, viz. eliminate λ from this equation and the equation $\frac{\partial F}{\partial \lambda} = 0$, we obtain a complex A. Then, considering *line*-complexes, it is seen that for any point, the ∞^1 planes of the complexes C touch the complex cone of the point for the complex A; for two consecutive linear complexes intersect in a linear congruence and the complex A may thus be regarded as composed of ∞^1 linear congruences of which two consecutive congruences belong to the same linear complex; considering therefore the lines of A which pass through any point, we see that two such *consecutive* lines always belong to the plane of one linear complex C.

If a D_{21}'' has a general first integral of the form $u = f(v)$, it must be a D_{11}; to the latter a line-complex corresponds, which is termed an Integral Complex. Each of the equations $u = \text{constant}$, $v = \text{constant}$ gives ∞^1 Integral complexes; they may be united in ∞^2 ways in pairs such as $u = u_a$, $v = v_b$; such a pair assigns to each point P of space one or several surface elements, viz. those of the common tangent planes of the complex cones of P in the two corresponding Integral complexes.

If we choose from the ∞^2 pairs a set of ∞^1 by any definite law $u_a = f(v_b)$, we thereby assign to each point P ∞^1 surface elements; they are the tangent planes of the complex cone of P in the Integral complex corresponding to $u = f(v)$.

Two consecutive pairs (u_a, v_b), $(u_a + \Delta u_a, v_b + \Delta v_b)$ assign to each point one or several directions which belong to an unlimited number of Integral complexes*; such directions therefore are common to all these complexes and hence belong to a line-congruence; by giving $\frac{\Delta u_a}{\Delta v_b}$ all values and keeping u_a and v_b constant we obtain ∞^1 such congruences forming a line-complex C; this complex is linear, or consists of several linear complexes; since its lines for each point lie in the plane (or planes) assigned to the point by $u = u_a$, $v = v_b$.

Now making $u_a = f(v_b)$, where f is some definite function, we obtain ∞^1 linear complexes C, whose planes for any point P are the tangent planes of the cone of P in the Integral complex corresponding to $u = f(v)$.

Hence, *if a* D_{21}'' *has a general first integral of the form* $u = f(v)$, *there correspond to it* ∞^2 *linear complexes* C *such that any* ∞^1 *complexes* C *have an envelope complex whose* D_{11} *is a particular first integral.*

Conversely, it can be shown that ∞^2 linear complexes determine a D_{21}'' (or a D_{22}''), with a general first integral. We notice in the first place that a *linear* line-complex determines ∞^3 surface elements, whose planes are its polar planes for the ∞^3 points of space; similarly a linear sphere-complex determines ∞^3 surface elements, for the spheres of such a complex cut a fixed sphere S at a constant angle, and therefore the ∞^3 surface elements along their trajectory circles have their points on S and intersect S at a constant angle; through each point of S there pass ∞^1 of these surface elements.

If therefore we consider ∞^2 linear sphere-complexes C, we obtain ∞^5 surface elements; each surface element of space belongs to one complex C (or a finite number of complexes C), and taking the intersection of each surface element of space with the "fundamental" sphere S of the complex C to which it belongs, we assign to each surface element two orthogonal directions which thus depend merely on the *position* of the surface element; this process therefore determines the D_{22}'' which performs the same definite assignment. The envelope of any set of ∞^1 linear complexes gives a complex which has a corresponding D_{12}; the surface elements of this D_{12} for any point are such that each is intersected

* Since there is an infinite number of functions f satisfying the conditions
$$u_a = f(v_b), \quad u_a + \Delta u_a = f(v_b + \Delta v_b).$$

along the direction assigned to it by the D_{22}'', by the consecutive surface element; hence this D_{12} is a particular integral of the D_{22}''.

306. D_{21}'' and D_{22}'' with two general first integrals. It has been seen that a D_{22}'' determines on every sphere S two sets of circles which are mutually orthogonal; hence all circles of one set pass through two points P_1 and P_2, and all of the other set through two points Q_1 and Q_2. The Integral complexes which contain S consist of two systems, viz. those whose trajectory circles pass through P_1 and P_2, and those whose trajectory circles pass through Q_1 and Q_2. It is obvious that all Integral complexes of the first system which contain S will also contain the two consecutive spheres S' and S'' which touch S respectively at P_1 and P_2; now the D_{22}'' determines on S' two points P_1' and P_2', and since P_1 lies on the trajectory circle of S' for all the Integral complexes considered, it is clear that P_1' coincides with P_1. Thus there are ∞^1 spheres touching at P_1, all of which belong to these complexes. Continuing this argument it is clear that all these Integral complexes have at least ∞^2 spheres in common which divide themselves into ∞^1 "pencils" of ∞^1 spheres (there cannot be ∞^3 spheres in common, for then the complexes would be identical).

Hence we have, for every sphere S, two sphere-congruences, one, which may be denoted by K, corresponding to the points P_1 and P_2; and another, denoted by L, corresponding to the points Q_1 and Q_2; every sphere of K belongs to each of two pencils of ∞^1 spheres in contact at the same point; similarly for L.

We have in correspondence in the space Λ, two line-congruences K_1 and L_1 which are disposed in ∞^1 plane pencils and so that each line of K_1 belongs to *two* of these pencils; but this is peculiar to *linear* congruences, therefore K_1 and L_1 are both linear congruences, and hence also K and L.

Now every sphere S gives rise to such a pair of linear congruences K and L, and since it is seen that the spheres of a congruence, *e.g.* K, form a closed system, *i.e.* the same congruence is obtained with whichever sphere of K we begin; it follows that there are, corresponding to the ∞^4 spheres of space, ∞^2 such congruences K and L.

Moreover, any pair of line-congruences K_1 and L_1 are such that their directrices form a twisted quadrilateral. For, a sphere S, common to K and L, touches the two fixed spheres S_1 and S_2 of the congruence K to which S belongs, at the points P_1 and P_2, and touches the two fixed spheres S_3 and S_4 of the other congruence L, at the points Q_1 and Q_2; and it has been seen that P_1, P_2 and Q_1, Q_2 are such that the circles through the one pair cut orthogonally the circles through the other pair. Hence, if we invert S from any point upon it, we obtain a plane π which touches the inverse spheres S_1', S_2'; S_3', S_4' at points having the same property for circles of the plane π. *It can be shown that S_1' and S_2' touch S_3' and S_4'.* For, taking π as the plane $z = 0$, and the lines joining the two pairs of points as the axes of coordinates in π, it is seen that if $P_1'P_2' = 2a$, the equations of S_1', S_2', S_3' and S_4' are respectively

$$(x-a)^2 + y^2 + (z - R_1)^2 = R_1^2,$$

$$(x+a)^2 + y^2 + (z - R_2)^2 = R_2^2,$$

$$x^2 + (y - ai)^2 + (z - R_3)^2 = R_3^2,$$

$$x^2 + (y + ai)^2 + (z - R_4)^2 = R_4^2.$$

Hence, since

$$a^2 - a^2 + (R_1 - R_3)^2 = (R_1 - R_3)^2, \text{ \&c.},$$

the result just stated follows at once.

Consequently, in the corresponding line-congruences K_1 and L_1, it is seen that the directrices of K_1 meet those of L_1.

Hence, each of the directrices of the ∞^2 congruences K_1 meets all directrices of the ∞^2 congruences L_1; and neither set of directrices can be ∞^2 in number. Hence the directrices of the congruences K_1 form a regulus ρ of which the directrices of L_1 form the complementary regulus ρ'.

Hence, *every D_{22}'' with two general first integrals gives rise to two sets of ∞^2 linear sphere-congruences K and L*; since the first integral corresponding to the congruences K contains an arbitrary function, the corresponding Integral complex can be determined so as to contain any ∞^1 spheres (belonging to different congruences K), so that any ∞^1 congruences K determine an Integral complex of the D_{22}''. Thus any correspondence established between pairs of lines of a given regulus ρ gives rise to a line-complex, whose corresponding D_{11} is a first integral of a D_{21}''; while any correspondence between the lines of the conjugate ρ' gives a first integral of a second species: and generally, two conjugate systems of three terms of linear complexes define the most general D_{21}'' or D_{22}'' with two general first integrals.

307. Application to the quadratic complex*. By introduction of the "elliptic" coordinates of a line (Art. 130), the condition $\Sigma dx_i^2 = 0$ for intersection of two consecutive lines assumes the form

$$d\mu_1^2 \cdot \frac{(\mu_1 - \mu_2)(\mu_1 - \mu_3)(\mu_1 - \mu_4)}{f(\mu_1)} + \ldots = 0,$$

where

$$f(\mu) \equiv (\lambda_1 + \mu)(\lambda_2 + \mu)(\lambda_3 + \mu)(\lambda_4 + \mu)(\lambda_5 + \mu)(\lambda_6 + \mu).$$

The partial differential equation $\Sigma \left(\frac{\partial \phi}{\partial x_i}\right)^2 = 0$ of a *special* complex (Art. 280) becomes in terms of the variables μ_i

$$\left(\frac{\partial \phi}{\partial \mu_1}\right)^2 \frac{f(\mu_1)}{(\mu_1 - \mu_2)(\mu_1 - \mu_3)(\mu_1 - \mu_4)} + \ldots = 0.$$

But of this partial differential equation a complete solution is known†, viz.

$$\phi = \int d\mu_1 \frac{\sqrt{(\mu_1 - a)(\mu_1 - b)}}{\sqrt{f(\mu_1)}} + \int d\mu_2 \frac{\sqrt{(\mu_2 - a)(\mu_2 - b)}}{\sqrt{f(\mu_2)}}$$
$$+ \int d\mu_3 \frac{\sqrt{(\mu_3 - a)(\mu_3 - b)}}{\sqrt{f(\mu_3)}} + \int d\mu_4 \frac{\sqrt{(\mu_4 - a)(\mu_4 - b)}}{\sqrt{f(\mu_4)}} + C,$$

in which a, b and C are arbitrary constants.

Attributing to a, b and C any definite values, we obtain a surface *whose two sets of principal tangents belong respectively to the cosingular quadratic complexes* $\mu = a$, $\mu = b$; to prove this we have merely to show (Art. 283) that the congruence $\phi = 0$, $\psi = 0$ is *special*, *i.e.* that the directrices of the linear congruence

$$\Sigma y_i \frac{\partial \phi}{\partial x_i} = 0, \quad \Sigma y_i \frac{\partial \psi}{\partial x_i} = 0$$

coincide, which occurs if

$$\Sigma \left(\frac{\partial \phi}{\partial x_i}\right)^2 \Sigma \left(\frac{\partial \psi}{\partial x_i}\right)^2 - \left(\Sigma \frac{\partial \phi}{\partial x_i} \cdot \frac{\partial \psi}{\partial x_i}\right)^2 \equiv 0,$$

where ϕ is the special complex considered, and ψ is either of the complexes $\mu_4 = a$, $\mu_4 = b$.

Now $\Sigma \left(\frac{\partial \phi}{\partial x_i}\right)^2$ is zero by hypothesis, hence we have to show that

$$\Sigma \frac{\partial \phi}{\partial x_i} \cdot \frac{\partial \psi}{\partial x_i} \equiv 0.$$

* See Klein, *Math. Ann.* v.

† See Jacobi's *Vorlesungen über Dynamik*.

The last condition becomes in terms of the coordinates μ_i

$$\frac{\partial\phi}{\partial\mu_1}\cdot\frac{\partial\psi}{\partial\mu_1}\cdot\frac{f(\mu_1)}{(\mu_1-\mu_2)(\mu_1-\mu_3)(\mu_1-\mu_4)}+\ldots\equiv 0,$$

which is satisfied, since

$$\frac{\partial\psi}{\partial\mu_1}=\frac{\partial\psi}{\partial\mu_2}=\frac{\partial\psi}{\partial\mu_3}\equiv 0,$$

and

$$\frac{\partial\phi}{\partial\mu_4}=\frac{\sqrt{(\mu_4-a)(\mu_4-b)}}{\sqrt{f(\mu_4)}}$$

which vanishes for $\mu_4=a$ or for $\mu_4=b$.

The value of the constant C has not come into consideration, hence *there is a singly infinite number of surfaces ϕ for each of which one set of principal tangents belongs to one, and the other set of principal tangents belongs to the other of two cosingular quadratic complexes.*

To the two quadratic complexes there correspond two partial differential equations D_{11}; these ∞^1 surfaces ϕ are common Integral surfaces of these two differential equations.

MISCELLANEOUS RESULTS AND EXERCISES.

1. If (xyz), $(x'y'z')$ are the Cartesian coordinates for rectangular axes of two points on a line, the Plücker coordinates of the line are

$$p_{41} = x' - x, \qquad p_{42} = y' - y, \qquad p_{43} = z' - z,$$
$$p_{23} = yz' - y'z, \quad p_{31} = zx' - z'x, \quad p_{12} = xy' - x'y.$$

The first three coordinates are proportional to the direction cosines l, m, n of the line; and since

$$p_{23} = y(z' - z) - z(y' - y), \quad p_{31} = z(x' - x) - x(z' - z),$$
$$p_{12} = x(y' - y) - y(x' - x),$$

it is clear that the Plücker coordinates are proportional to the quantities l, m, n; l', m', n' where l, m, n are the direction cosines of the line and

$$l' = yn - zm,$$
$$m' = zl - xn,$$
$$n' = xm - yl.$$

The relations satisfied by the coordinates are

$$l^2 + m^2 + n^2 = 1,$$
$$ll' + mm' + nn' = 0.$$

The form of the last two equations suggests that l', m', n' may be regarded as the derivatives of l, m, n with regard to a new variable t of which l, m, and n are such functions that

$$l^2 + m^2 + n^2 \equiv 1.$$

So that $$l' = \frac{dl}{dt}, \quad m' = \frac{dm}{dt}, \quad n' = \frac{dn}{dt}.$$

"Thus instead of taking six coordinates to represent a line we may take three variables and their three derivatives, and giving this a kinematical interpretation, we may represent a line by a point moving on a sphere of unit radius; the radius through the point is parallel to the line, and the three components of velocity of the point are the second set of three coordinates of the line*."

* Hudson, "A new method in line geometry," *Messenger of Mathematics*, 1902.

2. Denoting by M the "mutual moment" of two lines $(l, m, n;\ l', m', n')$ and $(\lambda, \mu, \nu;\ \lambda', \mu', \nu')$, *i.e.* the moment of unit force in one line about the other; show that if (xyz) is any point on the first line and $(x'y'z')$ any point on the other line,

$$M = \begin{vmatrix} x-x', & l, & \lambda \\ y-y', & m, & \mu \\ z-z', & n, & \nu \end{vmatrix} = l\lambda' + m\mu' + n\nu' + l'\lambda + m'\mu + n'\nu.$$

3. *The shortest distance of two lines $(lmnl'm'n')$, $(\lambda\mu\nu\lambda'\mu'\nu')$ is $(LMNL'M'N')$, where

$$L = m\nu - n\mu, \text{ etc.}; \quad L' = \frac{p\varpi L}{q^2} + m\nu' + m'\nu - n\mu' - n'\mu, \text{ etc.},$$

where

$$p = l\lambda + m\mu + n\nu, \quad \varpi = l\lambda' + l'\lambda + m\mu' + m'\mu + n\nu' + n'\nu, \quad q^2 = L^2 + M^2 + N^2.$$

4. In rectangular Cartesian coordinates

the line $(-l, m, n,\ l', -m', -n')$ is the reflexion of (l, m, n, l', m', n') in $x = 0$,

the line $(-l, m, n,\ -l', m', n')$ is the reflexion of (l, m, n, l', m', n') in $y = 0, z = 0$,

the line $(l, m, n,\ -l', -m', -n')$ is the reflexion of (l, m, n, l', m', n') in the origin.

5. The line (l, m, n, l', m', n') touches

$$ax^2 + by^2 + cz^2 + dw^2 = 0$$

if

$$(al^2 + bm^2 + cn^2)\, d + bcl'^2 + cam'^2 + abn'^2 = 0.$$

It touches

$$2axw + by^2 + cz^2 = 0, \text{ if } a(al^2 - 2bmn' + 2cnm') - bcl'^2 = 0.$$

6. The tangent planes drawn from the line to

$$ax^2 + by^2 + cz^2 + dw^2 = 0$$

are

$$P^2/a + Q^2/b + R^2/c + S^2/d = 0,$$

where

$$\dagger P \equiv ny - mz - l'w, \quad Q \equiv lz - nx - m'w,$$
$$R \equiv mx - ly - n'w, \quad S \equiv l'x + m'y + n'z.$$

[*Math. Trip.* 1896.]

To

$$2axw + by^2 + cz^2 = 0,$$

they are

$$2PS/a + Q^2/b + R^2/c = 0.$$

* Questions 3—6, 11—13 are due to Prof. Bromwich.

† $P = 0$, $Q = 0$, $R = 0$, $S = 0$ are planes through the line and the vertices of the tetrahedron of reference. The plane through the line and the point $(x_0 y_0 z_0 w_0)$ is

$$Px_0 + Qy_0 + Rz_0 + Sw_0 = 0.$$

To $(abcdfgha'b'c'\between xyzw)^2 = 0,$

they are

$$\begin{vmatrix} a & h & g & a' & P \\ h & b & f & b' & Q \\ g & f & c & c' & R \\ a' & b' & c' & d & S \\ P & Q & R & S & 0 \end{vmatrix} = 0.$$

7. The line is a generator of the general central quadric

$$(abcfgh\between xyz)^2 + d = 0,$$

if

$$al + hm + gn = l'(\Delta/d)^{\frac{1}{2}},$$
$$hl + bm + fn = m'(\Delta/d)^{\frac{1}{2}},$$
$$gl + fm + cn = n'(\Delta/d)^{\frac{1}{2}},$$

where

$$\Delta = abc + 2fgh - af^2 - bg^2 - ch^2.$$

[Bromwich, *Mess. Math.* 1900.]

8. The line is normal to $ax^2 + by^2 + cz^2 = 1$, if

$$\frac{ll'}{a} + \frac{mm'}{b} + \frac{nn'}{c} = 0,$$

$$(b-c)\,lm'n' + (c-a)\,l'mn' + (a-b)\,l'm'n = (b-c)(c-a)(a-b)\,lmn/abc.$$

It is normal to $2ax + by^2 + cz^2 = 0$, if

$$\frac{a}{c} - \frac{m'}{n} = \frac{a}{b} + \frac{n'}{m} = -\frac{a}{2l^2}\left(\frac{m^2}{b} + \frac{n^2}{c}\right).$$

9. If

$$all' + bmm' + cnn' = 0,$$

there is one and only one of the confocals

$$\frac{x^2}{a+\rho} + \frac{y^2}{b+\rho} + \frac{z^2}{c+\rho} = 1$$

to which the line is normal; this is given by

$$(b-c)(a+\rho)\,lm'n' + (c-a)(b+\rho)\,l'mn' + (a-b)(c+\rho)\,l'm'n$$
$$= (b-c)(c-a)(a-b)\,lmn.$$

10. Tangent planes are drawn to the confocals

$$\frac{x^2}{a+\rho} + \frac{y^2}{b+\rho} + \frac{z^2}{c+\rho} = 1$$

through the line (l, m, n, l', m', n'), prove that the locus of their points of contact is the twisted cubic given by

$$\frac{p}{x}[1 + (a-b)r^2 - (c-a)q^2] = \frac{q}{y}[1 + (b-c)r^2 - (a-b)p^2]$$
$$= \frac{r}{z}[1 + (c-a)p^2 - (b-c)q^2] = \frac{t^2 + l^2 + m^2 + n^2}{l'^2 + m'^2 + n'^2};$$

where p, q, r are linear functions of t such as

$$p = \frac{l't + mn' - m'n}{l'^2 + m'^2 + n'^2}, \text{ etc.} \quad \text{[H. F. Baker.]}$$

If normals are drawn at the points of contact of these tangent planes they generate the hyperbolic paraboloid

$$(lx+my+nz)[(b-c)\,m'n'x+(c-a)\,n'l'y+(a-b)\,l'm'z] + (b-c)(c-a)(a-b)\,lmn$$
$$=(b-c)\,lx\,[(c-a)nn'+(a-b)\,mm']+(c-a)\,my\,[(a-b)ll'+(b-c)nn'] + (a-b)\,nz[(b-c)mm'+(c-a)ll'].$$

If $all'+bmm'+cnn'=0$, the paraboloid becomes

$$(lx+my+nz-lmn/\theta)^2=0,$$

where $$\theta=\frac{ll'}{b-c}=\frac{mm'}{c-a}=\frac{nn'}{a-b}.$$ [Bromwich.]

11. The intercept on the line $(l,\ m,\ n,\ l',\ m',\ n')$ by the quadric

$$(abcfgh\between xyz)^2=1$$

is $$\frac{2\sqrt{l^2+m^2+n^2}\,[(abcfgh\between lmn)^2-(ABCFGH\between l'm'n')^2]^{\frac{1}{2}}}{(abcfgh\between lmn)^2}.$$

12. The polar line with respect to $(abcfgh\between xyz)^2=1$ is given by the equations

$$\frac{\lambda}{Al'+Hm'+Gn'}=\ldots=\ldots=\frac{-\lambda'}{al+hm+gn}=\ldots=\ldots.$$

13. Show that the polar of $(l,\ m,\ n,\ l',\ m',\ n')$ with regard to

$$\frac{x^2}{a}+\frac{y^2}{b}+\frac{z^2}{c}=1$$

is given by

$$(\lambda,\ \mu,\ \nu,\ \lambda',\ \mu',\ \nu')=(al',\ bm',\ cn',\ -bcl,\ -cam,\ -abn).$$

If $all'+bmm'+cnn'=0$, show that the polar lines with regard to quadrics confocal with the given quadric lie in the plane

$$lx+my+nz=-\frac{mn}{l'}(b-c)=\text{two similar expressions.}$$

14. If a line touches the quadric

$$ayz+bzx+cxy+abc=0,$$

show that

$$a^2l'^2+b^2m'^2+c^2n'^2-2bcm'n'-2can'l'-2abl'm'=4abc\,(amn+bnl+clm).$$

15. If the line $(lmnl'm'n')$ receives an infinitesimal rotation $d\theta$ on a screw of pitch ϖ whose axis is the line $(abca'b'c')$, it becomes

$$l+dl,\ \ldots,\ l'+dl',\ \ldots,\ \text{where, if } k^2=a^2+b^2+c^2,$$

$$k\frac{dl}{d\theta}=bn-cm,\quad k\frac{dm}{d\theta}=cl-an,\quad k\frac{dn}{d\theta}=am-bl;$$

$$k\frac{dl'}{d\theta}=bn'+b'n-cm'-c'm+\varpi(bn-cm);$$

$$k\frac{dm'}{d\theta}=cl'+c'l-an'-a'n+\varpi(cl-an);$$

$$k\frac{dn'}{d\theta}=am'+a'm-bl'-b'l+\varpi(am-bl).$$

Hence show that if a line $(l_1, m_1, n_1; l_1', m_1', n_1')$ is rotated through an angle θ about the line $(l, m, n; l', m', n')$, where θ is to be reckoned positive when given by a left-handed screw in the sense (lmn), then if $(l_2, m_2, n_2; l_2', m_2', n_2')$ are the coordinates of the line in its displaced position

$$l_2 = l_1 - (mn_1 - m_1 n)\sin\theta + (lk - l_1)\cos\theta, \text{ etc.};$$

$$l_2' = l_1' + \{n'm_1 - n_1'm + m'n_1 - m_1'n\}\sin\theta + (l'k - l_1' + l\varpi)(1-\cos\theta), \text{ etc.};$$

where
$$k = ll_1 + mm_1 + nn_1,$$
$$\varpi = ll_1' + mm_1' + nn_1' + l'l_1 + m'm_1 + n'n_1,$$

and (lmn), $(l_1m_1n_1)$ are the actual direction cosines of the lines*.

16. If the line $(lmnl'm'n')$ receive a small displacement given by (da, db, dc) parallel to the axes and $(d\phi_1, d\phi_2, d\phi_3)$ about them, find the consequent change in the coordinates of the line. Hence show that a line of the linear complex

$$A'l + B'm + C'n + Al' + Bm' + Cn' = 0$$

will be changed into a line of the same complex if

$$da\,(BC' - B'C) + db\,(CA' - C'A) + dc\,(AB' - A'B) = 0,$$
$$d\phi_1 : d\phi_2 : d\phi_3 = A : B : C,$$
$$(BC' - B'C)\,d\phi_1 = A\,(Cdb - Bdc), \text{ etc.}$$

17. With the same notation and coordinate system show that the equation of the axis of the complex

$$a'l + b'm + c'n + al' + bm' + cn' = 0$$

is
$$\frac{cy - bz + a'}{a} = \frac{az - cx + b'}{b} = \frac{bx - ay + c'}{c}.$$

The coordinates of the axis are given by the equations

$$\rho . l = a,\ \rho . m = b,\ \rho . n = c,\ \rho . l' = -a' + Ea,\ \rho . m' = -b' + Eb,$$
$$\rho . n' = -c' + Ec;\ E = (aa' + bb' + cc')/(a^2 + b^2 + c^2).$$

18. There is only one quadric of the type

$$ax^2 + by^2 + cz^2 + dw^2 = 0$$

of which a given line is generator. [Lie.]

It is
$$l'mnx^2 + lm'ny^2 + lmn'z^2 + l'm'n'w^2 = 0. \quad \text{[Bromwich.]}$$

19. If two quadrics S, S' do not touch, there are four generators of each system of S which touch S'; taking S as

$$x^2 + y^2 + z^2 + w^2 = 0,$$

and S' as
$$ax^2 + by^2 + cz^2 + dw^2 = 0,$$

* Bromwich, "The displacement of a given line by a motion on a given screw," *Mess. Math.* (1900).

the eight generators are determined by the equations

$$\left.\begin{aligned} &\frac{l'}{l} = \frac{m'}{m} = \frac{n'}{n} = \pm 1, \\ &\frac{l^2}{(b-c)(a-d)} = \frac{m^2}{(c-a)(b-d)} = \frac{n^2}{(a-b)(c-d)} \end{aligned}\right\}.$$

20. *If the line is a generator of the paraboloid

$$\frac{y^2}{b^2} - \frac{z^2}{c^2} = 2\frac{x}{a},$$

prove that

$$\frac{cm}{bn} = \frac{bm'}{cn'} = \frac{bcl}{al'} z = \pm 1.$$

21. The general quadric has $(lmnl'm'n')$ as a generator if

$$\frac{1}{l}\frac{\partial\phi}{\partial l'} = \frac{1}{m}\frac{\partial\phi}{\partial m'} = \frac{1}{n}\frac{\partial\phi}{\partial n'} = \frac{1}{l'}\frac{\partial\phi}{\partial l} = \frac{1}{m'}\frac{\partial\phi}{\partial m} = \frac{1}{n'}\frac{\partial\phi}{\partial n} = \pm 2\sqrt{\Delta};$$

where Δ is the discriminant of the quadric, and $\phi = 0$ is its complex equation, *i.e.* the condition that the line should touch the quadric, the coefficients of ϕ being therefore the second minors of Δ. Show also that if a is *any* line, b its polar line for the quadric, and g any generator,

$$\frac{\text{mutual moment of } (b, g)}{\text{mutual moment of } (a, g)} = \pm\sqrt{\Delta}. \quad \text{[Grace.]}$$

22. The regulus determined by three linear complexes being given by the equations

$$\begin{aligned} l' &= a_1 l + a_2 m + a_3 n, \\ m' &= b_1 l + b_2 m + b_3 n, \\ n' &= c_1 l + c_2 m + c_3 n; \end{aligned}$$

the quadric to which the regulus belongs is

$$(abcfgh \between x - \alpha w,\ y - \beta w,\ z - \gamma w)^2 + w^2 D = 0,$$

where

$$\begin{aligned} D &= abc + 2fgh - af^2 - bg^2 - ch^2, \\ a &= a_1,\quad b = b_2,\quad c = c_3, \\ 2f &= c_2 + b_3,\quad 2g = a_3 + c_1,\quad 2h = b_1 + a_2, \\ 2\alpha &= c_2 - b_3,\quad 2\beta = a_3 - c_1,\quad 2\gamma = b_1 - a_2. \end{aligned}$$

23. Show that any four points and their polar planes for a linear complex form two tetrahedra, of which each is inscribed and circumscribed to the other.

24. There is one pair of lines which are polar for a given linear complex and also for a given quadric.

25. The polars of a line l with reference to a pencil of linear complexes form a regulus, to which l itself and the directrices of the two special complexes of the pencil belong.

* Questions 19, 20 and 22 are due to Prof. Bromwich.

26. The polars of the lines of a linear complex C with reference to another C' form a third linear complex C'' which belongs to the pencil determined by C and C'.

27. If k_1 and k_2 are the chief parameters of two complexes A and B, d the shortest distance, and ϕ the inclination of their axes,

$$\sqrt{k_1 k_2}\frac{\Omega(a \mid b)}{\sqrt{\Omega(a)\,\Omega(b)}} = (k_1 + k_2)\cos\phi + d\sin\phi\,*.$$

28. Show that a linear complex may contain two lines of one regulus belonging to a given quadric and only one of the other. If the quadric is $\Sigma c_i x_i^2 = 0$, and the complex $\Sigma a_{ik} p_{ik} = 0$, show that in the case considered

$$4c_1 c_2 c_3 c_4 (a_{12}a_{34} + a_{13}a_{42} + a_{14}a_{23})^2 = (c_3 c_4 a_{12}^2 + c_1 c_2 a_{34}^2 + \ldots)^2.$$

[*Math. Tripos*, 1898.]

29. If six linear complexes are in mutual involution, three of them are right-handed and three left-handed.

30. Three linear complexes which are not all right-handed or left-handed intersect in a real regulus.

31. The two common intersectors of any line and its polars for three linear complexes belong to the regulus common to these complexes.

32. In the system of five terms determined by the complexes A, B, C, D, E, the directrices of the special complexes form the linear complex

$$| a_i, b_i, c_i, d_i, e_i, x_i | = 0.$$

33. The linear equation satisfied by the six coordinates of a line PQ which belongs to a linear complex being $f_1(PQ) = 0$, and similarly $f_2(PQ) = 0$, $f_3(PQ) = 0$, $f_4(PQ) = 0$ being analogous equations for three other linear complexes, show that the conditions that a plane ABC may contain *one* of the two lines common to the four complexes are

$$\left\| \begin{matrix} f_1(BC) & f_2(BC) & f_3(BC) & f_4(BC) \\ f_1(CA) & f_2(CA) & f_3(CA) & f_4(CA) \\ f_1(AB) & f_2(AB) & f_3(AB) & f_4(AB) \end{matrix} \right\| = 0.$$

34. The ∞^2 complexes $\lambda C_1 + \mu C_2 + \nu C_3 = 0$ being designated a *net* of complexes, where $C_1 = 0$, $C_2 = 0$, $C_3 = 0$ are three given linear complexes, show that there is one complex of the net which contains any given pencil, and that every point of a given plane determines one complex of the net for which the given point is the pole of the plane.

35. The polars of a line l with reference to the complexes of a net form a linear congruence whose directrices are the two generators of the regulus common to the complexes of the net which meet l.

The axes of the complexes of a net form a congruence (2, 3).

* See Segre, *Crelle*, Bd. 99.

36. The complex C_1 is said to be *orthogonal* to C when C_1 is in involution with C', where C' is the locus of lines polar to those of C with regard to a given quadric. [d'Ovidio.]

Prove that for each complex of a pencil of complexes there is one orthogonal complex belonging to the pencil.

37. If four tangents of a twisted cubic have their two intersectors coincident with one line p, then p belongs to the linear complex determined by the tangents of the cubic.

38. If two tetrahedra are inscribed in a twisted cubic their eight planes osculate another twisted cubic; and there are ∞^1 tetrahedra which are inscribed to one and circumscribed to the other tetrahedron. [Hurwitz.]

39. If P is the pole of the sphere-circle for the ray of a linear congruence in the plane at infinity, and p the ray through P, the rays which have a given inclination to p form a ruled quartic which has the directrices of the congruence as double directrices.

40. Two lines are said to be *conjugate* with respect to a quadric when each intersects the polar of the other. [Schur.]

If five lines meet a given line and are such that all but one of the ten pairs formed from them are conjugate with respect to a quadric, the quadric is uniquely determined and the remaining pair are also conjugate.

Let now a_6 be the first line and b_1, b_2, b_3, b_4, b_5 the meeting lines, then a_5 the other intersector of b_1, b_2, b_3, b_4 is the polar of b_5. So each of the lines a_1, a_2, a_3, a_4, a_5 is the polar of the corresponding b. But the five lines b all meet a_6, hence the lines $a_1 \dots a_5$ all meet the polar of a_6 which may be called b_6. The lines a and b thus form the *double sixer*

$$\left.\begin{matrix} a_1 a_2 a_3 a_4 a_5 a_6 \\ b_1 b_2 b_3 b_4 b_5 b_6 \end{matrix}\right\} \text{ of Schläfli.} \quad \text{[Grace.]}$$

It follows that opposite lines of a double sixer are polar lines with respect to a certain quadric. [Schur.]

41. Show that through any point three osculating planes can be drawn to the twisted cubic $y^2 = zx$, $z^2 = uy$, and that the three points of osculation are coplanar with the given point. Also show that each such point and plane are pole and polar plane for a linear complex which contains the four lines

$$z = 0,\ u = 0;\ \ y = 0,\ u = 0;\ \ x = 0,\ z = 0;\ \ x = 0,\ y = 0.$$

42. On every ruled surface whose generators belong to a given linear complex the family of principal tangent curves is found by a quadrature. [Lie.]

Denote by k the principal tangent curve of the surface whose tangents belong to the linear complex; any generator meets k in two points x and y, and any point z on this generator is $x+\mu y$; the coordinates x_i and y_i are functions of one parameter λ.

The differential equation of the principal tangent curves is

$$Pd\lambda^2+2Qd\lambda d\mu+Rd\mu^2=0,$$

where

$$P=\left| z_i, \frac{\partial z_i}{\partial\lambda}, \frac{\partial z_i}{\partial\mu}, \frac{\partial^2 z_i}{\partial\lambda^2} \right|, \quad Q=\left| z_i, \frac{\partial z_i}{\partial\lambda}, \frac{\partial z_i}{\partial\mu}, \frac{\partial^2 z_i}{\partial\lambda\partial\mu} \right|, \quad R=\left| z_i, \frac{\partial z_i}{\partial\lambda}, \frac{\partial z_i}{\partial\mu}, \frac{\partial^2 z_i}{\partial\mu^2} \right|.$$

Hence, since $R=0$, the equation of the principal tangent curves is

$$Pd\lambda+Qd\mu=0;$$

where $-P=| x_i, y_i, x_i'+\mu y_i', x_i''+\mu y_i'' |$, $-Q=| x_i, y_i, x_i', y_i' |$, the differentiations being with regard to λ. Hence the required differential equation is

$$\frac{d\mu}{d\lambda}+X_1\mu^2+X_2\mu+X_3=0,$$

where the X_i are functions of λ. Since $\mu=0$ and $\mu=\infty$ are solutions of this equation, we have $X_1=X_3=0$, and the equation reduces to

$$\frac{d\mu}{d\lambda}+X_2\mu=0.$$

43. The projection of every unicursal curve of degree m, whose tangents belong to a linear complex, on a plane perpendicular to the axis of the complex, has m points of inflexion at infinity. [Picard.]

44. A linear complex being given, any curve such that the polar plane of each of its points is normal to the curve at the point is a helix. [Picard.]

45. The lines of a quadratic complex C^2 determine upon any two lines l and l' a (2, 2) correspondence, the double ratio of the branch points on l being equal to the double ratio of the branch points on l'. Taking l to be any line and l' its polar line for C^2, show thereby that the locus of singular points is identical with the envelope of singular planes.

46. Each of the two lines common to any four of the fundamental complexes of C^2 is the polar of the other with regard to C^2; and there are no other pairs of lines so related except the 15 pairs of lines thus obtained. [Klein.]

47. Every quadratic complex through 16 given lines contains 16 other fixed lines.

For let $f_1(x)=0,\ f_2(x)=0,\ f_3(x)=0,\ f_4(x)=0$ be any four complexes through the 16 given lines, then *any* complex through these lines has an equation of the form

$$f_1(x)+\lambda f_2(x)+\mu f_3(x)+\nu f_4(x)=0,$$

since the constants λ, μ, ν can be determined so as to make it pass through any other three lines. But this complex contains *all* the lines of intersection of the four given complexes, which are 32 in number.

48. A Kummer surface is identical with its polar surface for a fundamental complex. The polars for C^2 of the lines of one of its fundamental complexes C_i belong to C_i.

49. Through two pencils of C^2 which have no common line one linear congruence passes; this congruence intersects C^2 in two other pencils each of which contains a line of the former two pencils; thus for any two skew pencils of C^2 there are two other skew pencils of C^2 having a common line with each of them.

50. If a line p is such that the points of intersection of p with the complex curve of C^2 in any plane whatever lie upon a second complex curve, the locus of p is a complex of the sixth degree. [Sturm.]

If C^2 is replaced by a tetrahedral complex $p(ABCD)=\lambda$, the complex of the sixth degree is replaced by the three complexes

$$p(ABCD)=\lambda^2, \quad p(ABCD)=(1-\lambda)^2, \quad p(ABCD)=\left(1-\frac{1}{\lambda}\right)^2.$$

[W. Stahl.]

51. By taking different values for the constants λ, μ, ν in the projective transformations

$$x_1=\lambda x, \quad y_1=\mu y, \quad z_1=\nu z$$

we obtain a set of ∞^3 transformations, connecting the point (xyz) with each point of space. If the line-element $(x, y, z; dx:dy:dz)$ is given, a definite line-element is assigned to each point $(x_1 y_1 z_1)$ of space, viz. that given by the equations

$$\frac{dx}{x}=\frac{dx_1}{x_1}, \quad \frac{dy}{y}=\frac{dy_1}{y_1}, \quad \frac{dz}{z}=\frac{dz_1}{z_1}.$$

These line-elements all belong to the same tetrahedral complex. [Lie.]

52. Every tetrahedral complex is invariant for reciprocation with regard to a quadric which has the tetrahedron of the given complex as a self-conjugate tetrahedron. [Lie.]

53. The normals of the quadric $(abcfgh \between xyz)^2=1$ belong to the complex

$$All'+Bmm'+Cnn'+F(mn'+m'n)+G(nl'+n'l)+H(lm'+l'm)=0.$$

54. Any line of the complex

$$all'+bmm'+cnn'=0$$

meets the quadric
$$\frac{x^2}{a}+\frac{y^2}{b}+\frac{z^2}{c}=1$$

in points P and Q the normals at which intersect each other. The coordinates of this point of intersection are given by the equations

$$x = \frac{mn}{\theta} \cdot \frac{cn'^2 + bm'^2 - bcl^2}{bcl^2 + cam^2 + abn^2}, \text{ etc.}$$

where $$\theta = \frac{ll'}{b-c} = \frac{mm'}{c-a} = \frac{nn'}{a-b}.$$

The line is a principal axis of the section of the quadric by the plane

$$al' \,.\, P + bm' \,.\, Q + cn' \,.\, R = 0,$$

where $P = ny - mz - l'$, $Q = lz - nx - m'$, $R = mx - ly - n'$.

The coordinates of the other principal axis are

$$asl', \; bsm', \; csn', \; \frac{lt}{a}, \; \frac{mt}{b}, \; \frac{nt}{c},$$

where $$t\left(\frac{l^2}{a} + \frac{m^2}{b} + \frac{n^2}{c}\right) + s\,(al'^2 + bm'^2 + cn'^2) = 0.$$

The lines of the complex which are in the plane

$$px + qy + rz + s = 0$$

touch a parabola whose directrix is in the plane

$$(b-c)\frac{x}{p} + (c-a)\frac{y}{q} + (a-b)\frac{z}{r} = 0.$$

As p, q, r and s vary, the directrix moves in the complex

$$(b-c)\frac{l}{l'} + (c-a)\frac{m}{m'} + (a-b)\frac{n}{n'} = 0.$$

The projection of the parabola on the plane $x = 0$ is

$$\sqrt{(a-b)\,qy} + \sqrt{(a-c)\,rz} = \sqrt{(b-c)\,s}. \quad \text{[Bromwich.]}$$

55. The lines of the complex

$$all' + bmm' + cnn' = 0$$

which meet the line $P_0 = 0$, $Q_0 = 0$, $R_0 = 0$, touch the surface

$$\sqrt{(b-c)\,xP_0} + \sqrt{(c-a)\,yQ_0} + \sqrt{(a-b)\,zR_0} = 0,$$

where $$P_0 = n_0y - m_0z - l_0', \text{ etc.}$$

This may be put in three other forms such as

$$\sqrt{(b-c)\,S_0} + \sqrt{(c-a)\,zR_0} + \sqrt{(a-b)\,yQ_0} = 0;$$

$$(S_0 \equiv l_0x + m_0y + n_0z). \quad \text{[H. F. Baker.]}$$

56. If a rigid body is turning about Oz, on a screw of pitch p, the lines of motion of its points belong to the complex

$$nn' = p\,(l^2 + m^2).$$

This complex belongs to the species [(22)(11)]. It cuts any plane in the lines of a parabola, and if the plane is

$$ax + by + cz + d = 0,$$

the directrix lies in the plane

$$bcx + cay + p\,(a^2 + b^2 + 2c^2) = 0. \quad \text{[Bromwich.]}$$

57. If l is any line of a complex C^2, then through l and its polar line for a fundamental complex there pass ∞^1 reguli of C^2 (not merely two).

58. Every regulus of a quadratic complex touches the singular surface four times.

59. Every regulus of C^2 which contains a given line l of C^2 belongs to a tangent linear complex of l.

60. There are ∞^1 quadratic complexes which contain a congruence (C^2, A) and have A as a fundamental complex; they have the same fundamental complexes.

61. If two (2, 2) congruences K_1 and K_2 are such that a complex for which K_1 is focal contains K_2, then a complex for which K_2 is focal will contain K_1. [Grace.]

62. A Plücker's complex surface is its own polar surface for each of four linear complexes which are mutually in involution. [Klein.]

63. If a line l describes a congruence (m, n), its polar l', with reference to any given quadratic complex, will describe a congruence

$$(2m + 3n,\ 3m + 2n).$$

64. The double tangents of a general complex surface form four congruences (2, 2); if the double line of the surface belongs to the complex the double tangents form three congruences (2, 2); if it is a singular line they form two congruences (2, 2). [Sturm.]

65. The singular lines of a harmonic complex H^2 are the intersections of H^2 with the tetrahedral complex formed by the lines whose polars with reference to f_1 and f_2 intersect each other; where f_1 and f_2 are a pair of quadrics which give rise to H^2.

66. If a quadratic complex contains a regulus and also its complementary regulus it is harmonic. [Schur.]

67. If a quadratic complex contains a plane system it has three double lines in the plane.

68. The axes of the complexes of *a system of four terms* form a quadratic complex.

69. Every congruence (m, n, r) possesses a ruled surface, of degree $4(mn - r) - 2(m + n)$, formed by rays with coincident focal points.

70. The six singular points of the congruence $(2, n)$ which lie in a singular plane are situated on a conic.

71. The tetrahedra of the 40 tetrahedral complexes which contain a given congruence (2, 2) can be arranged in 20 pairs so that the tetrahedra of a pair are inscribed and circumscribed to each other.

72. If a line moves with two of its points A, B in two fixed planes, it describes a complex of the fourth degree. If it has three of its points A, B, C in three fixed planes it describes a congruence (6, 2). Taking the two planes in the first case as $x=0$, $y=0$ the complex is

$$n'^2(l^2+m^2+n^2)=k^2l^2m^2$$

where $k=AB$; if the three planes in the second case are the coordinate planes, the congruence is contained in the tetrahedral complex

$$kmm'+k'nn'=0,$$

where $$k'=AC.$$

73. If $f=0$, $\phi=0$ are two quadrics, the complex of harmonic section determined by them is in general of the species [111111], having a tetrahedroid as its singular surface. When f and ϕ have certain projective relations to each other the complex is modified in form. By considering the elementary divisors of the discriminant of $f+\mu\phi$ we arrive at the following canonical forms to which f and ϕ can be reduced.

[1111], the general case; $\phi\equiv\sum_1^4 x_i^2$, $f\equiv\sum_1^4 a_i x_i^2$. The equation to determine the coefficients of H^2 is

$$\{\lambda^2-a(\alpha-a)\}\{\lambda^2-b(\alpha-b)\}\{\lambda^2-c(\alpha-c)\}=0,$$

where $\alpha=\Sigma a_i$, $a=a_1+a_2$, $b=a_1+a_3$, $c=a_1+a_4$.

[11(11)] $a_3=a_4$; the quadrics touch in two points and intersect in two conics. H^2 is [(11)(11)11].

[(11)(11)] $a_1=a_2$, $a_3=a_4$; the quadrics have four common generators forming a quadrilateral. H^2 is [(11)(11)11].

[(111)1] $a_1=a_2=a_3$; the quadrics touch along a conic. H^2 is [(111)(111)], *i.e.* consists of the tangents of a quadric.

If in [1111] we have $a_3+a_4=0$, an edge of the tetrahedron of reference belongs to H^2 which is [21111]. If $a_1+a_2=a_3+a_4=0$, two edges of the tetrahedron of reference belong to H^2 which is then [(11)1111]. In the case [(11)11] if either b or c is zero, *i.e.* if f and ϕ are harmonic with their pair of planes of intersection and either cone of the pencil $f+\mu\phi$, H^2 is [(22)11].

[112] $\phi\equiv x_1^2+x_2^2+2x_3x_4$, $f\equiv a_{11}x_1^2+a_{22}x_2^2+2a_{34}x_3x_4+x_3^2$. The quadrics touch each other and H^2 is [1122].

[(11)2] $a_{11}=a_{22}$; the quadrics intersect in a conic and two generators. H^2 is [(11)(11)11].

[1(12)] $a_{22}=a_{34}$; the quadrics intersect in two conics which touch. H^2 is [(12)(12)].

[(112)] $a_{11}=a_{22}=a_{34}$; the quadrics touch along two generators. H^2 is [(111)(111)].

If in [112] we have *either* $a_{11}+a_{34}=0$, *or* $a_{22}+a_{34}=0$, *i.e.* when two cones of the pencil $f+\mu\phi$ are harmonic to f and ϕ, H^2 is [411].

[13] $\phi \equiv x_1^2+x_3^2+2x_2x_4$, $f\equiv a_{11}x_1^2+a_{33}(x_3^2+2x_2x_4)+2x_2x_3$; the quadrics have *stationary* contact. H^2 is [33].

[(13)] $a_{11}=a_{33}$; the quadrics intersect in a conic and two generators which intersect on the conic. H^2 is [(111)(111)].

If in [13] we have $a_{11}+a_{33}=0$, *i.e.* if f and ϕ are harmonic with regard to the two cones of the pencil $f+\mu\phi=0$, H^2 is [6].

[22] $\phi\equiv x_1x_2+x_3x_4$, $f\equiv 2a_{12}x_1x_2+2a_{34}x_3x_4+x_1^2+x_3^2$; the quadrics have a common generator. H^2 is [(11)211].

[(22)] $a_{12}=a_{34}$; the quadrics touch along a generator. H^2 is [(111)(12)].

If in [22] we have $a_{12}+a_{34}=0$, H^2 is [(13)11].

[4] $\phi\equiv x_1x_4+x_2x_3$, $f\equiv 2a_{14}(x_1x_4+x_2x_3)+2x_1x_3+x_2^2$; the quadrics have a common generator which touches their residual cubic of intersection. H^2 is [(12)3]. [Segre and Loria.]

74. The harmonic complex for the quadric $\Sigma a_i x_i^2=0$ and the two planes $x_1=0$, $x_2=0$ is

$$a_3p_{13}p_{23}+a_4p_{14}p_{24}=0,$$

which is a tetrahedral complex.

75. The lines whose distances from two fixed lines l and l' have a constant ratio, form a complex of the fourth degree.

If l and l' have as their equations

$$\begin{cases} z=a, \\ y=x\tan\alpha, \end{cases} \qquad \begin{cases} z=-a, \\ y=-x\tan\alpha; \end{cases}$$

and if k is the ratio of the distances of a line p of the complex from l and l' respectively, the equation of the complex is

$$\frac{a(l\sin\alpha-m\cos\alpha)-(l'\cos\alpha+m'\sin\alpha)}{\sqrt{n^2+(l\sin\alpha-m\cos\alpha)^2}} = k\frac{a(l\sin\alpha+m\cos\alpha)-(l'\cos\alpha-m'\sin\alpha)}{\sqrt{n^2+(l\sin\alpha+m\cos\alpha)^2}}\,*.$$

76. Let there be two reciprocal systems Σ and Σ' of points and lines in a plane γ, and also two pencils (A, α), (B, β) having a common line; then if l is any line of Σ it determines a point L' of Σ'; through L' draw the line p which meets the same lines of the pencils (A, α), (B, β) as l; then the lines p form a congruence (3, 2). [W. Stahl.]

* See Schoute, *Ann. de l'école polytechnique de Delft*, T. III. (1887).

77. Show that any point of the surface

$$\sqrt{(b-c)\,x\,(l'+mz-ny)}+\sqrt{(c-a)\,y\,(m'+nx-lz)}$$
$$+\sqrt{(a-b)\,z\,(n'+ly-mx)}=0$$

can be represented in terms of the parameters t and u as follows:

$$\beta\gamma x=(t+a)^2/(l'u-p),\quad \gamma\alpha y=(t+b)^2/(m'u-q),\quad \alpha\beta z=(t+c)^2/(n'u-r),$$

where $\alpha=b-c$, etc.; $p=(mn'-m'n)/(l'^2+m'^2+n'^2)$, etc. (Compare Question 55.)

78. If the congruence $x_i=f_i(u,v)$ is such that a linear complex can be found which contains a line x of the congruence and all lines consecutive to it neglecting terms of the *third* order, the coordinates x_i satisfy an equation of the form

$$A\frac{\partial^2 x_i}{\partial u^2}+2B\frac{\partial^2 x_i}{\partial u\partial v}+C\frac{\partial^2 x_i}{\partial v^2}+D\frac{\partial x_i}{\partial u}+E\frac{\partial x_i}{\partial v}+Fx_i=0.\quad \text{[Koenigs.]}$$

79. If l, m, n, l', m', n' are functions of a parameter t and represent the coordinates of a generator of a developable, then

$$\frac{dl}{dt}\frac{dl'}{dt}+\frac{dm}{dt}\frac{dm'}{dt}+\frac{dn}{dt}\frac{dn'}{dt}=0.$$

The tangent plane through the generator is

$$P\frac{dl}{dt}+Q\frac{dm}{dt}+R\frac{dn}{dt}=0 \text{ (see No. 6)};$$

and the generator cuts the cuspidal edge in the point given by

$$x=\left(n'\frac{dm'}{dt}-m'\frac{dn'}{dt}\right)\Big/\left(l\frac{dl'}{dt}+m\frac{dm'}{dt}+n\frac{dn'}{dt}\right),\text{ etc.}$$

If the surface is skew instead of developable, the first condition does not hold, and the generator cuts the line of striction in the point given by

$$\frac{x-mn'+m'n}{l}=\frac{y-nl'+n'l}{m}=\frac{z-lm'+l'm}{n}$$

$$=\begin{vmatrix} l & m & n \\ \frac{dl}{dt} & \frac{dm}{dt} & \frac{dn}{dt} \\ \frac{dl'}{dt} & \frac{dm'}{dt} & \frac{dn'}{dt}\end{vmatrix}\div\left[\left(\frac{dl}{dt}\right)^2+\left(\frac{dm}{dt}\right)^2+\left(\frac{dn}{dt}\right)^2\right],$$

where $l^2+m^2+n^2=1.$

The direction cosines of the normal to the surface at this point are proportional to

$$\frac{dl}{dt},\quad \frac{dm}{dt},\quad \frac{dn}{dt}.^*$$

* See also Larmor's method, *Quarterly Journal*, vol. XIX.

The line being a generator of

$$ax^2 + by^2 + cz^2 + d = 0$$

we have $\theta . l' = al,\ \theta . m' = am,\ \theta . n' = an,\ \theta^2 = abc/d;$

and this generator cuts the line of striction in the point given by

$$\theta . x = -\frac{(b-c)\,mn}{l^2+m^2+n^2} - \frac{(b-c)(c-a)(a-b)}{mn\left[(b-c)^2/l^2 + (c-a)^2/m^2 + (a-b)^2/n^2\right]}.$$

The cone joining the line of striction to the origin is

$$\frac{(b-c)^2}{ax^2} + \frac{(c-a)^2}{by^2} + \frac{(a-b)^2}{cz^2} = 0. \quad \text{[Bromwich.]}$$

80. The general reciprocity being defined by the equations

$$(\lambda - \lambda_1)\,u_1 = x_4',\ (\lambda - \lambda_3)\,u_2 = x_3',\ (\lambda + \lambda_3)\,u_3 = x_2',\ (\lambda + \lambda_1)\,u_4 = x_1',$$

if a line p_{ik}' joining two points x_i', y_i', intersects the line p_{ik} joining the corresponding planes u_i, v_i, then p_{ik} belongs to the harmonic complex

$$\lambda^2\,(p_{14}^2 + p_{23}^2 + 2p_{12}p_{34} - 2p_{13}p_{42}) = (\lambda_3 p_{14} + \lambda_1 p_{23})^2.$$

[Lindemann.]

81. Show that the line-complexes which are unaltered by an infinitesimal rotation about the axis of z are those of the form

$$\Psi\,(r^2 + s^2,\ \rho^2 + \sigma^2,\ s\rho - r\sigma) = 0. \quad \text{[Lie.]}$$

INDEX.

The numbers refer to the Articles.

Printed in the United States
By Bookmasters